FROM CELL TO PHILOSOPHER

FROM

TO

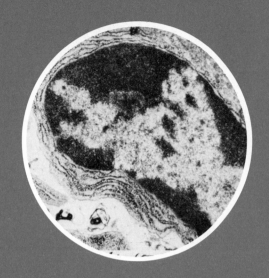

CELL

PHILOSOPHER

Michael D. Nicklanovich

PRENTICE-HALL, INC.

Englewood Cliffs, New Jersey

PRENTICE-HALL BIOLOGICAL SCIENCE SERIES
William D. McElroy and Carl P. Swanson, editors

FROM CELL TO PHILOSOPHER

by *Michael D. Nicklanovich*

Library of Congress Catalog Card No. 72–5460

ISBN: 0–13–331553–3

10   9   8   7   6   5   4   3   2   1

Printed in the United States of America

PRENTICE-HALL INTERNATIONAL, INC., *London*
PRENTICE-HALL OF AUSTRALIA, PTY. LTD., *Sydney*
PRENTICE-HALL OF CANADA, LTD., *Toronto*
PRENTICE-HALL OF INDIA PRIVATE LIMITED, *New Delhi*
PRENTICE-HALL OF JAPAN, INC., *Tokyo*

to Andrew, Mary, LuAnne, Mara, and Anna

CONTENTS

9

RISEN APES OR FALLEN ANGELS   473

10

BIOPHILOSOPHY   515

PREFACE

THIS BOOK has been written primarily for the nonscience major: the students of the humanities, liberal arts, or other areas. It is intended to serve as a text for a one-term biology course. Each of the ten chapters has been written as a nearly independent essay. The order of topics is suggested, but the text should lend itself well to either a modular approach or another sequence of presentation by the instructor. There are basically four units: chapters 1–3; chapters 4–5; chapters 6–7; chapters 8–10. A successful student should be able to perform the tasks requested of him in the objectives. It is suggested that he consult them before, after, and during the reading of the chapters.

The writing of this text was a pleasure to me. The task prodded me to organize and synthesize my philosophy. If it does the same for the student, I will have succeeded.

MICHAEL D. NICKLANOVICH

FROM CELL TO PHILOSOPHER

OBJECTIVES

1. List and define four primary properties of life.

2. State the Cell Theory.

3. List the five major categories of biochemicals.

4. Cite three examples of biologic macromolecular polymers and give their monomeric components.

5. Name three compounds employed in the construction of cell membranes.

6. List four structural polysaccharides found in plant and animal tissues.

7. Briefly state the relationship between nucleic acids, proteins, and metabolism.

8. Explain the role of vitamins in biochemistry.

9. Describe the role of enzymes in biochemical reactions.

10. Contrast biochemical and inorganic reactions.

11. Define catabolism and anabolism; give two pathways exemplifying each.

Life

and Its

Chemicals

DEFINITIONS OF LIFE

LIFE IS DIFFICULT TO DEFINE. Few attempt the task. In the words of Harvard's great biology teacher and Nobel laureate, George Wald,

Biologists long ago became convinced that it was not useful to define life. The trouble with any such definition is that one can always construct a model that satisfies the definition, yet clearly is not alive. And of course we do not ever measure life. We can measure many of its manifestations accurately; and we combine those with others that we observe, but perhaps cannot measure, to make up our concept of what it means to be alive. The life itself is neither observed nor measured. It is a summary of and judgment upon our measurements and observations.

What biologists do about life is to recognize it. If that seems a slipshod procedure, I beg the reader, try to define your wife. You have no trouble recognizing her; I think you will grant the operation to be accurate and unequivocal. Well, that's the way it is with biologists and life.[1]

Dictionaries certainly assume that one also knows what is meant by life. Here's a sample definition of life from Webster's *New World Dictionary:* "the quality that distinguishes a living animal or plant organism from inorganic matter or a dead organism."

Perhaps life can be defined in terms of its opposite, death. Professor

[1]George Wald, "Innovation in Biology," *Scientific American,* Vol. 199, No. 3 (Sept., 1958), p. 113.

Hans Jonas has written "That life is mortal may be its basic self-contradiction, but it belongs to its nature and cannot be separated from it even in thought: life carries death in itself, not in spite of, but because of, its being life, for of such revocable, measured kind is the relation of form and matter upon which it rests." [2]

The most credible modern hypothesis of the origin of life contends that life arose from nonliving matter. Even today, the energy and matter that flow into life ultimately come from the nonliving world and are finally returned to it, following death and decay. Viruses are often said to "straddle" the boundary between the living and nonliving realms.

The Anglo-Saxon word "life" is akin to the German word for body, *Leib*. (The German verb *leben* means "to live.") The term "organism," which is used to refer to "any living thing," is closely related to the Greek *organon*, which means "instrument," "implement," "machine." An organism is a machine of life. So far it has not been possible to separate life from its form.

Describing the characteristics of life that distinguish it from nonlife is an attempt at definition, perhaps the best attempt we can make. The primary properties of life are *reproduction, metabolism, sensitivity,* and *cellular organization.*

The ability to reproduce probably distinguished the first "living" molecules from the dead. To reproduce is to make again. In a sense, life makes itself again. Reproduction is the final criterion of the survival of the species. Asexual reproduction is mere duplication, but sexual reproduction results in new individuals rather than copies of the old. All living organisms contain nucleic acids, the hereditary chemicals, and most contain DNA, which "autoduplicates" or self-duplicates in the process of reproduction.

Metabolism is a term used to refer to the total sum of biochemical reactions. The two major subdivisions of metabolism are *anabolism*, synthesis, and *catabolism*, breakdown. The major biosyntheses are photosynthesis and protein synthesis; reproduction is basically a synthesis. The major catabolic reaction sequences are respiration and digestion. In photosynthesis, inorganic compounds, water, nitrate, phosphate, and other salts, along with the simple organic compound carbon dioxide, are converted into the complex organic molecules of the biochemicals. The energy of sunlight is transformed into the chemical bond energies of the biochemicals. In respiration, the bonds between the atoms in biochemicals are broken, and the energy released is used to do work. Photosynthesis and respiration are processes of energy transformation. Living organisms can fix energy in the form of chemical bonds, then liberate and utilize it to do work.

---

[2] Hans Jonas, *The Phenomenon of Life* (New York, N. Y.: Dell Publishing Co., Inc., 1968), p. 5.

All living organisms are sensitive, irritable, capable of making *adaptive response,* or self-regulating. A stimulus is a change in the external or internal environments of an organism that evokes an adjustive response. The organism can move to withdraw from an unsuitable environment or seek a favorable one. The property of sensitivity makes possible the adaptation or adjustment of the organism to its environment. Failure to respond adaptively to a changing environment can result in death for the organism, or the species, since evolution is really the adaptive response of generations to gradually changing environments. The living organism is somehow like a gyroscope or a thermostat, self-righting. There are control systems within the cell that regulate metabolic rates and directions, as well as organ systems within the multicellular organism that are devoted to coordinating, integrating, and controlling the interactions of the other systems with the environments.

All living organisms except viruses are composed of cells and their products. Even viruses are comparable to certain parts of the cell: they have been called "naked genes." The *Cell Theory* states that all living organisms are composed of cells and their products or are at least cellular analogues.

BIOCHEMICALS

The structure and function of life in the final analysis depend upon the organization and properties of the chemicals of life, the biochemicals. Water is the medium of the reactions of life, and it accounts for most of the percent by weight composition of organisms. Inorganic salts are found in both cellular and body fluids, and they play no small role in biochemistry. Yet, water and inorganic salts do not a living organism make; it is only in their interactions with the biochemicals that they play a role in the reactions of life. The modern theory of the origin of life on earth holds that chemical evolution preceded organismal evolution: it is believed that "biochemicals" formed before life appeared. There are five major categories of biochemicals: (1) *proteins,* (2) *nucleic acids,* (3) *lipids,* (4) *carbohydrates,* and (5) *vitamins.*

The word "protein" is derived from the Greek *proteios,* meaning "prime" or "chief." Proteins were so named because they are the primary constituents of living organisms. Not too long ago they were even thought to be the genes, the biochemicals of heredity. This was not too illogical. The proteinaceous *enzymes* control almost all the reactions of metabolism. Much of the structure of living organisms and the cells of which they are composed is protein in nature. No known state of organization ever exhibited the properties of life without proteins. Like

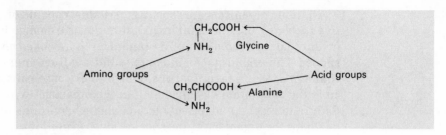

Fig. 1-1   **The two simplest amino acids. There are about 20 others.**

many biologic molecules, proteins are *macromolecules,* giant molecules, and *polymers,* made up of many similar repeating units, the *monomers.* The monomers of protein polymers are *amino acids,* of which there are a common 23 (Fig. 1-1). The bond between amino acids is called a *peptide bond,* and proteins are *polypeptides* of 50 amino acid units or more (Fig. 1-2). Many proteins contain hundreds and even thousands of amino acid units. Several proteins contain more than one chain, and the chains themselves are spiraled and folded (Fig. 1-3). When 23 different

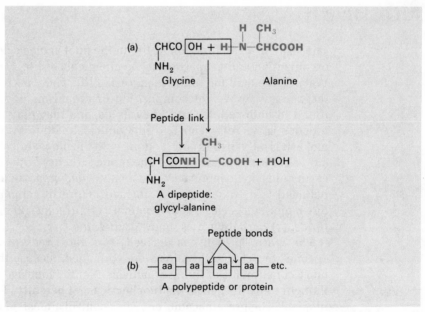

Fig. 1-2   **(a) Peptide bond formation and (b) a diagram representing amino acids (aa) in polypeptide chain.**

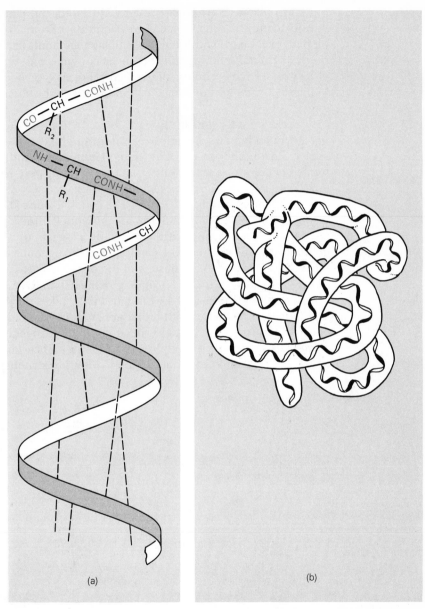

(a)

(b)

Fig. 1-3   (a) The spiral pattern of a polypeptide chain (the alpha helix). The peptide bonds are shown in color. Hydrogen bonds tie the loops together. The R's represent the remainder of the various amino acid molecules. (b) A drawing of a protein molecule, showing the polypeptide chain folded and knotted. The polypeptide rope of this knot has the helix within it.

units are taken in groups of several hundred, the possible number of
various combinations seems almost infinite. (From 26 letters of the alphabet, millions of combinations and arrangements can be made.) Since
they controlled the reactions of life as enzymes and were so various, it
is easy to see why scholars of even 20 years ago considered the proteins
to be the key compounds of life, and therefore the chemicals of inheritance.

It has been only within the last 30 years that the nucleic acids have
been proven beyond a shadow of a doubt to be the biochemicals of heredity. The nucleic acids DNA (deoxyribonucleic acid) and RNA (ribonucleic acid) are, like proteins, macromolecular polymers. In fact, they
are the *largest* biological molecules; proteins are second in size. Nucleic
acids determine protein structure. The genetic code of DNA is a code for
protein synthesis. Nucleic acids are *polynucleotides;* their monomers
are *nucleotides.* A nucleotide molecule consists of a 5-carbon sugar
compounded with phosphoric acid and a nitrogenous base (Fig. 1-4). In
DNA, the sugar is deoxyribose, and the possible bases are four: adenine
(A), thymine (T), guanine (G), and cytosine (C). In RNA, the sugar ribose
replaces deoxyribose, and the base uracil (U) takes the place of thymine.
The letters of the genetic alphabet are A, T, G, and C. Phosphate molecules join the nucleotide sugars in chains, and nitrogenous bases project
off to the sides. In DNA, there are two chains, and certain nitrogen bases
pair (A with T and G with C) to form a double structure like a ladder,

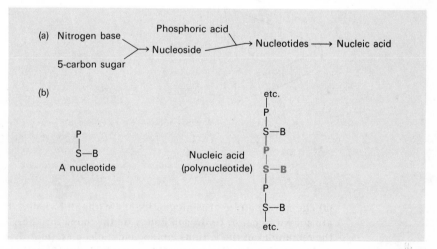

Fig. 1-4   **(a) The sequence of synthesis in nucleic acid construction. (b) Nucleotide and
nucleic acid diagrams; B represents nitrogen bases; S indicates 5-carbon sugar;
and P stands for phosphate.**

Fig. 1-5   **The double helix of DNA, represented three ways.** TOP: **the "twisted ladder," or spiral "staircase."** MIDDLE: **a more detailed representation: phosphate (P), sugar (S); A, T, G, and C stand for the four nitrogen bases: adenine, thymine, guanine, and cytosine. Sugar and phosphate make up the sides of the ladder, and hydrogen-bonded nitrogen bases make up the cross rungs.** BOTTOM: **a molecular model showing the atoms of carbon (C), oxygen (O), hydrogen (H), and phosphorus (P).**

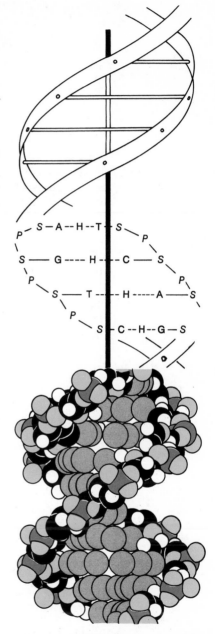

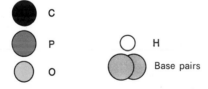

which is twisted to form a double helix (Fig. 1-5). It is the sequence of bases that codes for the various amino acids and the proteins they compose. In one sense, the earlier biologists were right: proteins mirror the genetic code. A sequence of three bases (a triplet) codes for each amino

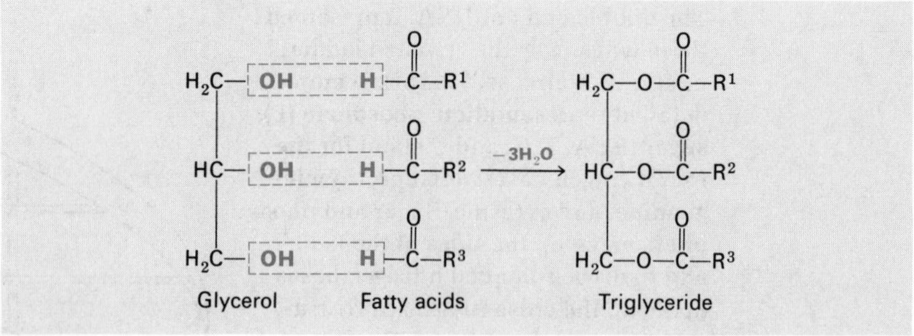

Fig. 1-6

The synthesis of a fat. Fats are composed of molecules of triglycerides, which are compounds of glycerol with three molecules of fatty acids. The R groups can be identical or different long carbon chains of fatty acids.

acid. With few exceptions, RNA seems to be an intermediary of DNA, a director of protein construction according to the instructions of DNA. DNA has been compared to a computer, programmed to direct the reactions of life. The proteins are the "executors" of the "will of life," DNA. In most cases, a form of RNA is the messenger of DNA.

Lipids are biochemicals "insoluble in water," the fats, oils, waxes, and

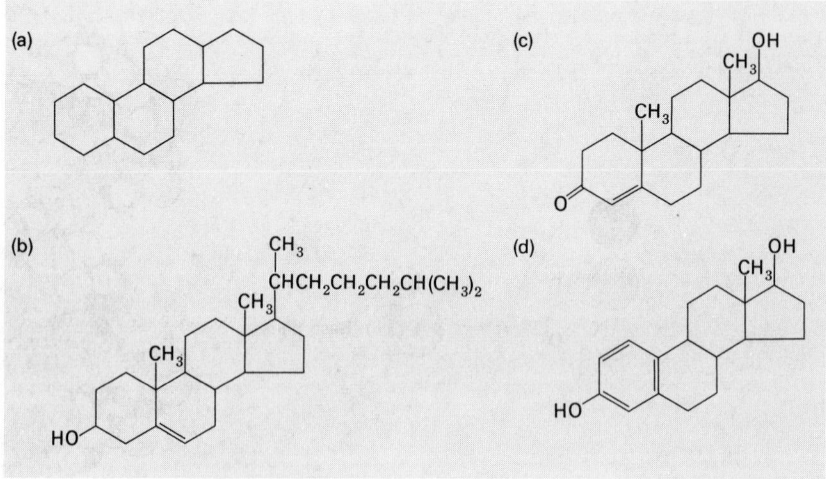

Fig. 1-7

Steroids. (a) Steroid nucleus. (b) Cholesterol. (c) Testosterone, the male sex hormone. (d) Estradiol, the female sex hormone. Note the tiny difference in the hormones of the opposite sexes. A change of a few atoms converts one to the other; yet, they are metabolic antagonists.

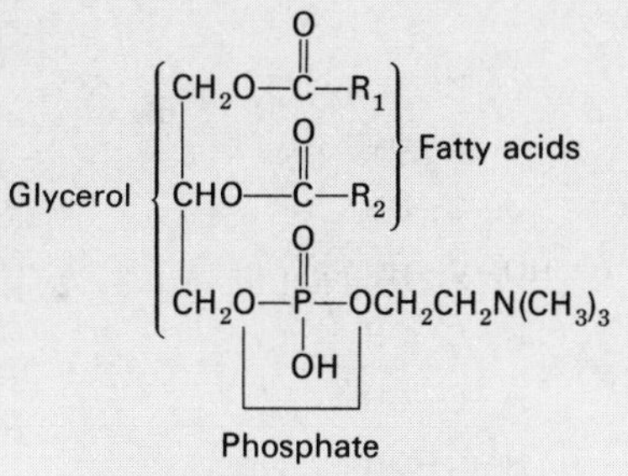

Fig. 1-8

A phospholipid. The lower end of the molecule is water soluble, whereas the upper end is oil soluble. Such molecules form films and are most important constituents of the filmlike membranes employed in the construction of the cell. Note also the presence of the element nitrogen (N) in this common variety of lipid.

sterols. *Fats* are solids at room temperature and primary "food" storage compounds. (There is no storage form of proteins.) Fats yield the highest energy per weight unit of any biochemical, when they are respired for energy purposes. They are composed of the alcohol glycerol combined with three molecules of fatty acid (Fig. 1-6). *Oils* are like fats, except that they are liquid at room temperature. Oils also contain less hydrogen per acid molecule, and their fatty acids are said to be "unsaturated." *Waxes* are composed of long carbon chain alcohols and fatty acids. The *sterols* are joined ring compounds whose only similarity to the fats and oils is their water insolubility. Cholesterol is an important lipid. Vitamin D and the sex hormones are sterols (Fig. 1-7). Bile salts are steroids that function as soaps or detergents, playing an important role in fat digestion.

The phospholipids are basically like fats and oils, but they have a phosphoric acid molecule substituted for one of the fatty acid molecules (Fig. 1-8). Phospholipids and certain sterols can form transition layers between water and oil interfaces. Phospholipids, with proteins, form the membranes that are so important in cell construction. Cholesterol is another important lipid in cell membranes. Thus, lipids as well as proteins are important structural biochemicals for life.

The word "carbohydrate" implies that these biochemicals are water compounds of carbon, and their ratios of $C:H:O$ are usually nearly $1:2:1$, respectively. Carbohydrates have become known as the "fuels of life," but fats and even proteins can be respired for energy purposes. Nevertheless, carbohydrates are the most readily available form of en-

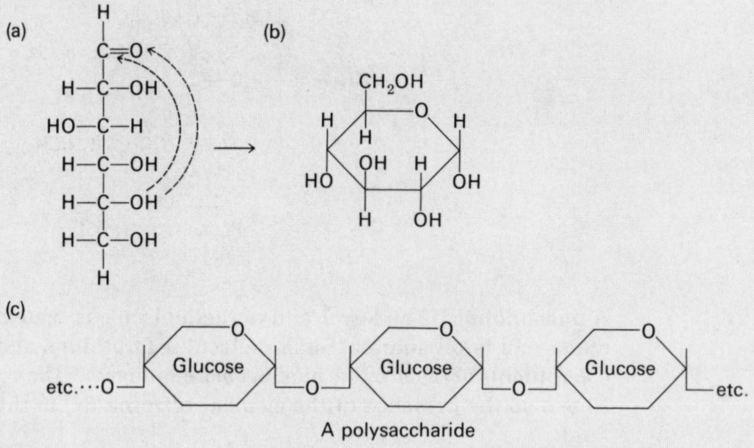

Fig. 1-9 **Carbohydrates. (a) Straight chain formula for glucose. (b) Ring formula for glucose. (c) Diagram of starch molecule.**

ergy. Simple sugars, the *monosaccharides,* such as glucose (Fig. 1-9), fructose, and galactose, are carbohydrates. The *polysaccharides,* starch and glycogen (animal starch), are glucose storage compounds. In their giant molecules, glucose units are strung together.

Some polysaccharides are not food storage forms at all. Cellulose is a structural polysaccharide forming the walls of plant cells. It is usually stated that the primary elements of carbohydrates are carbon, hydrogen, and oxygen, but nitrogen is found in many structural and functional polysaccharides as well. The outer skeleton of insects and crustaceans is composed of chitin, which is largely a compound of a nitrogenous polysaccharide, polyglucosamine. The interwall substance of plants, pectin, is a polysaccharide, and the major extracellular compounds of animals are the acid mucopolysaccharides (AMP's), which are nitrogen derivatives. The AMP's are not food storage forms. They bind tissue water and mineral ions. Nutrients as well as wastes must percolate through the AMP's to and from the tissue cells and the blood; thus, they are most important physiological (functional) compounds. Most AMP's are jellylike in consistency, but firm AMP's are found in cartilage (Fig. 1-10) and other supporting tissues.

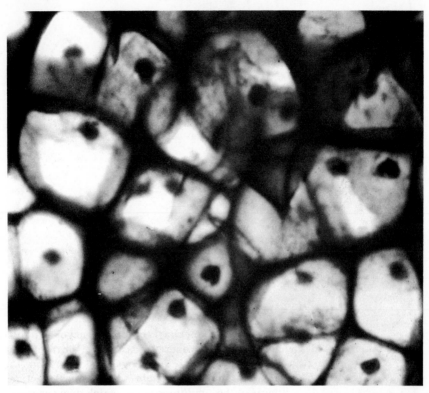

Fig. 1-10   **Animal carbohydrate extracellular substance (ECS). In this photomicrograph of cartilage, the cells occupy the white spaces and synthesize and secrete the acid mucopolysaccharides (AMP'S), which together with a few small protein fibers make up the darkly stained latticework surrounding the cells.**
From M. J. Zbar and M. D. Nicklanovich, *Cells and Tissues,* Saturn Scientific, Inc., Fort Lauderdale, Fla., 1971.

Most enzymes contain a nonprotein portion in addition to the protein component. The whole enzyme molecule, "holoenzyme," is composed of protein, the *apoenzyme,* and the nonproteinaceous *coenzyme.* Most vitamins serve as coenzymes. Although vitamin deficiency diseases frequently exhibit specific symptoms, the effects are usually more general or widespread than it seems, since most of the reactions of life (metabolism) are controlled by enzymes, whose apoenzymes are often functionless in the absence of vitamin coenzymes.

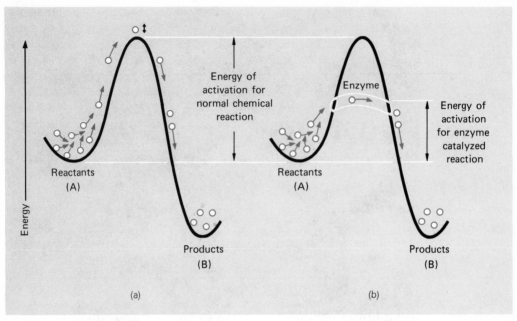

Fig. 1-11   **Energy diagrams of (a) uncatalyzed and (b) enzyme-catalyzed reactions. Enzymes somehow lower the energy of activation for reactions so that the rate is greatly accelerated. It is as if the enzyme makes a "tunnel in the energy barrier." This explains the ease with which biochemical reactions proceed, as well as the mildness of the conditions under which they occur.**

CONDITIONS OF
BIOCHEMICAL REACTIONS

Many inorganic reactions take place in one step, and occur only under violent conditions of great heat, pressure, and extremely alkaline or acidic conditions. The burning of glucose, for instance, requires the application of heat until the ignition point is reached; then the combustion proceeds until all the glucose is converted to CO_2 and H_2O.

In the biologic "burning" of glucose, no heat is applied, and the complete oxidation of glucose to carbon dioxide and water requires many steps, each of which involves small energy changes and is enzymatically controlled. *Catalysts* promote reactions by lowering the energy of activation required of the reactant molecules (Fig. 1-11). Although catalysts participate in the reactions, they emerge unchanged. Enzymes are catalysts, among the most efficient known. They are the agents that permit the reactions of metabolism to take place under mild conditions of temperature, pressure, and acidity.

DISCUSSION QUESTIONS

1. *Why is it incorrect to say that living organisms are composed of cells alone?*

2. *What are the uses to which the various biochemicals are put?*

3. *How do many otherwise difficult reactions proceed in the cell under very mild conditions?*

4. *What order do some of the biochemicals exhibit, and what is the consequence of this arrangement?*

5. *How does the living differ from the nonliving?*

REFERENCES

ASIMOV, ISAAC, *Chemicals of Life*, Abelard-Schuman Limited, New York, 1954.

AWAPARA, J., *Introduction to Biological Chemistry*, Prentice-Hall, Inc., Englewood Cliffs, N. J., 1968.

STEINER, R. F., and H. EDELHOCK, *Molecules and Life*, D. Van Nostrand Co., Inc., Princeton, N. J., 1965.

WALD, GEORGE, "Innovation in Biology," *Sci. Am.*, Nov., 1958.

OBJECTIVES

1. List and define three levels of organization between cells and organisms.

2. Enumerate four steps in the preparation of tissue for microscopic study.

3. Compare the construction, function, and resolving powers of the light and electron microscopes.

4. Contrast the cellular detail seen with light and electron microscopy.

5. Name ten cellular organelles, describe their structure, and give their functions.

6. Draw and label a diagram of the fine structure of a typical cell.

7. Contrast the fixed cell and the living cell as observed in tissue culture with phase microscopy and time lapse photography.

8. Diagram Danielli's model of the cell membrane.

9. List five ways things get into cells.

10. Contrast diffusion and active transport.

11. Specify those organelles highly developed in transport and protein secretory cells; correlate structure and function.

12. Write the summary equations for photosynthesis and respiration.

13. Contrast the end products and energy yields of anaerobic and aerobic respiration.

14. Summarize the major events of the light and dark reactions of photosynthesis; explain the relationship of the two phases.

15. Contrast procaryotic and eucaryotic cell structure.

16. Describe the bioblast theory of the origin of mitochondria and chloroplasts. Cite several lines of evidence tending to support the theory.

17. List four major stages in the evolution of higher cells.

18. Define the steady state.

A
Microcosmic
Steady State

THE SIMPLEST ORGANIZATION OF LIVING MATTER capable of independently carrying out the basic activities of life is the cell. There is no life without cells. All plants and animals and living things that are neither plant nor animal, sometimes in between, are composed of single cells or communities of cells and their products. One cell can be an *organism,* a living "thing." Cell seems like a poor name: too anatomical. It sounds like a "container of life." The first units called cells were actually empty rooms, the cell walls constructed by plant cells that had died. As far as is scientifically known, it is impossible to separate life from its machinery, the living cell. Just as an atom is the final subdivision of inorganic matter that exhibits an element's chemical identity, defined in terms of reactivity, so the cell is the last fraction of living matter that displays the concerted properties of life, reproduction, directed synthesis, energy transformation, and adaptive response. In the analysis of life it is impossible to separate form and function. This is the biological principle of complementarity. The primary form of life is the cell and life is basically cellular function! It has been said that when we completely understand the cell, we will understand life itself.

Life has written many variations on the theme of the single cell or some derivation of it. There is an incredible variety of microscopic unicellular organisms. This can only mean that this design was very successful and has persisted long after the first communities of cells, the multicellular organisms, began their long twisted marches to other ends, not that nature is ever "satisfied" with her creations. With the evolution of multicellularity, the independence of single cells was sacrificed for specialization, and, therefore, interdependence appeared. The individual cells of many-celled organisms are no longer capable of independent life. In multicellular plants and animals, the *level of organization* above the cell is the *tissue,* a collection of similar specialized cells and their products having a common function. *Organs* are composed of several tissues, and the complex animal is, in turn, organized into a number of interacting *systems* of organs. The cell consists of a number of subcellular organelles, the "little organs of the cell."

It took man more than a thousand years to learn his body's gross anatomy, visible to the unaided eye. In the last hundred years, particularly the last quarter-century, he has discovered the anatomy of the cell. Much of the knowledge of cell structure has come from the study of the cell corpse.

MICROSCOPES, KNIVES, AND STAINS

The earliest microscopists trained their powerful magnifying glasses upon the cut surfaces of fresh organs, such as the kidney, spleen, and liver, or thin translucent organs, such as tadpoles' tails and frogs' lungs. The greatest feat of the seventeenth-century microanatomists was the discovery of the capillaries, those microtubes that carry blood from the arteries to the veins. Animal tissues are soft and almost impossible to slice thinly. Fibrous tissues like muscle could be teased apart. Stiff plant tissues, however, lent themselves well to freehand thin sectioning with pocketknives. Sunlight reflected from the surface of objects under study was the major light source of the early microscopists. Only fresh tissues were used for investigation, and except for blood and muscle most tissues were colorless. Although the compound microscope (with a combination of lenses and greater power) was known in the seventeenth century, the advance of cell biology awaited the development of new techniques of tissue preservation, preparation, and staining.

Alcohol, weak acids, and metal salts had been used to preserve food and whole biological specimens for centuries before they came into general use as tissue fixatives in the nineteenth century. The microscopic appearance of tissue fixed by these agents differed little from that of fresh tissue, although fixed tissue was hardened and less perishable. Such fixed tissues were infiltrated with molten wax and cast in blocks. This technique of paraffin imbedding provided support for soft animal tissues. Thus, they could be thinly sliced or sectioned. Microtomes ("small cutters"), instruments for cutting extremely thin sections of tissue, were invented. They used mechanical devices such as screws, and later gears, to advance the tissue block gradually toward the knife's edge. Thin sections of tissue were mounted on slides and stained with dyes to increase the contrast within the relatively transparent cells. Compound microscopes were equipped with stages, and light was transmitted through the section rather than reflected from a sample surface.

Technological advance often precedes the progress of science and society. With the described tools, late nineteenth-century biologists furiously described most of histology, the kinds and arrangements of tissue cells in organs. Pathology, or the study of disease, moved to the tissue level. The greatest achievement in cell biology was the discovery of the events of cell division, or mitosis, but the foundations of cellular anatomy were also laid.

THE LIGHT MICROSCOPIST'S CELL

Although cell nuclei had been described early in the nineteenth century, staining showed the nucleus to be a constant feature of all cells. Figure 2-1 is a diagram of a typical animal cell as revealed by the light microscope. The cell had been a blob of jelly to early biologists, but the students of fixed and stained cells saw membranes around the cell as well as its nucleus. The region of the cell between the *nuclear membrane* and the *cell membrane* was called the *cytoplasm*. Specific stains were developed, which revealed oil droplets, starch granules, and protein *secretory granules* in the cytoplasm of various cells. Rod-shaped and spherical bodies like bacteria were also detected in the cytoplasm. The function of these *mitochondria* ("thread granules") remained unknown. Colored bodies, the chromoplastids, particularly the *chloroplast,* were obvious even in fresh unstained plant cell cytoplasm. Bubblelike clear areas of the cytoplasm, *vacuoles,* were found to be highly developed in plant cells. In contrast to the proteinaceous cytoplasmic fluid, which usually stained, vacuoles apparently contained a watery nonprotein fluid, since they remained unstained.

Near the nucleus of animal cells, two small particles, the *centrioles,* were located. It was observed that fibrous threads radiated from them in the animal cell mitotic apparatus, and they formed the poles of the dividing cell. Within the nucleus were one or two dark staining spheres,

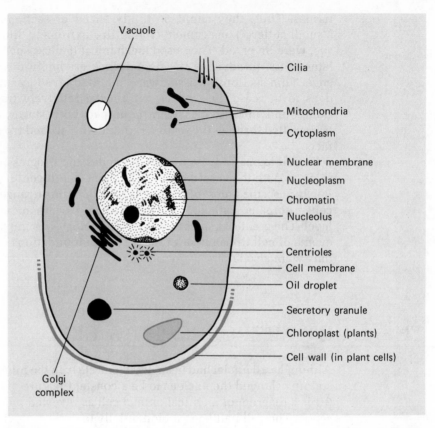

Vacuole

Cilia

Mitochondria

Cytoplasm

Nuclear membrane

Nucleoplasm

Chromatin

Nucleolus

Centrioles

Cell membrane

Oil droplet

Secretory granule

Chloroplast (plants)

Cell wall (in plant cells)

Golgi complex

Fig. 2-1   **Drawing of a typical cell as revealed by the light microscope. This was nearly all the structure that could be made out until about 1945.**

the *nucleoli,* which disappeared during cell division. Other nuclear contents included clumps or strands of staining material, the *chromatin,* which was packaged into the staining bodies of the *chromosomes,* which appeared during division, doubled, and separated only to be dispersed as chromatin in the nuclei of daughter cells (see Fig. 2-2).

The enclosed nucleus, in the heart of the cell, logically seemed to be the control center. When it was partitioned into a halved cell, the half with it survived. The portion of the cytoplasm without it died. In the first decade of the twentieth century, it was noted that the behavior of the chromosomes during meiosis (sex cell divisions by means of which the chromosome number is halved) paralleled the segregation of the paired genetic factors, which separated in sex cell formation. So it was proposed that the hereditary particles, the genes, were carried on the chro-

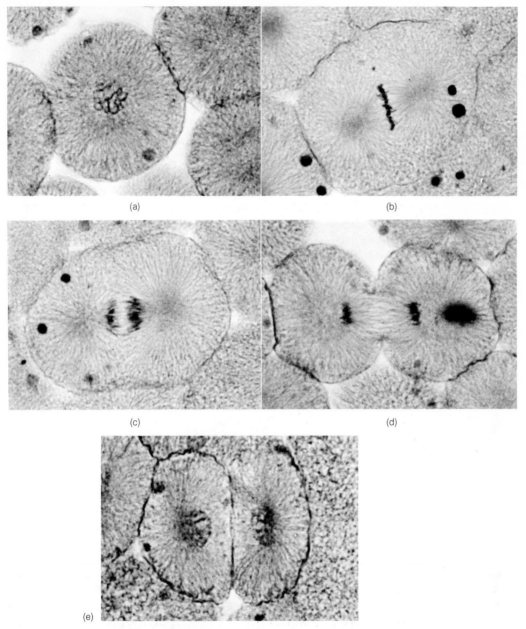

Fig. 2-2   **Stages of cell division: (a) prophase, (b) metaphase, (c) anaphase, (d) telophase, and (e) daughter cells. The chromosomes appear in prophase, line up along the equator in metaphase, and migrate to the poles in anaphase. In telophase, the cytoplasm is divided, and the nuclei are reformed. Note the spindle-shaped mitotic apparatus in (b).**

From M. J. Zbar and M. D. Nicklanovich, *Biology of Reproduction,* Saturn Scientific, Inc., Fort Lauderdale, Fla., 1971.

mosomes and thus resided in the nucleus. The *nucleoplasm,* the fluid of the nucleus, remained as much of a mystery in terms of structure and function as the sea of cytoplasm.

The term "protoplasm" ("first fluid") was given to the substance of the cell, "the stuff of life," early in the nineteenth century. As cell structure was more clearly delineated in the late 1900's and distinct particles were described, the vagueness of the word "protoplasm" was transferred to the unknown cytoplasm. Protoplasm had meant at first the whole cell; protoplast still refers to the plant cell within the wall. As details of cellular anatomy were described, particularly the structures of the nucleus during cell division, protoplasm came to mean more and more the complex fluid jelly of the cell without visible structure, the cytoplasm. The nucleoplasm was neglected in view of the strikingly distinct chromosomes of division. Protoplasm died a slow death, and even then we were hesitant to bury it. The word served a purpose in its time to focus attention upon the material of the cell. Other words have died, but the demise of this one, which was so religiously taught even in the 1940's, seemed to sound the death knell over the mystical wish to find an unanalyzable substance of life, really the desire to separate form and function, the anatomical cell and life. The subtleties of the cytoplasm awaited the coming of the electron microscope and the birth of biochemistry.

A LIVE MICROANATOMY

Not all studies of the light microscopists were limited to dead cells, as is the case in electron microscopy. Early in this century biologists began to culture tissues and cells in nutrient fluids in glass chambers outside the animal or plant body. The shortcomings of the direct light microscope in viewing nearly transparent, unstained cells were largely overcome through the use of the phase microscope, which increased the contrast between the various portions of the cell, making many structures visible without staining (see Fig. 2-3). In the twenties and thirties, time lapse motion pictures were made of cells in *tissue culture,* and even the scientific world was astonished by the activities of the living cell so recorded. The community of biological scholars was accustomed to the frozen postures of fixed and stained cells. In fact, the closer actual preparations came to the idealized painted forms, the more pleased their makers were. It was as though the world had forgotten that there was another dimension, time, and that the primary quality of life from moment to moment and generation to generation was change. Life is dy-

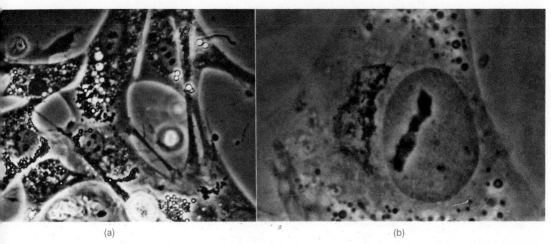

(a)   (b)

Fig. 2-3   **Connective tissue cells from tissue culture. (a) Living, unstained cells. Nuclei**
are the central, grey structures with interior black dots (nucleoli). The com-
pacted, black networks are Golgi complexes, and the white spots are bubbles
of protein solution taken in from the surrounding culture medium. (b) A dead
cell, fixed and stained, showing the same structures.
Courtesy of Dr. Alexis L. Burton.

namic; preserved death is static. Before we proceed to the submicroscopic (beyond the powers of the light microscope) anatomy of the
cell as revealed by the electron microscope, let us pause to consider for
a moment what can be seen of living cellular anatomy.

With time lapse cinematography, it is possible to see in seconds what
occurs in hours. Thus, what is seen is time distorted, but it is the best we
can do to see that which the fourth dimension, time, would have otherwise hid from our impatient eyes. Man has always cheated a bit. If we
had depended upon our naked eyes or our limited sense of time (although the greatest of all animals), we would be ignorant indeed. Even
in tissue culture, the visible changes of the cell are almost imperceptible
to an exhaustingly patient man, excepting only those naturally crucial
and quick events. Still these are slow, requiring as much as several continuous observational hours by the observer.

The cell membrane flows back and forth like a rapid tide, the seeming
tide of life. The cytoplasm flows out into microfingers of the membrane
at the same time that other areas of it inpocket and pinch off as miniature
bubbles in the cytoplasm. The "slime animalcule," the amoeba, and the
phagocytes ("eating cells") of multicellular animals extend projections
of cytoplasm and membrane (*pseudopodia,* "false feet") on the sides of

Fig. 2-4

Diagram of phagocytosis ("cell eating"). The microorganism becomes enclosed in a food vacuole, whose walls are made up of the old cell membrane, where the latter infolds and pinches off between the advancing cytoplasmic arms, the pseudopodia ("false feet").

particles, such as bacteria (see Fig. 2-4). The projections approach each other, and the inner membranes of each fuse to form a *food vacuole,* enclosing the bacterium in a fluid-filled, membrane-bound bubble. This process of engulfment is called *phagocytosis* ("cell eating"). Most cells in tissue culture, bathed in protein solution, form minute bubbles or *vesicles* in the clefts of smaller projections. This process, first observed in tissue culture, was called *pinocytosis,* or "cell drinking," since fluid rather than particles was ingested (review Fig. 2-3). Phagocytosis and pinocytosis are essentially similar. A vesicle is a very small vacuole. The surface of most cells in tissue culture froths and bubbles, seethes in time lapse cinematography. It is a far cry from the still cell membrane of the "fixed" cell. Mitochondria move about like worms in the cytoplasm and sometimes pinch in half. Other particles jitter to and fro with Brownian movement, due to the unequal bombardment of water molecules on opposite surfaces, but they do not exhibit the prolonged directional movements of the mitochondria.

There is nothing chaotic or haphazard about the precise events of cell division in tissue culture. The nuclear membrane dissolves, and the nucleoli disappear as the murky chromatin organizes into distinct doubled chromosomes. One of each moves to the dividing cell's poles, formed by the two centrioles, which play out seeming anchor lines to the chromosomes. When the chromosomes reach their polar destinations, they disperse into the nearly formless chromatin again, as the nuclear membrane reforms and the nucleoli reappear. The cytoplasm of the mother cell either pinches in half or is divided by newly formed membranes. Each centriole duplicates, and the cell returns to the resting stage.

NEW EYES AND INVISIBLE "LIGHT"

The maximal magnification obtainable with the light microscope does not exceed 2000. More importantly, it is impossible to see objects smaller than 0.17 micron (μ) (1 μ = $\frac{1}{1000}$ mm). This limitation upon what can be seen with an optical instrument is called the *resolving power*. If a particle or a line is thinner than 0.17 μ, it cannot be visualized with the light microscope. If two points or two lines are closer than 0.17 μ, they will appear as one point or one line under the light microscope. If the points or lines are separated by 0.17 μ or more, they can be seen as two.

The resolving power of an optical instrument is inversely related to the wavelength of the illumination—or, the shorter the wavelength, the greater the resolving power. The light microscope is limited in its resolving ability by the wavelengths of visible light (0.4 to 0.7 μ). The shortest wavelength of visible light (violet) is about 0.4 μ, much longer than invisible ultraviolet, X rays, and electron waves. Electron beams can be produced with a wavelength of 0.000005 μ, which is 1/100,000 of the average wavelength of visible light (0.5 μ). Since the resolving power is greater with shorter wavelengths, an electron microscope would theoretically have a resolving ability 100,000 times as great as the light microscope. It would be possible to see things invisible to the light microscope, 100,000 times smaller than the smallest objects visible with the light microscope. In reality, because of mechanical factors we shall not consider here, the practical resolving power of the electron microscope is only about 0.001 μ; nevertheless, it is approximately 200 times greater than the ability of the light microscope (0.17 μ). A more useful unit of measurement in this range is the angstrom (Å). One micron equals 10,000 Å, or 1 Å equals 1/10,000 μ (see Fig. 2-5). The resolving power of the electron microscope (0.001 μ) is thus about 10 Å. This figure approaches the size of atomic structure. Large molecules have been seen with the electron microscope.

It is possible to take a light micrograph (photomicrograph) and enlarge the picture 100 times, but the additional magnification will be nearly meaningless, an "empty magnification," because no more detail can be made out than the resolving power of the instrument permits. Although magnifications of 100,000 can be made with the electron microscope and these enlarged 10 times, no more detail will be revealed than the resolution allows.

The lenses of the electron microscope that focus the electron rays are electromagnetic coils. The image is focused upon a fluorescent screen or photographic plate. The source of "light" is an electron gun, and since electron beams can travel only through a vacuum, the tube of the scope

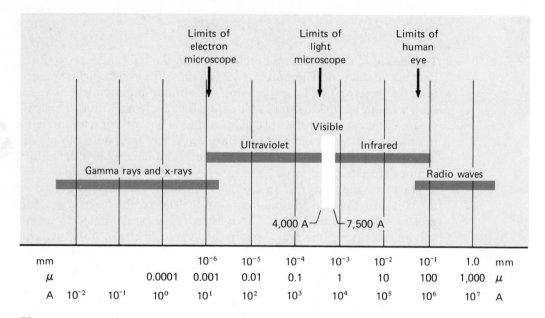

Fig. 2-5   The electromagnetic spectrum, including visible light, ultraviolet, infrared, radio, and X-ray waves. Their wavelengths are given, and the limits of resolution, (the smallest things that can be seen) of the various optical instruments are shown above in comparison. One millimeter equals 1000 microns (μ) and 1 micron (μ) equals 10,000 angstroms (A). The resolving power of the light microscope is 1000 times greater than the naked eye, and the resolving power of the electron microscope is several hundred times greater than that of the light microscope.

is evacuated. Air molecules would deflect electron rays. Compare and contrast the human eye, the light and electron microscope (Fig. 2-6).

The section of tissue must be extremely thin to allow the electrons to pass through. The development of a technique of very thin sectioning was a major obstacle to the application of the electron microscope to tissue anatomy. Thinner sections than could be obtained with the steel microtomes of light microscopy were required. Sections of 0.06 to 0.5 μ were needed. A pioneer of electron microscopy, the Swede Sjostrand, took the thinnest sections that could be produced on a high-speed microtome (about 1 μ thick) and reimbedded them and cut again, hoping to cleave an even thinner section by chance. Finally, it was discovered that the broken edges of glass cut thinner slices than the finest honed steel blade. Wax was too soft, crumbling at these thicknesses. Soft plastics came into use for imbedding. Then the hardest of all plastic glues, the epoxy resins, were used to support the tissues, and they dulled the

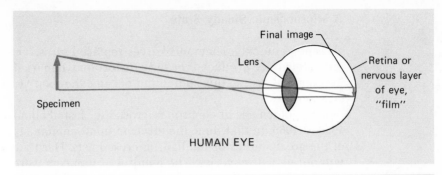

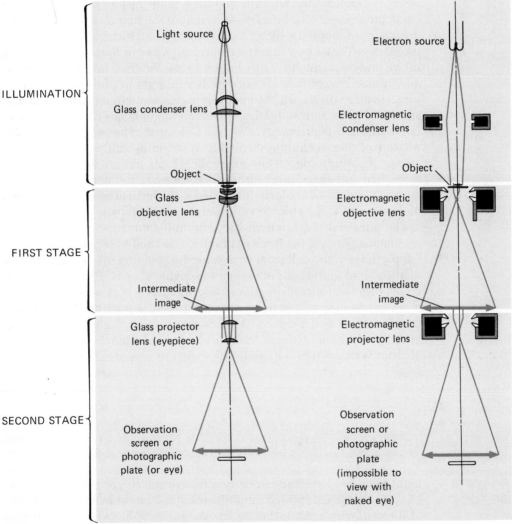

Fig. 2-6   The optical systems of the eye, the light microscope, and the electron microscope. The electron beam is focused by varying the voltage of the current flowing through the electromagnetic coils, which function as "lenses" here.

glass edge quickly. Diamond knives replaced glass. The vapors of osmium tetroxide produced the least distortion when used to fix live cells in tissue culture, so this compound became the fixative of electron microscopists.

The "golden age of electron microscopy" lasted about 15 years, from 1950 to 1965. In that time, the electron microanatomists repeated most of the great studies of the light microscopists. Hardly a month passed without new discoveries. The handful of pioneers were bound upon a voyage of exploration with few parallels in the history of science. Their excitement was every bit as great as that of the first microscopists, who turned lenses upon the living world and found structures unseen by the naked eye. They knew that the resolving power of their instrument lay in the molecular realm. They hoped to see the final machinery of life, and in many ways they succeeded. It was a grand philosophic adventure. History will record their achievements, but will it record their passion? Today's electron microscopists rely upon teams of technicians to maintain their instruments. The early electron microscopists were the masters of their techniques, capable of stripping and cleaning their instruments blindfolded. They knew their tools like soldiers know their rifles. The advances in electron microscopy correlated with the concurrent revelations in biochemistry led to the birth of a movement called molecular biology, which revolutionized all of biology.

The ultracentrifuge, a high-speed spinning machine capable of generating many times the force of gravity, was used to separate the different particles of the cell from each other by spinning the debris from the disruption of tissue cells at higher and higher speeds. This technique is called *differential centrifugation,* and by means of it cellular particles were isolated in various fractions, and it became possible to assign certain functions and reactions to different cellular organelles. Electron micrographs of centrifugal fractions were made; biochemical analyses of them were carried out, and thus structure was identified with function.

THE BOUNDARY OF LIFE

Light microscopists had performed microsurgery upon the living cell. Microneedles encountered a definite resistance at the external boundary of the cell, which, nevertheless, broke under slight pressure. Although they could see no definite structure at the cell's edge, they hypothesized that a porous membrane existed there, separating the extracellular and intracellular fluid (cytoplasm), whose compositions differed sharply. Analyses of red blood cell membranes revealed a high concentration of lipids and proteins, and it was discovered that molecules such as oils with a high solubility in fat solvents passed through the cell membrane with the greatest of ease. The amount of lipid present in the cell membrane was twice that necessary to cover the surface area of a red blood cell with a layer one molecule thick. So it was proposed that there was a double layer of lipid molecules in the cell membrane.

The lipid molecules found in membrane analyses differed from ordinary, neutral fat or oil molecules in that they were charged phospholipids. Phospholipid molecules are partially soluble in water, whereas purely lipid molecules are not soluble in water in the least. Like detergents and soaps, phospholipids can dissolve in both water and oil to form films such as soap bubbles, where water and air meet. In 1941, J. F. Danielli of King's College, London, predicted on the basis of molecular size that the cell membrane would be 80 Å thick, two layers of protein (on the outside and inside) sandwiching a double layer of phospholipid between. He also theorized that there would be purely protein-lined holes or pores in the membrane to allow the passage of water and water soluble molecules, which entered the cell easily (see Fig. 2-7).

Electron microscopic studies revealed a three-layered membrane, 75 Å thick, very close to Danielli's prediction. In electron micrographs, the outer and inner layers appeared dark, indicating the obstruction of electron passage, and the intermediate area was light, allowing electron rays to pass as a lipid layer would (see Fig. 2-8). This was the visualization of several of the predictions of the theory. No pores were seen, however. Perhaps they appeared and disappeared intermittently in the living membrane. This theory remains the best available explanation of membrane properties and structure.

The cell membrane seemed to have many of the characteristics of a dead, artificial membrane, such as cellophane, with a limited pore size, allowing some molecules to pass and preventing the passage of others. Such a membrane is said to be *semipermeable*. The molecules of a gas or liquid (a major state for the reactions of the matter of life) pass from a high concentration to a low concentration, until they are evenly spread. The constant, random, molecular motion and collision favors the equally random distribution of particles. A concentration is like a mountain that wears away rapidly to form a flatland of low-lying particles. This movement of gaseous and liquid particles from a high to a low concentration is called *diffusion,* and the incline from the high to the low is called the *diffusion* or *concentration gradient.*

Water moves across the cell membrane, into the cell, along a diffusion gradient. As long as there is more water outside the cell than inside,

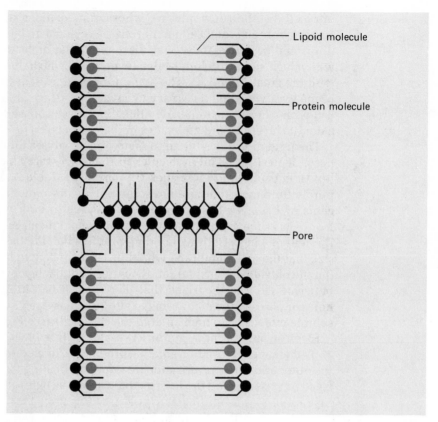

Fig. 2-7

Theoretical molecular model of the cell membrane. The spheres represent water soluble portions of the molecules, and the lines represent the water insoluble or oil soluble portions of the molecules.
From J. F. Danielli, *Colston Papers,* 7:1, 1954.

water molecules move into the cell. When there is more water in the cell than outside it, water molecules leave the cell. Water molecules move back and forth across the membrane at equal rates, when the concentrations on the inside and the outside are the same. The diffusion of water across a semipermeable membrane is called *osmosis* (Fig. 2-9).

When red blood cells are placed in pure water, they burst, since the net entry of water molecules swells the cell and strains the cell membrane to the point of rupture. Plant cells will likewise take up water when placed in distilled water, but they do not burst, for the cell wall restrains the cell membrane. If a plant cell is immersed in salt water, the cell will shrink away from the wall, due to the exit of water from the cell

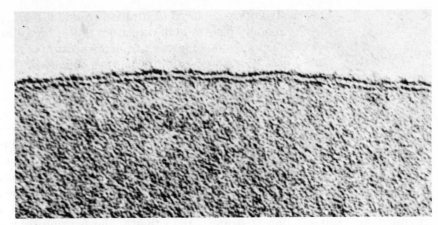

Fig. 2-8   The cell membrane under the electron microscope. It is of the size predicted in Danielli's hypothesis, and its structure also seems to agree with the preceding model. The two dark lines probably represent the layers of protein that sandwich a light, intermediate lipid layer between them.

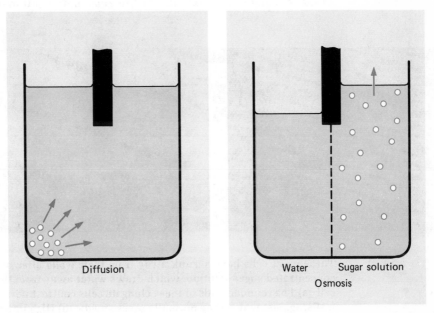

Fig. 2-9   **Diffusion and osmosis.**

following the diffusion or concentration gradient (Fig. 2-10). Molecules of water actually pass in both directions, but more so in one than the other. The net movement is away from the greatest concentration.

When comparing the osmotic properties of two salt solutions, the one with the most salt and least water is said to be *hyperosmotic* to the one with least salt and most water. A concentrated sugar solution will take water from a more dilute solution and therefore rise in a tube. The rising fluid has an *osmotic pressure* (Fig. 2-11). The osmotic pressure of a strong salt solution is greater than that of a weak one. The solution with the least salt, most water, and lowest osmotic pressure is said to be *hypo-osmotic* to a stronger salt solution. If a salt solution bathing a cell is stronger than the intracellular fluid (ICF), it is said to be *hypertonic* to the cell. A dilute cell bath is said to be *hypotonic* to the cell if the ICF salt solution is stronger than that of the bathing solution. If the ex-tracellular fluid is the same strength as the ICF, it is said to be *isotonic* with the cell. Two solutions with the same osmotic pressures are said to be *isosmotic*.

Many of the properties of the cell membrane can be visualized by representing it as a semipermeable membrane with a limited pore size allowing the free diffusion of molecules smaller than the hole diameter. Large molecules do not diffuse across the cell membrane, yet many small molecules and ions (charged atoms) do not seem to cross the cell membrane to a significant extent. For this reason, the cell membrane is said

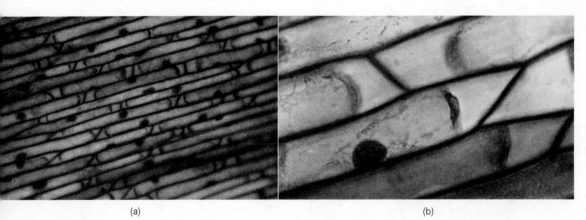

(a)   (b)

Fig. 2-10   **These plant cells have shrunk away from their walls after being bathed in a concentrated sugar solution, which draws water away from the interior of the cell. (a) The rounded ends of these elongate cells contrast with the straight end walls, against which they normally press snugly. (b) High power.**
Courtesy of Saturn Scientific, Inc., Fort Lauderdale, Florida.

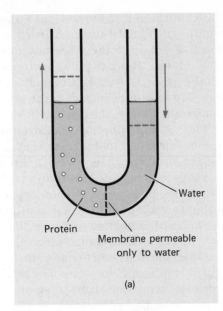 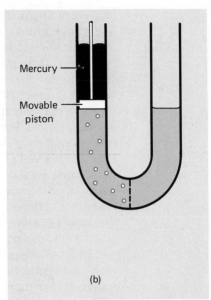

Fig. 2-11   **If a protein solution and water are separated from each other by a semipermeable membrane whose pore size does not allow the passage of protein molecules, more water will enter than leaves the protein solution per unit of time.
As a result, the liquid in the arm of the U-tube that contains the protein solution will rise, and the level of water in the other arm will drop (a). The protein
solution is said to have an *osmotic pressure*, which can be measured as in (b).**

to be *selectively permeable* rather than semipermeable, active rather
than passive. It is almost as though the membrane of the living cell decides what shall pass, and the physical membrane that we have studied
so far does not account for the total transport properties of it. An
alcohol-fixed or boiled cell's membrane does act completely as a passive, semipermeable membrane, however. The protein molecules are
irreversibly altered by these treatments. Perhaps the characteristics of
the live membrane depend upon the activities of intact protein molecules.

Most living cells contain high concentrations of potassium ions and
extremely low numbers of sodium ions, whereas the extracellular fluid
is rich in sodium and poor in potassium. The living cell maintains these
concentration differences. Is the cell defying the physicochemical laws
of diffusion, which argue for equal distribution? Because the concentration of sodium ions is higher outside the cell than in, sodium should
move into the cell until the concentrations on both sides of the mem-

brane are equal. Potassium should likewise diffuse outward from its high concentration within the cell until an equal distribution is reached on either side of the cell membrane. Yet the cell seems to exclude sodium and concentrate potassium ions. There is little difference in the size or chemical properties of the two ions. Radioactive sodium does enter the cell, but it does not build up an appreciable concentration there. These facts can only mean that sodium enters the cell, but is removed, and that potassium does leave the cell, but is brought back within at a greater rate. To maintain the concentration difference, the cell must pump out sodium ions that diffuse inward. To accomplish this directed transport, work must be done, energy expended. Respiration is the process by which the cell provides energy to do work. Respiratory poisons destroy the power of the cell to concentrate or dilute potassium and sodium ions. This sort of *osmotic work* is called *active transport,* in contrast to the passive "transport" of diffusion. In active transport, substances are moved from a region of low concentration to one of high concentration, with the expenditure of energy. In diffusion or passive "transport," molecules or ions move of their own accord from a region of high concentration to one of low concentration, without the expenditure of energy. Active transport and diffusion are exact opposites.

The typical electron microscopic anatomy of a *transport cell* reveals numerous, fingerlike convolutions of the cell surface, the *microvilli,* which presumably increase the surface area for the purpose of absorption, and abundant mitochondria in the cytoplasm (see Fig. 2-12). We shall see that the mitochondria are the major energy suppliers of the cell. Their high degree of development in transport cells is a reflection of the energy requirements of active transport, a major category of cell work. The fine structural features discussed are exhibited by the intestinal epithelial cells, having the primary task of absorbing digested nutrients. Certain cells in the kidney tubules are also engaged in absorption, and their anatomy shows an advance over that of the intestinal epithelial cells. The kidney transport cells possess basal infoldings as well as microvilli. The interrelationship of the structure of these cells and their function is an excellent example of the principle of *complementarity,* the inseparability of form and function.

It has been theorized that an energy-requiring *sodium pump* consisting of *carrier enzymes* transports the indiffusing sodium ions out of the cell. Since enzymes are largely protein in nature, this idea would explain why the alcohol or heat-treated membranes (with altered proteins) no longer actively transport. The differential distribution of metallic ions is essential to the functions of muscle and nerve.

When a portion of a cell membrane pinches off to form a pinocytotic vesicle or phagocytotic vacuole, the contents have not really entered the

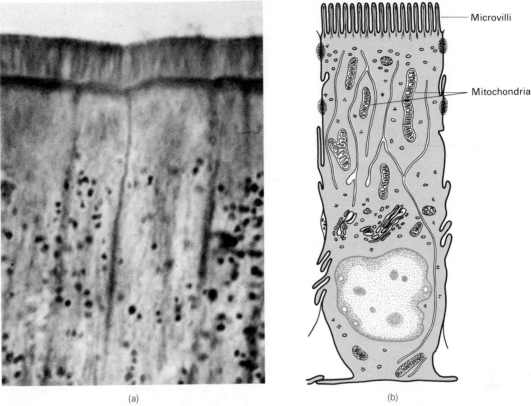

(a)   (b)

Fig. 2-12   **Transport cell: intestinal cell with tiny fingers of microvilli increasing the surface area and numerous mitochondria ("powerhouses of the cell") to provide
energy to do the cellular work of active transport. This cell is specialized for
absorption. Its structure reflects its function and illustrates well the principle
of complementarity: the wedding of form and function. (a) Roundworm intestinal cells showing microvilli above and granular mitochondria below, as
seen under the light microscope. (b) Drawing of the electron microscopic anatomy of a mouse intestinal cell.**
Photo from M. J. Zbar and M. D. Nicklanovich, *Cells and Tissues*, Saturn Scientific, Inc.,
Fort Lauderdale, Fla., 1971. Drawing from H. Zetterquist, *Thesis*, University of Stockholm,
Stockholm, 1956.

cell, for they still face a portion of the old cell membrane, which must
be traversed before they can enter the cytoplasm. A protein molecule
or a bacterium must first be digested before the products can cross the
cell membrane. The amino acids resulting from protein digestion are all
actively transported across the membrane into the cell cytoplasm.

Glucose, common "blood sugar," almost everywhere in the plant and animal world, cannot simply diffuse across most tissue cell membranes in higher animals. The molecule is large. The hormone insulin seems somehow to regulate the permeability of the cell membrane to glucose. Glucose absorption does not require the spending of respiratory energy. Insulin is said to "facilitate" the diffusion of glucose. There may be another type of transport as yet undescribed.

There are scientists who dispute the very idea of a membrane. After all, the structure of the cell membrane observed under the electron microscope is the anatomy of a fixed or dead cell. We are almost at a loss for words to describe the nature of the membrane of the living cell. It has been called a "liquid crystal." (Recall our discussion of the behavior of the cell membrane in tissue culture.) It seems to have definite structure (like a solid crystal) at the same time that it has the fluidity of a liquid.

Man tries to visualize that which he does not know in terms of that which he has seen. In the process, he usually evolves new ideas. So it has been with the cell membrane. We know more than we knew, but many questions remain. The quest of science has too often been to explain what we did not know in terms of what we did, a technique in itself illogical. Fortunately, somehow, we have gropingly discovered the new— imperfectly to be sure, but better than not having known.

The cell membrane is the boundary between the living and the dead. Beyond it lies the sea of life, yet the membrane itself is not dead, or passive. Nature has experimented with a "new" form of the organization of matter at the cell surface. Nevertheless, the laws of physics are obeyed. To "disobey" these laws, work must be done. To concentrate or dilute the naturally tending random, uniform distribution of molecules, energy must be expended. This expenditure of energy is a prime property of life, which is said to be "wasteful of energy." It seems a small price to pay for life in the natural world and the fifth dimension.

THE MEMBRANE SYSTEM
OF THE CELL

It has been noted that several particles floated in the light microscopists' cytoplasm: mitochondria, secretory granules, oil droplets, starch grains, etc. The first great triumph of electron microscopy was the revelation that the structureless cytoplasm of the light microscopist indeed contained structure, a "cytoskeleton," a cell framework of membranous channels coursing throughout the cytoplasm. In 1945, before the development of much electron microscopic technique, Keith Porter of the Rockefeller Institute discovered this cytoplasmic membrane system in flattened mouse cells that had been grown in tissue culture and then examined live under the phase contrast microscope. In similar but fixed cells, the electron microscope revealed the same definite structure. He called this system of cytoplasmic tubes an *endoplasmic reticulum* (er), since the tubules were best developed in the internal cytoplasm and constituted an interlacing network in three dimensions. In the first 10 years after its discovery, continuities between the cell membrane and the endoplasmic reticulum were religiously sought. A few were found, but far fewer than would be necessary to assume that the endoplasmic reticulum was a definite frequent avenue of communication between the interior and exterior of the cell. So the idea of the cell as a Venetian city with a network of canals to the open sea of extracellular fluid remains seriously open to question. Perhaps, like the pores of the external membrane, the openings of the canal system to the exterior appear and disappear in the living cell with some sort of rhythm. The structure of the endoplasmic reticular membranes is quite similar to that of the cell membrane, a fact that argues for their identity. It may be that the process of fixation destroys their connection. These are speculations. Meanwhile, more definite functions have been assigned to this intracellular membrane system.

In cells specialized for extensive protein synthesis for extracellular purposes (secretory cells with secretory granules), a highly developed endoplasmic reticulum studded with minute particles is the predominant feature of the cytoplasm (see Fig. 2-13). Such *rough endoplasmic reticulum* (rer) is a constant anatomical characteristic of the cytoplasm of these cells. Synthesized protein first appears as granules within dilations of the *rer,* before it is packaged into smooth membrane-bound secretory granules. The major function of the pancreas is the synthesis of extracellular digestive enzymes, which are protein in nature, like all enzymes. The pancreatic secretory cells and many salivary gland cells are loaded with rough endoplasmic reticulum.

Light microscopists had noted that the cytoplasm of secretory cells was more acid than the average and therefore stained intensely with basic dyes. When he noted that RNA (ribonucleic acid) was also high in the cytoplasm of such cells, the Belgian Jean Brachet hypothesized that RNA was pivotal to protein synthesis as early as 1940. George Palade of the Rockefeller Institute observed in electron micrographs that the microsomal fraction (left after the nuclei and mitochondrial fractions had been spun down from cellular debris) contained minute particles and

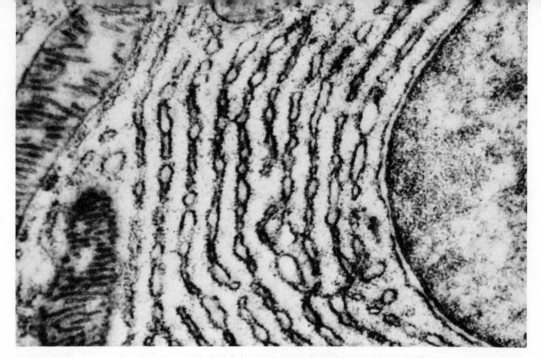

Fig. 2-13   **Pancreatic secretory cell. This cell manufactures and exports digestive enzymes, which are protein in composition. A nucleus lies to the right and mitochondria are seen to the left. The cytoplasm is loaded with concentric layers of rough endoplasmic reticulum (rer), the cellular organelle that is highly developed in cells specialized for protein synthesis and secretion. The tiny granules of the ribosomes, which stud the surface of roughened endoplasmic reticulum, are quite visible. The ribosomes are the tools of protein synthesis.**
Courtesy of Dr. George E. Palade.

fragments of torn er bearing such particles. He broke down the membranes of er with a detergent and obtained a purely particulate fraction, whose analysis showed RNA and protein. These ribonucleoprotein particles came to be known as *ribosomes* (see Fig. 2-14). They were shown to be the cytoplasmic sites of protein synthesis, since it is only in the microsomal fraction that significant quantities of radioactive amino acids are incorporated into protein structure. Inasmuch as the cytoplasm of protein secretory cells had been seen to be loaded with layer upon layer of er speckled with ribosomes, it was only logical to conclude that the function of the rer was protein synthesis. The plasma cell, which manufactures the protein antibodies of gamma globulin, is filled with rer (see Fig. 2-15). There are no secretory granules, however. It may be that the openings of the canal system to the surface release protein here. The cytoplasm of developing red blood cells is loaded with aggregates of ribosomes, but lacks er. This cell is specialized for extensive intracellular protein (hemoglobin) synthesis. Studies of these cells have indicated that ribosomes function in groups, as polyribosomes or polysomes. No

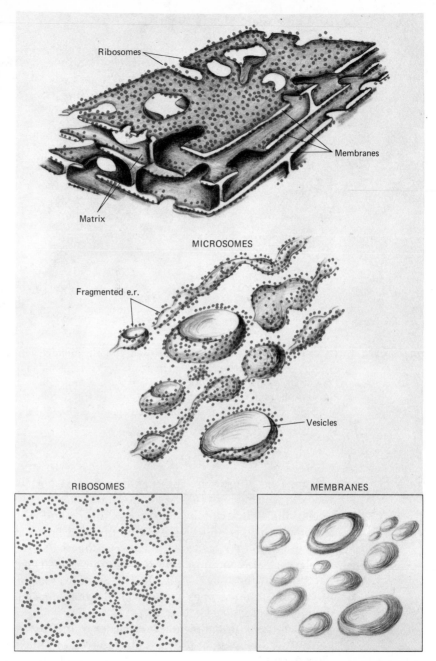

Ribosomes

Membranes

Matrix

MICROSOMES

Fragmented e.r.

Vesicles

RIBOSOMES

MEMBRANES

Fig. 2-14   TOP: **The three-dimensional arrangement of rough endoplasmic reticulum (rer)
with windows and cross connections between layers of membranes.** MIDDLE:
The fragmentation of the rer to form the microsomes. BOTTOM: **isolated ribosomes and membranes.**

From E. D. P. DeRobertis, W. W. Nowinski, and F. A. Saez, *Cell Biology*, W. B. Saunders
Co., Philadelphia, 1965.

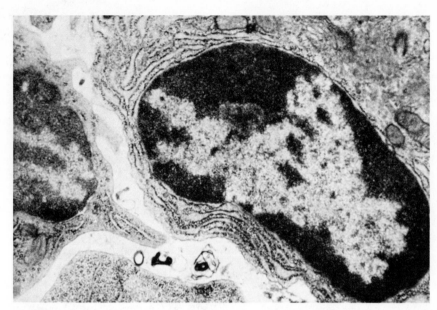

Fig. 2-15   **A protein synthetic and secretory cell, the plasma cell. This cell is specialized
                 for the manufacture and export of protein, as is reflected in the highly developed
                 rer in its cytoplasm. The plasma cell makes and releases the proteinaceous
                 antibodies, the immune chemicals that are the body's third line of defense.**
                 By permission from K. R. Porter and M. A. Bonneville, *Fine Structure of Cells and Tissues,*
                 Lea & Febiger, Philadelphia, 1965.

cell engaged in protein synthesis for secretion has ever been observed
that did not have a well-developed *rer.* The electron microscope revealed that the *nucleolus,* the dark staining body within the nucleus,
was made up of granules whose structure and size were nearly identical
with those of the ribosomes of the *rer.* Chemical tests showed that nucleoli were also composed of RNA and protein. Their granules are apparently proribosomes. Nucleoli arise in close association with portions
of certain chromosomes. In some cells, the ribosomes lie free in the
cytoplasm, and the *er* is smooth.

   In keeping with the principle of order in the living system, enzymes
controlling the sequential reactions of a type of lipid (steroid) synthesis
have been located in the membranes of *smooth endoplasmic reticulum
(ser)* of certain cells. It is hardly possible to separate the enzymes engaged in male hormone (testosterone) synthesis from the *ser* membranes
of testicular cells (see Fig. 2-16). Lipid manufacturing cells of the ovary
and liver also have well-developed smooth endoplasmic reticulum.

In 1898, the Italian microanatomist Camillo Golgi described a reticular apparatus surrounding the nuclei of nerve cells and made visible with silver stains. The *Golgi complex* was discovered in other cells. In the first half of the twentieth century, more than 2000 papers were written about it. Many light microscopists questioned its existence in the living cell. In 1954, intact Golgi apparati were isolated from disrupted cells. This settled the question of the complex's existence as a definite subcellular organelle. Electron micrographs showed that the Golgi complex was—like the cell membrane and *er*—constructed of three-layered membranes, organized into flattened sacs and vesicles, frequently adjacent to the nucleus (see Fig. 2-17). Numerous Golgi bodies are scattered in the cytoplasm of plant cells. Electron microscopic studies of glandular secretory cells seem to indicate that the secretory protein first appears within the spaces of the *er*, then makes its way to the Golgi complex, where it is packaged in smooth Golgi membranes as secretory

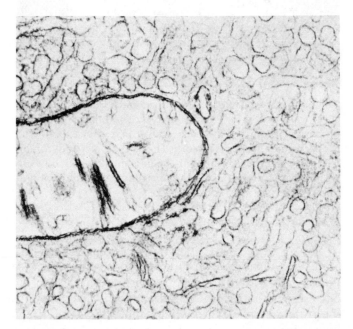

Fig. 2-16   **This lipid hormone manufacturing cell from the testis shows numerous bubbles and channels of smooth endoplasmic reticulum (ser), which is believed to be involved in lipid synthesis. The large structure projecting into the center from the left is a mitochondrion.**
Courtesy of Dr. D. W. Fawcett.

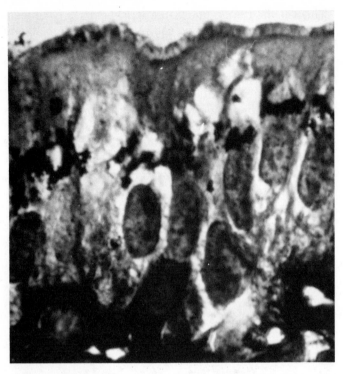

Fig. 2-17   **Golgi apparatus. The nuclei of these intestinal cells are the
gray ovals just above the cell bases. The microvillous free
borders of these cells is seen above. Just above the nuclei
is an irregular black meshwork. This is the Golgi body, and it
is involved in the secretion of the pale mucus, which lies in
white masses above the Golgi bodies and below the microvillous borders.**
From M. J. Zbar and M. D. Nicklanovich, *Cells and Tissues*, Saturn
Scientific, Inc., Fort Lauderdale, Fla., 1971.

granules. Mucus is a carbohydrate-protein complex secreted by cells
lining the digestive and respiratory passages. Radiotracer experiments
have shown that the carbohydrate portion of the molecule is added to
the protein in the Golgi complex and that mucous droplets are likewise
bound for secretion in Golgi membranes. Therefore, it is well established that the Golgi complex plays a role in secretion. It was once proposed that the Golgi region was a center of membrane synthesis, but no
proof was provided. In plant cell division, Golgi complexes seem to form
the new membranes separating the daughter cells. It is difficult to distinguish between Golgi elements and *ser;* Golgi membranes are always
smooth, but continuities with *rer* have been observed.

The nuclear membrane of the light microscopist was seen to be a *nuclear envelope* under the electron microscope. It consisted of two three-layered membranes separated by a space comparable in width to the diameter of the channels of *er*. The outer membrane of the two was often studded with ribosomes like the rough endoplasmic reticulum. In places the two membranes fused, interrupting the continuity of the envelope and forming pores through which the nucleoplasm and the cytoplasm communicated. Since the nucleus was known to ultimately direct and regulate cytoplasmic reactions as well as to be influenced by them, it seemed logical that there would be definite lines of communication between the cytoplasm and the nucleus. In plant cells these breaks in the envelope occur frequently and are evidently completely free passageways (see Fig. 2-18). In animal cells, short cylinders project through the pores like gun barrels. The function of these pore tubes is unknown. Continuities between the nuclear membrane and *er* have been observed as well as fusions of Golgi and nuclear membranes. Perhaps the nuclear

Fig. 2-18   **The lowest membranous semicircle in this electron micrograph is the nuclear membrane or envelope. Note its pores, and also that it communicates by way of a bridge near the center with a horizontal channel of endoplasmic reticulum. A Golgi body and several round mitochondria are visible to the left.**
Courtesy of Dr. G. Whaley.

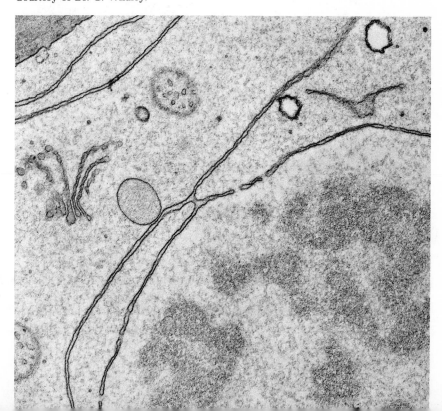

membrane is not so unique, but rather a portion of a continuum, the membrane system of the cell.

Because of the structural similarity of the cell membrane, the endoplasmic reticulum, the Golgi membranes, and the nuclear membrane and their observed continuities at times, it has been proposed that they should all be considered regional portions of a *unit membrane* system of the cell (see Fig. 2-19). They are all three-layered (trilaminar) membranes, and this unity of structure suggests a unity of origin. Even the plant cell vacuolar membrane is structurally similar in electron micrographs. The scientist looks for universal generalizations in the traditional mode of thought. Such generalizations are prized as ultimate truths of science, scientific laws. Yet biochemical tests show regional enzymatic differences in these membranes, which is a great puzzle. Sci-

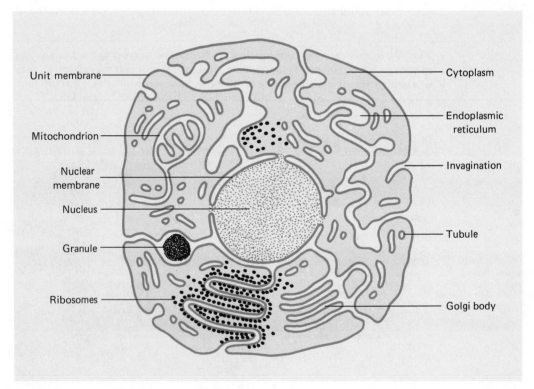

Fig. 2-19   **Possible interrelations of the various external and internal membrane systems of the cell. The diagram suggests that all membranes have a common structure and a common origin.**
Redrawn from Dr. J. D. Robertson.

ence would be inhuman without such frustrated attempts to generalize. How does the living cell construct membranes that are structurally alike but have different enzymes? We have a tendency to oversimplify. Half-truths are better than no knowledge. The unit membrane concept is useful. We can think of the cell as a polyphasic system. The fluids of the cell, the cytoplasm and nucleoplasm, are one phase, the membranes are another. Their faces are often the sites of reactions. For some reason, the cell has partitioned the contents of the endoplasmic reticulum from the cytoplasm. Similarly, the intracellular digestive enzymes are segregated from the rest of the cell in membrane-bound packages, the *lysosomes*. For summary of the anatomy of the cell as recorded by the electron microscope, study Fig. 2-20.

ENERGY TRANSFORMERS OF THE CELL

Bacteria were the first cells fixed and stained. These small, rodlike and spherical germs were just barely visible under the highest powers of the light microscope. In 1886, in the great era of bacteriology, Altman called cytoplasmic spheres and rods "bioblasts" (living germs) and compared them with bacteria, since they were of similar size, shape, and staining reactions. He did not consider these particles to be intracellular parasites, but rather symbionts, organisms that had become dependent upon the cell and established a mutually beneficial relationship with it. This idea was forgotten until recently when the *bioblast theory* was revived. In 1903, these particles were named mitochondria. It may be that the mitochondria were originally independent cells that took up residence within the cytoplasm of other cells in the course of evolution. No higher plant or animal cell is without mitochondria, and mitochondria cannot survive long outside the cell. Cells die following the blockage of mitochondrial function.

In the 1930's, mitochondria were isolated from tissue cell debris by differential centrifugation, and analyzed. They were found to be high in protein and phospholipids. In 1946, enzymes known to be of crucial importance in the process of *respiration* or biological oxidation were identified in their fraction.

Cellular respiration has been compared with burning, but it differs in that the process takes place under mild conditions in many steps with small release of energy, rather than a sudden burst of heat. Not all the energy derived from the oxidation of food molecules appears as useless heat. A substantial portion of the energy of respiration is transformed

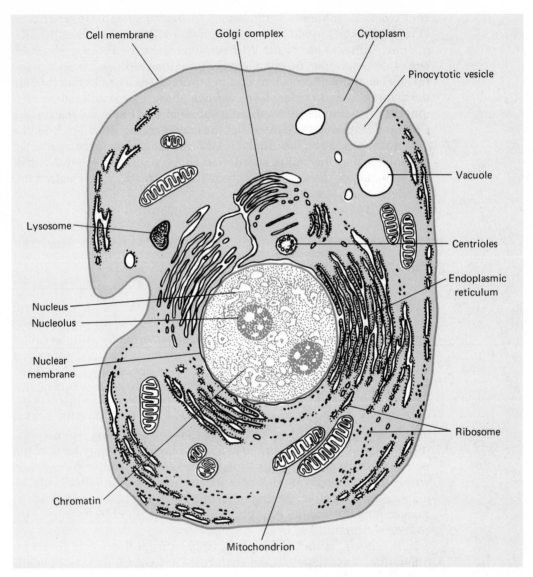

Cell membrane

Golgi complex

Cytoplasm

Pinocytotic vesicle

Vacuole

Lysosome

Centrioles

Endoplasmic
reticulum

Nucleus

Nucleolus

Nuclear
membrane

Ribosome

Chromatin

Mitochondrion

Fig. 2-20   **Modern diagram of a typical cell as revealed by the electron microscope.**
Modified from Jean Brachet, "The Living Cell," *Sci. Am.,* Sept., 1961.

into the chemical bond energy of adenosine triphosphate (ATP), the
molecule that has been called "the energy currency of the cell," since its
breakdown and consequent energy release are inevitably linked or coupled to every type of cell work. The greatest ATP-producing phase of

respiration occurs in the mitochondria. Thus, these particles have been described as "the powerhouses of the cell."

Respiration begins in the cytoplasm. Here float the enzymes, food molecules, and their fragments that are engaged in the 11 reactions of the first phase of respiration. This phase of respiration is anaerobic, not requiring oxygen. If glucose is the starting fuel, only two molecules of ATP are produced by the reaction sequence of *anaerobic respiration,* and the end products of this process are still potent fuels (Fig. 2-21). We evaluate the process in terms of the molecules of ATP produced because it is the goal and purpose of respiration, the production of an immediately available energy, the number of ATP molecules formed. The anaerobic respiration of yeast cells breaks glucose molecules down into alcohol, which certainly can be further oxidized or burned. Complete biological oxidation of glucose produces CO_2, a molecule whose carbon cannot be further oxidized, and water. The summary equation for the overall process of complete respiration of glucose is

$$\underset{\text{(glucose)}}{C_6H_{12}O_6} + 6O_2 \longrightarrow 6CO_2 + 6H_2O + \text{energy}$$

The hydrogen atoms that appear in the water molecules have been transported from the hydrocarbon molecule of glucose to oxygen. The carbon atoms of the carbon dioxide molecules have been cleaved from the food molecule.

The essential step of bio-oxidation is the removal of hydrogen atoms

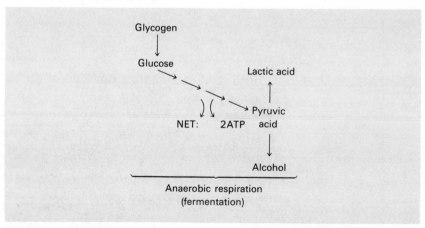

Fig. 2-21   **Summary of anaerobic respiration. Only two molecules of ATP are formed, and the end products, lactic acid or alcohol, are far from completely oxidized and still contain considerable energy.**

from the fuel molecule. The enzymes that catalyze such biological oxidations are proteins, and they function with nonprotein *coenzymes* that act as hydrogen acceptors initially and carriers subsequently. At the heart of a coenzyme molecule lies a vitamin, such as niacin. Following the dehydrogenation reaction, the hydrogen-loaded coenzyme molecule dissociates from the protein enzyme. In anaerobic respiration, the hydrogens are returned to the incompletely oxidized fragments of the food molecule. In those organisms that complete the oxidation, the two-carbon fragments of the glucose molecule are fed into the common metabolic mill, the Krebs cycle, where oxidation is completed.

The Krebs cycle and its interlocking process of electron transport take place within the mitochondrion, which is the seat of the aerobic (oxygen-requiring) phase of respiration. The end products of *aerobic respiration* are CO_2, the water of biological oxidation, ATP, and heat (Fig. 2-22). The aerobic respiration of the fragments of the glucose molecule within the mitochondrion produces 36 molecules of ATP, an eighteenfold increase over the two resulting from the anaerobic respiratory phase.

Under the electron microscope, the mitochondrion was seen to consist of two membranes, an outer membrane and an inner membrane thrown into numerous folds or crests (see Fig. 2-23). Within the innermost membrane lies a protein-rich fluid. The proteins contained therein are the enzymes of the Krebs cycle.

This cycle is also called the citric acid cycle, since a 2-carbon fragment of glucose is joined to a 4-carbon acid to form the 6-carbon acid, citric acid. Then by a series of eight reactions, in which CO_2 is split off and

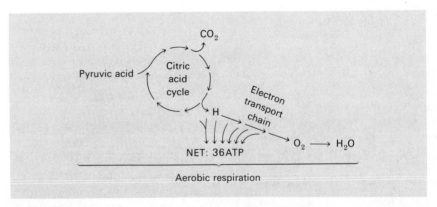

Fig. 2-22   **Aerobic phase of respiration summarized. Note that the number of molecules of ATP yielded here is many times greater than in anaerobic respiration, and that the carbon in carbon dioxide (CO_2) is completely oxidized.**

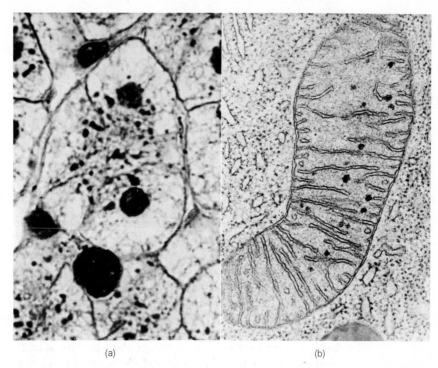

(a)   (b)

Fig. 2-23   **The mitochondrion. In the photomicrograph (a), the large dark spheres are the nuclei of liver cells, and the small dark rods and spheres are mitochondria in their cytoplasms. Nothing of the internal structure of these small organelles can be seen. (b) In the electron micrograph, the inner and outer membranes and crests of the mitochondrion are visible. The crests are formed by folds of the inner membrane. The "dusty" interior is the matrix. Profiles of rer surround the sausage-shaped organelle.**

(a) From M. J. Zbar and M. D. Nicklanovich, *Cells and Tissues,* Saturn Scientific, Inc., Fort Lauderdale, Fla., 1971. (b) From Guidance Associates, *Biology 500,* Harcourt Brace Jovanovich, Inc., New York, 1968.

hydrogen removed, the 6-carbon acid becomes once again the 4-carbon acid, which began the cycle by condensing with the 2-carbon fragment of anaerobic respiration. Most important, hydrogen removal occurs in several places along the circular route, and this hydrogen is not given back to fragments of the fuel molecule (Fig. 2-24). The endless cleavage of molecules would in itself be meaningless.

The hydrogen carriers, the coenzymes, separate from the Krebs cycle enzymes and carry the hydrogen atoms to the flavoproteins, which contain the vitamin riboflavin. The flavoproteins separate the electrons

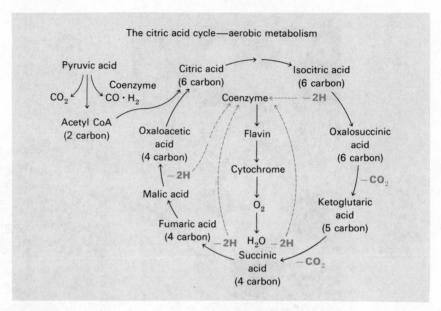

The citric acid cycle—aerobic metabolism

Fig. 2-24

Active acetic acid (2C) condenses with oxaloacetic acid (4C) to form citric acid (6C). Citric acid then goes through a series of oxidations and carbon cleavages, finally leading back to the reformation of oxaloacetic acid. The net effect is the complete combustion of pyruvic acid via acetate to carbon dioxide and water. The hydrogen removed from the fuel molecules are transported to oxygen by means of the electron transport chain. All these reactions occur in the mitochondria.

from their hydrogen atoms. The protons (H^+) so formed are released into the fluid. Then the electrons are transported from the flavoproteins to oxygen (the final electron acceptor) by a chain of respiratory enzymes, the *cytochromes* (the cell pigments). These iron-containing, pigmented enzymes constitute a respiratory assembly, or an *electron transport system*. The iron ions within them capture electrons and pass them on to the next cytochrome in the series.

As the electrons move from one respiratory enzyme to the next, they drop each time in energy level, releasing energy in the process. Electron transport has been compared to a waterfall. The water at the top of the fall (the electrons of the loaded coenzymes) has the greatest potential energy that can be used to do work, to turn the dynamos of a hydroelectric plant (or to provide the energy for ATP synthesis). The water at the bottom of the fall (the electrons firmly bound to oxygen) has lost its usefulness (see Fig. 2-25). As electrons drop in energy level from cyto-

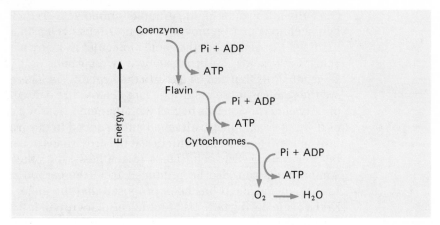

Fig. 2-25 **The "electron waterfall." ATP is formed in conjunction with the transfer of electrons over the electron transport chain (oxidative phosphorylation).**

chrome to cytochrome, packets of energy are released. A portion of this released energy is captured in the forming bonds of the ATP molecules. At three places along the electron transport route, ATP molecules are synthesized. For every pair of electrons transported, three molecules of ATP are generated.

Coenzymes I & II $\longrightarrow$ flavoprotein $\longrightarrow$ coenzyme Q, 1 ATP formed
Coenzyme Q $\longrightarrow$ cytochrome c $\longrightarrow$ cytochrome c, 1 ATP formed
Cytochrome c $\longrightarrow$ cytochrome oxidase $\longrightarrow$ O_2, 1 ATP formed

An electric current consists of a flow of electrons. The quantity of current flowing is measured in amperes (amps). It has been calculated that the number of electrons flowing from foodstuff to dehydrogenases, coenzymes, and cytochromes to oxygen per minute in a human body at rest amounts to 76 amps. This is an "electricity of life."

The oxygen that receives the transported electrons combines with the protons ($2H^+$) released into the medium to form the "water of biological oxidation." The unloaded coenzymes (with their hydrogens removed) return to their enzymatic partners to accept hydrogen from fuel molecules again. The Krebs cycle cannot turn unless the electron transport system functions. The Krebs cycle and the respiratory assembly (electron transport) are inseparable. Why? What are the connecting links between the two? Before we leave the Krebs cycle, let us look at another side of it.

Thus far we have only discussed the biochemical breakdown (respiration) of carbohydrate (glucose). Proteins and fats can also be respired.

Once the nitrogen or amino group of amino acids is removed, their carbon skeletons can be moved into the Krebs cycle, since in their semioxidized forms they are the same molecules as many of the Krebs cycle acids. Fatty acids are broken down two carbons at a time to provide 2-carbon units that can be fed into the Krebs cycle. These 2-carbon units are the same as those resulting from glucose breakdown. The synthesis of fatty acids and sterols begins with the same 2-carbon compound. One of them after another is attached and reduced in the growing chains of fatty acids. The 2-carbon unit lies at the crossroads of carbohydrate and fat metabolism. One road leads to synthesis, the other to respiration. Amino acids can also be produced from the carbon skeletons of the Krebs cycle acids. It has been proposed that the major function of the Krebs cycle during periods of rapid cellular growth is to provide carbon skeletons for amino acid (and thus protein) biosynthesis. Amino acids can be converted into glucose. The pathways of respiration converge upon the Krebs cycle, and routes of synthesis diverge from it. The Krebs cycle is the greatest biochemical crossroad (Fig. 2-26). There are no

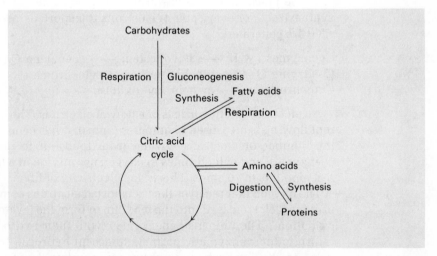

Fig. 2-26   **Crossroads of metabolism.** Not only are biochemicals broken down and respired via the citric acid cycle, but also the carbon skeletons of the cycle itself can serve as building materials for protein and lipid construction. Everyone is well aware of the fact that sugar and starch can be converted into fat, but few realize that under certain conditions sugars such as glucose can be produced from the amino acids of proteins. The formation of glucose from these other biochemicals is called gluconeogenesis ("new-born glucose"), and it illustrates one aspect of the interconvertability of biochemicals.

snarls of molecular traffic here; it is an incredible freeway exchange. Yet the price of life is paid; the tolls are hydrogens lost or gained, coming and going.

The enzymes of the Krebs cycle are suspended within the mitochondrial fluid. The enzymes of electron transport, the cytochromes (respiratory enzymes), are built into assemblies in the inner mitochondrial membrane. "Elementary particles," resembling golf balls on tees, cover the crests of the inner membrane, and they are thought to contain the respiratory assemblies. At first it was proposed that the oxidative enzymes were fastened upon the inner membrane in order, like enzyme mosaics. George Palade said that these multienzyme systems were "built in, or woven into, the texture of the mitochondrial membranes in the same manner as repeated decorative patterns are woven into a sheet of damask." The mitochondrion is a membranous organelle. The outer and inner membranes are three-layered and lipoprotein in composition, like most other cell membranes. They provide an excellent example of one of the major functions of cell membranes, which is to provide a surface for enzymatic reactions. Adherents of the unit membrane hypothesis proposed that the mitochondria arose from infoldings of the cell membrane.

The other energy transformer of the cell is the chloroplast. Mitochondria oxidize ready-made fuels; chloroplasts take the products of biological burning, carbon dioxide and water, and make biochemicals with the energy of the sun.

Both chloroplasts and mitochondria contain DNA (the hereditary chemical) and are self-replicating. The chloroplast is somewhat similar to the mitochondrion in construction and function. It is larger, about 5μ long and 3μ wide, with roughly an egg shape. The chloroplast, too, is a membranous cytoplasmic organelle, and its membranes have the typical electron microscopic structure of lipoprotein membranes. Two triple-layered membranes surround its contents, and a series of stacked, flattened, membranous sacs extend throughout the length of its interior. These internal membranes are thickened in places to form what look like stacks of coins in profile view, the *grana* (see Fig. 2-27). Between the membrane sacs lies a protein-rich fluid resembling the inner mitochondrial fluid in density. It is called the *stroma*, because it is the matrix or framework in which the membranes are imbedded. The "coins" of the grana contain chlorophyll, the accessory pigments of photosynthesis, and the enzymes in charge of the light reactions of photosynthesis in addition to the phospholipid and protein components of typical membranes. The grana are something like the cells of a solar battery. Surface views of sectioned grana show a number of particles, the *quantasomes*, which are similar to the elementary particles along the mitochondrial crests (see Fig. 2-28).

(a)   (b)

Fig. 2-27   **The chloroplast. In photomicrograph (a), the rectangular leaf cells specialized for photosynthesis are filled with round or oval chloroplasts. The electron micrograph (b) reveals an internal structure of stacks of membranous coins, the grana, and a protein-rich fluid, the stroma. Compare this structure with the mitochondrion.**

(a) Courtesy of Saturn Scientific, Inc., Fort Lauderdale, Fla. (b) Courtesy of Dr. Elliot Weier.

The summary equation for the multitude of reactions constituting photosynthesis is

$$6CO_2 + 12H_2O \longrightarrow C_6H_{12}O_6 + 6O_2 + 6H_2O$$

which is roughly the reverse of summarized respiration. Carbon dioxide is fixed and reduced (hydrogenated) to form a hydrocarbon like glucose in photosynthesis, whereas glucose and other metabolic fuels are cleaved and oxidized to CO_2 in respiration. Since the biochemical details of the respiratory reactions were discovered first, it was only logical that it would be proposed that the stepwise reactions of photosynthesis would be an exact reverse of respiration. This did not turn out to be the case. Mitochondria cannot photosynthesize, nor can chloroplasts re-

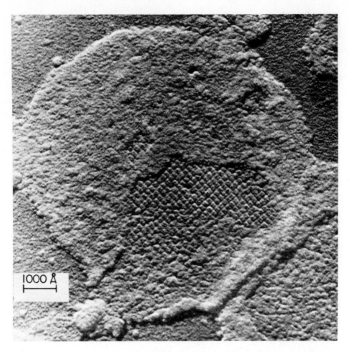

Fig. 2-28

This electron micrograph of the inner surface of one wafer of a granum shows numerous spherical particles, called quantosomes, which may well be the smallest units that can carry out the photochemical reactions. They are reminiscent of the particles found upon the internal crests of mitochondria, which are believed to contain the electron transport assembly lines of enzymes.
Courtesy of Dr. R. B. Park.

spire; however, the raw materials of one process are the end products of the other. The two processes complement each other in a drama of balance in nature. They do have similarities and a meeting place; plant cells respire as well as photosynthesize.

Photosynthesis can be divided into the *light reactions* and the *dark reactions*. The two sets of reactions are firmly linked. The light reactions provide ATP for the work of synthesis and hydrogen-loaded coenzymes for the reduction of the carbon dioxide fixed in dark reactions of photosynthesis. The major events of the dark reactions are the fixation of carbon (CO_2) and its reduction (hydrogenation).

In one set of light reactions, packets of light energy (quanta) in the far red portion of visible light are absorbed by chlorophyll molecules. In consequence, electrons of the chlorophyll molecule are raised to such a higher energy level that they are actually ejected to an iron compound and then transported along a chain of cytochromes back to the chlorophyll molecule (see Fig. 2-29). In the process of electron transport (similar to that in the mitochondrion), ATP molecules are formed with the energy released by the falls in energy levels of the electron in its circular path. This process is called cyclic photophosphorylation, since the transport is circular and the energy for phosphorylating adenosine diphosphate (ADP + P → ATP) originally comes from the wavelengths of absorbed light. (The process of ATP formation in aerobic respiration is called oxidative phosphorylation, because the production of ATP entails electron transport to oxygen.)

There is another type of photophosphorylation, in which the electrons ejected from one chlorophyll molecule do not return, but are picked up by a coenzyme. These lost electrons are eventually replaced by two from a split water molecule, after they have traversed a cytochrome system and generated ATP (see Fig. 2-30). Another type of chlorophyll is involved along the way. The coenzyme also accepts hydrogen ions (protons: H^+) from the split water molecule to form a typical hydrogen and electron-loaded coenzyme with reducing power. The remainder of the water molecule rearranges to form free oxygen, a by-product of photosynthesis and new water molecules. This process is called noncyclic

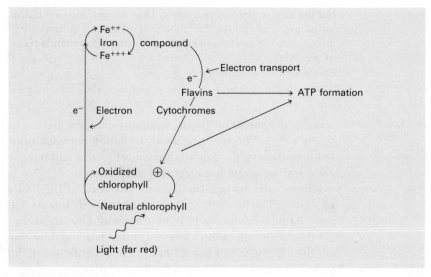

Fig. 2-29   **A light reaction of photosynthesis, cyclic photophosphorylation. The ATP so formed is used in the dark reaction of carbon fixation.**

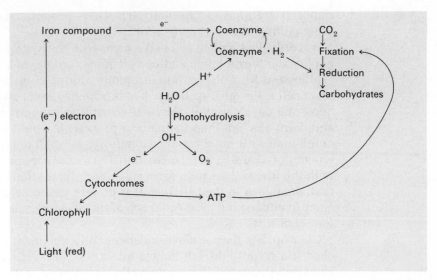

Fig. 2-30   **Photosynthesis. Noncyclic photophosphorylation and the dark reactions. The ATP from the cyclic light reactions is also used in carbon fixation.**

photophosphorylation, since the electrons of the original chlorophyll molecule do not return, but are replaced by electrons from a split water molecule via another chlorophyll molecule and an ATP-generating cytochrome system. The second chlorophyll molecule absorbs light in the blue region of the spectrum. Taken together, the chlorophylls absorb blue and red light but not green. Why are most plants green?

The dark reactions of photosynthesis take place within the chloroplastid stroma, where the enzymes controlling these reactions are suspended. In the dark reactions of photosynthesis, the ATP and hydrogen-loaded coenzymes from the light reactions are used to fix or bind carbon (CO_2) and reduce it to some equivalent of carbohydrate (CH_2O). Carbon dioxide is fixed to a 5-carbon sugar phosphate forming a 6-carbon compound, which is split into two 3-carbon fragments, that are reduced (hydrogenated) to form 3-carbon molecules of phosphoglyceraldehyde (PGAL). Phosphoglyceraldehyde molecules can be built into glucose. This model of carbon fixation was discovered by Melvin Calvin and his colleagues at the University of California at Berkeley, and its clarification ranks alongside Sir Hans Krebs' unveiling of the cycle named after him as two of the greatest achievements of biochemistry. Calvin and coworkers exposed photosynthesizing leaves to radioactive carbon dioxide and progressively identified the successive compounds in which radioactivity appeared. Recently, it has been proposed that there may be other pathways of carbon fixation.

Not only does a plant cell make sugar with the hydrocarbon products of photosynthesis, but in addition all other varieties of biochemicals are synthesized from the carbon skeletons produced in the process. The plant cell is the most accomplished biochemist in the living world. Perhaps, one day, man will carry out an artificial photosynthesis, simpler and more efficient than the natural process. It would be a greater triumph than splitting the atom. Hopefully, we shall not lose sight of the wonder of nature. A plant takes air (CO_2) and water and combines them with the fire of the sun to form sugar and the myriad of biochemicals that, organized in the cell, live. It is no less awesome that man should come to understand the mechanism of this seeming magic of the Greeks' four elements.

Chloroplasts do not move independently about in the cell as mitochondria seem to do, but the cytoplasm of plant cells streams around and around, carrying the photosynthetic factories in its currents.

THE EVOLUTION OF THE CELL

The cellular anatomy we have considered so far is characteristic of higher plant and animal cells. Bacteria and blue-green algae seem to consist of primitive cells. Their cells lack endoplasmic reticulum, Golgi complexes, mitochondria, and chloroplasts; their chromosomes are composed of nucleic acid (DNA) alone. The chromosomes of higher cells contain protein in addition to nucleic acid. Since bacteria and blue-green algae lack a nuclear membrane as well as chromosomes typical of higher cells, they have been called *procaryotic* (prenuclear) cells, whereas cells with true nuclei have been termed *eucaryotic* (true nuclear). Procaryotic cells do possess, however, functional nuclear equivalents, and *protocell* ("primitive cell") may better describe their anatomy (higher cells are called *eucells*, "true cells"). Protocells do possess the protein synthetic machinery of ribosomes, and the blue-green algae do contain chlorophyll in internal membranes, the *photosynthetic lamellae*. Study Fig. 2-31 carefully.

The oxidative enzymes of protocells are frequently bound up with the cell membrane itself or inpocketings of it, or membranous whorls in the cytoplasm, seemingly derived from it. A theory of mitochondrial evolution derives them from infoldings of the cell membrane, and these buried membranous whorls would thus represent an intermediate stage. The photosynthetic lamellae of blue-green algae seem to have originated

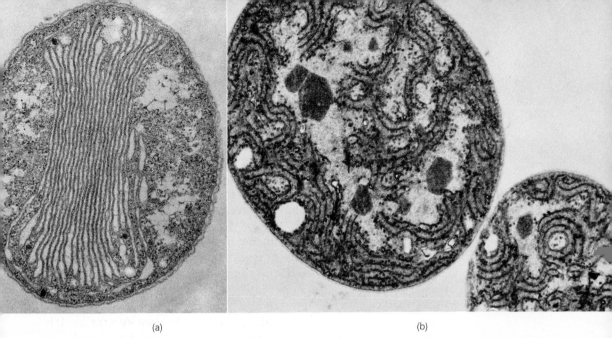

(a)   (b)

Fig. 2-31   **Electron micrographs of bacteria and blue-green algal cells. (a) A marine
bacterium. (b) A blue-green algal cell. Notice the similarities between the
structures of these cells and those of mitochondria and chloroplasts. The
particles in the cytoplasm of (b) are ribosomes.**

(a) Courtesy of Dr. S. Watson. (b) Courtesy of Dr. Norma J. Lang, University of California,
Davis, California.

by inpocketings of the cell membrane. This observation led to the speculation that chloroplasts have evolved like mitochondria from buried
portions of the cell membrane.

However, mitochondria and chloroplasts both possess DNA and ribonucleoprotein (like ribosomes) and are self-duplicating structures like
cells themselves. The mitochondrial "chromosome" is similar to the
bacterial chromosome. The blue-green algal cell looks like a primitive
chloroplast with a cell wall. (The cell wall of protocells is not composed
of cellulose fibrils as is the case with higher plant cell walls, but rather
consists of carbohydrate-protein spheres.) The size of the mitochondria
is quite close to that of bacteria, and chloroplasts are not much
smaller than blue-green algal cells. It does not seem at all unlikely that
these subcellular organelles were at one time independent cellular
organisms.

Even today, green and brown unicellular algae often live in the tissues
of flatworms, jellyfish, and coral animals, finding therein a CO_2-rich environment for photosynthesis and nitrogen (NH_3)-rich surroundings for
protein synthesis. The relationship of these unicellular plants and their
animal hosts is called *symbiosis,* or, more specifically, *mutualism,* since
both organisms derive benefits from the intimate relationship. The ani-

mal tissue cells receive the oxygen by-product of photosynthesis and sometimes digest the algae as food. It is something like having a vegetable garden within the tissues of your own body. Indeed, some flatworms have lost their digestive tract and depend wholly upon their internal algae for food manufacture.

It is believed that hydrocarbons and biochemicals appeared before life itself and that the first organisms broke down the sea of fuels. These first cells are supposed to have been *heterotrophs,* depending upon the degradation of compounds already formed, not photosynthetic organisms or *autotrophs,* manufacturing their own fuels and biochemicals. Oxygen was quite likely absent from the atmosphere before the evolution of photosynthetic life forms, and so the heterotrophic cells were limited to the meager energy yields of anaerobic respiration, such as certain bacteria today, the *strict* or *obligate anaerobes.* When photosynthetic organisms evolved and released free oxygen into the air, it became possible to oxidize foodstuffs completely. Before the appearance of oxygen in the atmosphere, incomplete respiration (anaerobic) occurred, and its end products, alcohol and lactic acid, accumulated, transforming the sea of life into a dilute sour wine. When aerobic respiration evolved, it took up where anaerobic respiration (fermentation) had left off. Perhaps, in certain cells, aerobic respiration was tacked onto anaerobic. Some cells might have become completely dependent upon oxygen-requiring respiration, starting with the end products of the ancient respiration (fermentation), the alcohols and acids that had accumulated. A few bacteria today are strict aerobes, perishing in the absence of oxygen, the final electron and hydrogen acceptor. In the transition period from anaerobic to aerobic respiration, it may have been that some cells tried to compromise and became *facultative microorganisms,* capable of surviving with or without oxygen, such as many bacteria today. Facultative cells would undoubtedly have been less efficient than symbiotic associations of strictly aerobic and obligatorily anaerobic cells, which were specialized for the two phases and thus formed a respiratory partnership. Was the mitochondrion a strict aerobe? Is it a degenerate facultative organism, having lost its anaerobic ability? Or was it derived from the cell membrane of anaerobes evolving into aerobes? We shall never know for sure, but we can see how it might have happened.

It is not so difficult to visualize the evolution of chloroplasts. It is easy to see how it would have been definitely advantageous for a heterotrophic cell (dependent upon the meager income of a dilute sea) to harbor an autotrophic blue-green algal cell. Over a period of time, the unicellular "plant" might have lost its ability to respire, as well as its cell wall. Somewhere, sometime, a combination of all three types of nutritional organisms must have evolved. A strict anaerobe (the cell minus mitochondria) came to host strict aerobes (the mitochondria) and degenerate photosynthetic autotrophs (chloroplasts) which had lost their ability to respire.

The cellular fluid is the location of what seem to be the most ancient reactions. Anaerobic respiration or fermentation occurs in the nucleoplasm as well as the cytoplasm. (There are no mitochondria in the nucleus.) The nucleoplasm (aside from its genetic chemicals) is thus quite like cytoplasm. Before the organization of the nuclei of higher cells, the cell fluid was a combined "nucleocytoplasm." It is almost impossible to conceive of a cell without membranes. Most of the very ordered, high-energy-yielding reactions take place upon membrane surfaces. In the studies into the origin of life, too much emphasis has been placed upon proteins and nucleic acids. Yet lipids are essential to the construction of all known cells. How were the first lipoprotein membranes formed? Were they generated spontaneously in the ancient seas by the association of proteins with lipid films? How are membranes formed today? Was there ever a naked nucleocytoplasm? It is easy to imagine primordial cells consisting of a nucleocytoplasm surrounded by a membrane. The evolution of higher cells from these would involve: (1) an expansion of the membrane; (2) an infolding of it to produce primitive endoplasmic reticulum; (3) the specialization of portions of the internal membrane system; (4) the acquisitions of mitochondria and chloroplasts (see Fig. 2-32). Viruses do not conform to cellular construction, yet they are, in many ways, biochemically and functionally comparable to the hereditary factors of the cell. For this reason, viruses have been called "naked genes." No known virus is capable of independent existence. All are obligatory parasites, capable of carrying out their single life function,

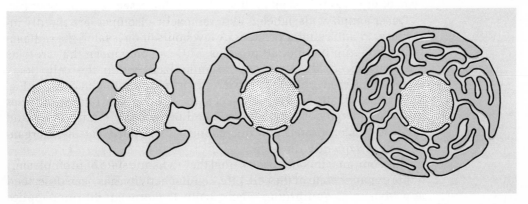

Fig. 2-32   **Evolution of the cell. It has been suggested that the membrane system of the cell evolved in the following steps, from left to right.**
Redrawn from Dr. J. D. Robertson.

reproduction, only within other cells. A major hypothesis of life origin holds that the first living organisms were independent viruses. This hypothesis will be discussed at length in the chapter on evolution. Viruses are at least comparable to parts of cells. Viruses which lost their cell-destroying ability and became attached to the genes of the host cell may have become in time functional genes of the cell itself. Except for viruses, all known organisms are composed of cells and their products. Although this generalization is called the Cell Theory, it is at least a terrestrial law and may apply beyond the earth as well. In construction, viruses are nucleoprotein, or composed of nucleic acids and protein. These are the biochemicals of the chromosomes.

THE STEADY STATE OF THE CELL

The second law of thermodynamics states that all systems in the universe continually tend to run down, go to lower energy states and greater disorganization. The cell consists of an intricate organization of matter beyond any construction in the inorganic world. At first, scientists thought that life disobeyed the second law before they realized that the cell expended a vast amount of energy to maintain the integrity of its complex functional architecture. This is the price of life.

We tend to think of two types of cellular molecules: the "permanent" molecules making up cell organelles, and the perishable molecules of "fuel." The cell makes little such distinction. The nucleic acid DNA seems to be the only biologic molecule conserved; otherwise, the composition of the cell changes from moment to moment. Only a functional state of organization is maintained. This is the essential activity of life.

Tracer amino acids, labeled with radioactive isotopes, are rapidly incorporated into cellular proteins. A few hours or days later, the radioactivity shows up in different proteins. This can only mean that proteins are synthesized, digested, and resynthesized within the cell. Some amino acids are respired; others are transformed into glucose (see Fig. 2-26). Labeled fatty acids are subject to the same diverse fates. Molecules of RNA undergo constant synthesis and breakdown. Amino acids and lipids are incorporated into mitochondrial structure, yet these organelles are frequently digested.

This continual flux has been called the "dynamic state of protoplasm," or the dynamic state of the cell. Life, cellular activity, has been described as a "dynamic equilibrium." Technically, the term equilibrium applies only to closed systems. The cell is an open system. It engages in constant energy and molecular traffic with the inorganic environment. The highly organized yet fluid state of the living cell is more properly called a *steady state,* since it is kept by a constant flow of energy into the cell, whether in the form of sunlight (plant cells) or the chemical bond energy of biochemicals already formed (animal cells). An equilibrium does not require an input of energy. The dynamic state is really synonymous with what we call metabolism, the total sum of biochemical reactions occurring in the cell.

The systems of advanced multicellular animals are said to have as their common goal *homeostasis,* "the maintenance of the constancy of the internal environment." Homeostasis is just another word for the steady state. The concepts of the steady state and homeostasis, originally intended to apply to the cellular and multicellular organismal levels, have been expanded to include the highest levels of biologic organization—populations and communities. Can we achieve a homeostatic or steady state for our social world, a state of stability yet change, dedicated to the maintenance of life? The cell is a steady state. Life is the steady state.

DISCUSSION QUESTIONS

1. *What is the principle of complementarity in biology?*

2. *How many times greater is the power of the electron microscope than that of the light microscope?*

3. *How were fine structure and biochemical function correlated?*

4. *What is meant by the phrase "the electricity of life?"*

5. *How do the dark reactions depend upon the light reactions?*

6. *Which came first: photosynthesis or respiration, aerobic or anaerobic respiration?*

7. *Are bacteria primitive or degenerate cells?*

8. *How does the steady state differ from an equilibrium?*

REFERENCES

BRACHET, JEAN, "The Living Cell," *Sci. Amer.,* Sept., 1961.

DE ROBERTIS, E. D., W. N. NOWINSKI, and F. A. SAEZ, *Cell Biology,* W. B. Saunders Co., Philadelphia, 1970.

GERARD, R. W., *Unresting Cells,* Harper & Row Publishers, New York, 1940.

LOEWE, A. G., and P. SIEKEVITZ, *Cell Structure and Function,* Holt, Rinehart & Winston, Inc., New York, 1969.

McELROY, W. D., and C. P. SWANSON, *Modern Cell Biology,* Prentice-Hall, Inc., Englewood Cliffs, N. J., 1968.

OBJECTIVES

1. Define homeostasis and cybernetics.

2. Explain how feedback is involved in homeostasis.

3. Outline the mechanisms of the hormonal control of basal metabolism and blood sugar.

4. List ten hormones, their sources, their targets, and their effects.

5. Give three theories of the level of hormone action.

6. Diagram and label a typical nerve cell.

7. Describe the events in the generation and propagation of a nerve impulse.

8. Diagram and label the components of a simple spinal reflex arc.

9. List the major subdivisions of the human brain and briefly describe their functions.

10. Compare the anatomy of amphibian, reptilian, bird, and human brains.

11. Explain how the nervous system controls the endocrine to a large extent.

12. List four homeostatic functions of the kidney.

13. Diagram and label the parts of a human nephron.

14. Trace the steps in the formation of urine from blood.

15. Describe the countercurrent and hormonal mechanisms of urine concentration.

16. Compare and contrast the structure and function of the nephron in the various classes of vertebrates.

17. Contrast excretion and the maintenance of salt and water balance in marine, freshwater, and terrestrial vertebrates.

18. Compare the human brain with computers.

The Even Keel of Life

HOMEOSTASIS

"La Fixité du milieu interieur est la condition de la vie libre." (The constancy of the internal sea is the prerequisite of independent life.) So said the father of modern physiology, the Frenchman Claude Bernard. He had been a failure as a playwright, but he became an incredibly successful and poetic scientist. This simple statement of principle probably outweighs all of Bernard's numerous other and significant discoveries in the science of biologic function, *physiology*. "The constancy of the internal sea" became the motto of the infant science. English physiologist and soldier of fortune, J. B. S. Haldane, regarded these words as the most "pregnant sentence . . . ever framed by a physiologist." American biologist Walter Cannon coined the word *homeostasis* for Bernard's "maintenance of the constancy of the internal environment."

Just as the independent, unicellular organism maintains a steady state, so the dependent specialized cells of the tissues, organs, and systems of the advanced multicellular organism have as their common goal this homeostasis (steady state). Although different words were chosen by the cell biologist and the physiologist, it was soon realized that the terms "steady state" and "homeostasis" were really one and the same. In fact, the students of the highest levels of living organization, the ecologists, have applied these same terms to biologic communities in the last decade. It seems that scientists used the words too narrowly at first, underestimating the influence of the external environment, considering the individual organism too independently. Since the steady state of life is dynamic, and the term homeostasis literally means "a steady standing," it has been suggested that the term "homeokinesis" ("steady motion") might be more appropriate, but this term has never won favor.

The "internal sea" originally referred to the *tissue fluid,* or the *extracellular fluid (ECF),* that solution which bathes the many cells of the higher multicellular animal and through which all exchanges (respiratory, excretory, and nutritive) take place. The acidity (pH), salt and water composition, oxygenation, and, in many cases, nutrient (glucose, amino acids, etc.) levels must be kept within narrow bounds if life is to continue. Indeed, death has been defined as a failure of homeostasis, an inability to maintain "the constancy of the internal environment." Homeostasis is the fragile—yet tenacious—balance of life. Since life is a dynamic process, the even keel of homeostasis comes very close to defining it; if death is homeostatic failure, then life is homeostasis.

If life is a machine, then it is a wondrous mechanism, self-regulating to an unparalleled degree. It is this autoregulation that has led many biologists to conclude that "The whole is greater than its parts." Life is its own helmsman, and the word for the science of *cybernetics,* which began as an attempt to explain the nature of living control systems by analogizing them with electronic calculating machines, is derived from the Greek *kybernetes,* which means "helmsman," "pilot," or "steerer." The comparison of mechanical models with living mechanisms has been a most rewarding endeavor, providing an invaluable viewpoint of investigation. To a certain extent, life is programmed like a computer. Stimuli, changes in the external and internal environments, are detected by receptors, which are connected with integration centers that interpret the stimuli and call into play effectors, which "effect" an appropriate response, adjusting the organism to environmental alteration. This is what is meant by "adaptive response." Life resists change, strives to maintain its "equilibrium" (homeostasis), or to return to it when it is disturbed. There are several mechanical devices whose function (like life's) is to maintain a steady condition.

Basically, the gyroscope consists of a wheel mounted in a ring so that its axis is free to turn in any direction. When the wheel is spun rapidly, it will keep its original plane of rotation no matter which way the ring is turned. Gyroscopes are used in gyrocompasses and to keep moving airplanes, ships, etc., level. An automatic pilot is a "gyropilot." The thermostat is more often compared with living control systems.

A thermostat is a mechanical apparatus for automatically regulating room temperature. It controls the operation of a heating unit, usually some type of furnace. The thermostat is essentially two strips of different metals, having contrasting rates of expansion. At a certain temperature setting, when the temperature drops too low, one strip contracts more than the other and the device bends, closing the circuit and activating the heating element. As the temperature rises, one strip expands more than the other, bending the device the opposite way, breaking the circuit, and thus deactivating the heater. The thermostat is at once the detector, integrator, and activator, according to the "program" of the temperature setting. The heater is the effector, switched on and off by the thermostat. The rise and fall of temperature are like stimuli, changes in the environment. Temperature drop is a positive stimulus, resulting in the turning on of the heater. This is stimulation. The rise of temperature above the setting is a negative stimulus, or inhibition, producing a negative *feedback,* the switching off of the furnace (Fig. 3-1). Once the temperature is set, the thermostat is independent. It seems to possess an intellect, perceiving differences and acting accordingly. Considering it, one can feel the inkling of the cybernetician that the living organism so replete with control systems could be usefully compared with servomechanisms, or regulatory devices. The cryostat is a similar device except that it turns on a cooler and is therefore a regulator for maintaining a nearly constant, low temperature. The "warm-blooded" animals, the homeotherms (constant body temperature), contain a temperature regulator, "thermostat," in their brain bases. The center consists of specialized nerve cells that respond to slight increases or decreases in blood temperature by turning on or off skin circulation and sweat flow (see Fig. 3-2).

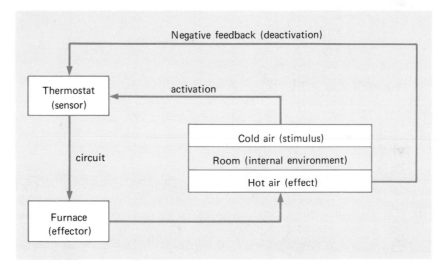

Fig. 3-1   **The sensing device (thermostat) activates or deactivates the furnace, thus changing the room air temperature, which, in turn, influences the thermostat.**

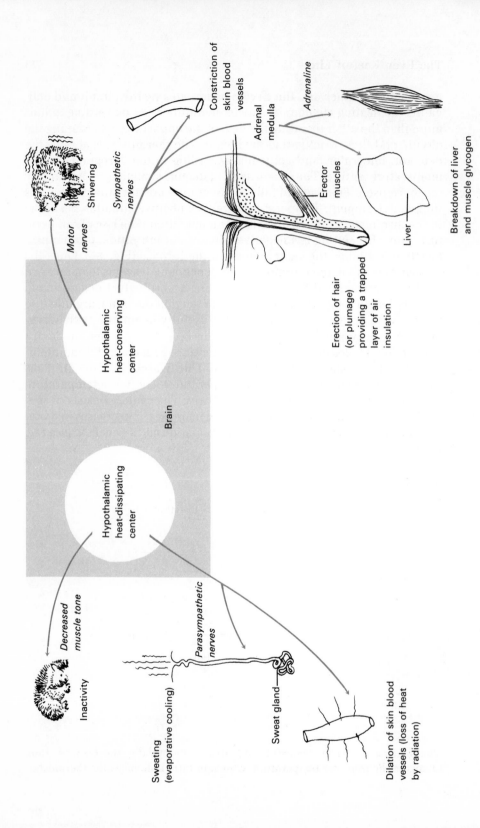

Fig. 3-2 **Thermoregulation. Birds and mammals (the animal class of man) maintain constant body temperatures through the above adaptive responses. Thermostatic control centers are located in the region of the brain base called the hypothalamus. The muscular contractions of shivering produce heat as a by-product.**

Constriction of skin blood vessels

Shivering

Sympathetic nerves

Adrenal medulla

Adrenaline

Erector muscles

Breakdown of liver and muscle glycogen

Liver

Erection of hair (or plumage) providing a trapped layer of air insulation

Motor nerves

Hypothalamic heat-conserving center

Brain

Hypothalamic heat-dissipating center

Decreased muscle tone

Inactivity

Sweating (evaporative cooling)

Parasympathetic nerves

Sweat gland

Dilation of skin blood vessels (loss of heat by radiation)

In the living "machine," stimulation alternates with inhibition, so that the curve of life rises and falls by turns, or oscillates, rather than maintaining a flat line. In homeostasis, the values of the rates of activity fluctuate about a norm. A straight line, the average between crests and troughs, represents the mark narrowly but continually overshot and undershot. The waves of some precise living control systems are so slight that their lines are nearly but never quite horizontal. Other functions, such as reproduction, are designed to be cyclic, with peaks and lows occurring periodically in harmony with optimal and minimal circumstances of the internal and external environments. In the advanced organism, the curves of many systems are being described simultaneously. All interact and life results. The curves rise and fall in the lifetimes of individuals and species.

There are intercellular as well as intracellular control systems, and the *intracellular fluid* (ICF) traffics with the *extracellular* (ECF). Although the organ systems of the body are somewhat separate, none is independent. Biologists refer to the organ systems as a matter of analytic convenience. In the outline of control, communication is the crux of regulation. Communication is the function of the nervous system. The nervous and hormonal systems are often cited as *the* two major control systems. Nevertheless, the other organ systems play important roles in homeostasis. The hormonal or endocrine system is said to be the "system of chemical control," and in its operation are found excellent examples of homeostasis.

HORMONAL CONTROL OF METABOLISM

The basal metabolic rate of all tissue cells in higher animal organisms is controlled by the hormone *thyroxine,* secreted by the bilobed thyroid gland (Fig. 3-3), which lies beneath the voicebox and embraces the windpipe. Normally, the rate of secretion of this biochemical control agent is inversely proportional to the metabolic rate. When the metabolic rate is high, virtually none of this hormone is released. Conversely, quantities of this stimulatory chemical are secreted into the bloodstream, when the metabolic rate is low.

The thyroid gland itself does not respond directly to metabolic stimuli. The thyroid is controlled by the *pituitary* gland, an anatomically small but functionally crucial gland, suspended from the base of the brain. The pituitary was called the "master gland," since most of the hormones it releases have as their target organs other endocrine (hor-

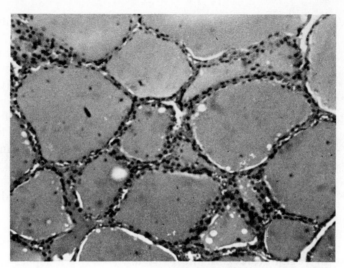

Fig. 3-3

Thyroid follicles. The spherical spaces are filled with a protein-bound form of the hormone thyroxine, which controls the basal metabolic rate of tissue cells. This hormone is synthesized and secreted by the cells lining the follicles.
From M. J. Zbar, *Introduction to Human Anatomy,* Saturn Scientific, Inc., Fort Lauderdale, Fla., 1968.

monal) glands. (Hormones that regulate the activities of other endocrine organs are called *trophic hormones,* or "tropins.") Then it was learned that a portion of the brain base, the *hypothalamus,* was the "master of the master gland," and, further, that stimuli from the internal and external environments were the "masters of the master of the master gland." The circle of control was complete. The two major control systems, the nervous and hormonal, met not only with themselves but with the environments as well. The organism was not environmentally independent.

A drop in metabolic rate beneath the optimum is the change in the internal environment (stimulus) which causes sensitive cells in the hypothalamus to secrete a *thyrotropin releasing factor,* (TRF), that stimulates the pituitary to release *thyrotropin,* or the thyroid stimulating hormone (TSH). TSH stimulates the thyroid to secrete thyroid hormone (thyroxine) into the bloodstream. After about 24 hours, the metabolic rate of tissue cells rises not only to normal but slightly above. A response has been effected. The response (a rise in metabolism) negates or abolishes the stimulus (a lowered metabolic rate), and the hypothalamus ceases to liberate TRF. The pituitary stops secreting TSH, and the thyroid gland halts the release of thyroxine. This is negative feedback (Fig.

3-4). It was said that the thyroid hormone fed back negatively against its stimulatory hormone. As thyroxine rose, TSH fell. This is true. More directly, however, the rise in metabolic rate inhibits (shuts off) the release of TRF. Thus, the "fires of life" are perpetually stoked and damped.

In the homeothermic (constant temperature) organisms (birds and mammals), exposure to cold produces a "calorigenic" (heat-producing) response by activating the same system. Normally, more than 50% of the energy released by respiratory metabolism appears as heat. Less than 50% is stored in the bonds of ATP. Thyroxine partially uncouples oxidation from phosphorylation, ATP formation, and 70 to 80% of the energy liberated in aerobic (oxidative) respiration appears as heat. Thus, this system is not isolated from the external environment. An adaptive response is made. Heat is produced.

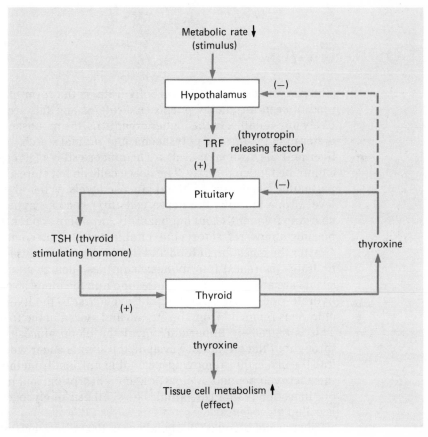

Fig. 3-4   **Control of metabolism by thyroid secretion.**

Many hours are required for thyroxine to exert its effect. The neuro-muscular response of shivering, which yields heat as a by-product of muscular contraction, is much quicker and more effective. The nervous control of the rate and volume of blood flow through the skin vessels is carefully regulated and constitutes the major mechanism of controlling heat loss and gain. In general, nervous control is rapid and hormonal regulation is slow.

In "cold-blooded" (inconstant temperature) vertebrates (reptiles, am-phibians, and fish), the body temperature and metabolic rate fluctuate to a great extent with the conditions of the surrounding environment. Their thyroxine has no calorigenic (heat-producing) effect. In many am-phibians, the thyroid hormone plays a key role in transforming the tad-pole into the adult. Without homeostatic body temperature, homeo-thermy, the lower vertebrates are less "free."

BLOOD SUGAR

Glucose, the simple sugar carbohydrate, is the form of foodstuff most readily used for energy purposes. Proteins and fats are normally rela-tively unavailable. In the higher organisms, the monosaccharide glucose is stored in the form of polysaccharide "animal starch," *glycogen,* in the liver and the skeletal muscles. The glucose stores (glycogen) of muscle cannot be tapped directly. The nerve cells do not store glycogen and are particularly sensitive to blood glucose levels. When the blood glucose level drops too low, the nerve cells first become hyperexcitable, dis-charging frequently and haphazardly, producing convulsions, and then become almost refractory, inexcitable, resulting in coma.

After the ingestion of food and the digestion of polysaccharide carbo-hydrates (starches) into monosaccharides such as glucose, the simple sugars are absorbed from the intestine into the blood stream. The carbo-hydrate-rich blood from the gut travels first to the liver, where excess glucose is removed from the stream and stored in the form of glycogen. This is extremely important, for, if the blood glucose level was high throughout the circulatory system, the simple sugar would "spill over" into the urine, thus removing a valuable nutrient from the body, wasting the energy expended in digestion and absorption, and making continu-ous ingestion (eating) essential. The synthesis of glycogen from glucose is called *glycogenesis.*

The normal blood level of glucose is 100 milligrams (mg) percent [per 100 milliliters (ml) of blood]. After the digestion of a meal, the blood sugar is elevated, which is somehow a stimulus to certain cells in the pancreas (Fig. 3-5) to release the hormone *insulin* into the bloodstream. Insulin is thought to alter the permeability of tissue cell membrane to glucose, "facilitating" its inward diffusion, and perhaps to catalyze the catalysts (enzymes) that transform the glucose molecule and lead it into metabolism, synthesis, or degradation. After a meal, the blood glucose level can rise to 140 mg %. Between meals, it can fall to 90 mg %. The average level maintained is about 100 mg %. A buffer is an agent that prevents the development of extremes, a padding device that absorbs shock. The liver is said to be the "blood glucose buffering organ," for, when the blood glucose drops, the liver cells break down glycogen (glycogenolysis) and release glucose into the bloodstream, raising its levels. When the blood glucose levels rise too high, the liver cells remove the sugar from the blood and store it in the form of glycogen. Glycogenesis alternates with glycogenolysis.

Several agents are antagonistic to insulin. *Glucagon* is a hormone from another variety of pancreatic cells that promotes glycogenolysis and therefore elevates blood glucose levels, the opposite to the effect of

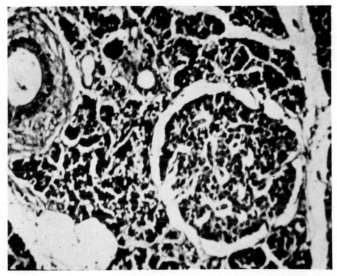

Fig. 3-5

Pancreas. The knot of tissue encircled by white space is an "islet" of insulin-secreting cells. The islet consists of cords of secretory cells interlaced with capillaries. In many diabetics, the islets are degenerate.

From M. J. Zbar, *Introduction to Human Anatomy,* Saturn Scientific, Inc., Fort Lauderdale, Fla., 1968.

insulin. The neurohormone adrenaline (from nervous tissue) also raises blood sugar, as do certain hormones called the glucocorticoids. Hunger begins when the blood glucose level drops too low. There are glucose-sensitive cells in the hunger center of the hypothalamus.

Lack of insulin produces the disease "sugar diabetes," in which glucose levels rise so high that sugar is lost in the urine and takes body water along with it. At the same time, proteins and fats are broken down for energy purposes, since carbohydrates can no longer be utilized. *Diabetes* literally means a "passing through." In the first years of the twentieth century, the untreatable diabetics ate and drank voraciously to no avail. Voluminous urination and wasting continued, and they died the seeming victims of starvation and dehydration. When it was discovered that the pancreas of diabetics was defective, and when a means of extracting the hormone insulin from normal animal pancreases for injection was perfected, it became possible to treat—but not cure—the disease by replacement therapy. The survival of experimental dogs following the removal of their pituitaries after the removal of the pancreas suggests that the "master gland" plays some role in this homeostatic process.

The rise and fall of hormones in response to internal and external environmental stimuli are "chemical reflexes," analagous to nervous reflexes. The word "hormone" is derived from the Greek *hormaein,* which means "to excite," and most hormones are stimulatory. The effects they bring about usually require several hours to even days and months. It has been noted that nervous responses are rapidly effected, often in milliseconds. Although nerve impulse pathways are frequently described as "physical," there are breaks in the chains where chemical transmitters are released. The neurohormones released by the hypothalamus will be discussed subsequently. Sex and adrenal hormones will be described in the chapters on reproduction and ecology, Chapters 4 and 7, respectively. Table 3-1 lists the mammalian hormones, their formation, sites, and actions.

THE NERVOUS SYSTEM

The evolution of higher nervous systems led to the "freest" and most independent life, the greatest range of adaptability. Can it be said that the human brain is the ultimate homeostatic organ thus far evolved?

"To integrate" means to make whole by bringing the parts together so that harmony results. Homeostasis is equivalent to harmony. The nerv-

Table 3-1

Mammalian hormones

Hormone	Site of Formation	Action
Somatotrophin release factor (STRF)	Hypothalamus	Stimulates adenohypophysis to secrete somatotrophin
Corticotrophin release factor (CTRF)	Hypothalamus	Stimulates adenohypophysis to secrete adrenocorticotrophic hormone
Thyrotrophin release factor (TTRF)	Hypothalamus	Stimulates adenohypophysis to secrete thyrotrophin
Gonadotrophin release factors (FSH-RF and LH-RF)	Hypothalamus	Stimulate adenohypophysis to secrete gonadotrophins (FSH and LH)
Prolactin inhibiting factor (PIF)	Hypothalamus	Inhibits adenohypophysis from secreting prolactin
Melatonin	Pineal	Regulates gonadal annual cycle (?)
Antidiuretic hormone (ADH)	Hypothalamus (via neurohypophysis)	Promotes water retention in kidneys
Oxytocin	Hypothalamus (via neurohypophysis)	Promotes milk release, uterine contractions (?)
Melanocyte-stimulating hormone (MSH)	Pars intermedia of pituitary	Stimulates melanin synthesis
Somatotrophin (growth hormone)	Adenohypophysis (anterior pituitary)	Promotes somatic (especially skeletal) growth
Adrenocorticotrophin (ACTH)	Adenohypophysis	Stimulates adrenal cortex to secrete glucocorticoids
Thyrotrophin (TTH)	Adenohypophysis	Stimulates release of thyroid hormones
Follicle-stimulating hormone (FSH)	Adenohypophysis	Promotes maturation of ovarian follicles, secretion of sex hormones
Luteinizing hormone (LH) (same as ICSH)	Adenohypophysis	Promotes growth of corpus luteum, secretion of sex hormones
Prolactin	Adenohypophysis	Promotes milk production, parental behavior
Thyroxine	Thyroid	Stimulates oxidative metabolism, promotes somatic growth
Thyrocalcitonin	Thyroid	Lowers blood Ca^{2+} level
Parathormone	Parathyroid	Elevates blood Ca^{2+} level
Glucocorticoids (corticosterone and the like)	Adrenal cortex	Regulate carbohydrate metabolism
Mineralocorticoids (aldosterone and the like)	Adrenal cortex	Regulate excretion and retention of Na^+ and K^+
Adrenaline and noradrenaline	Adrenal medulla	Reinforce action of sympathetic nervous system
Insulin	Pancreatic islets (Islets of Langerhans)	Promotes passage of glucose into cells, lowers blood sugar level
Glucagon	Pancreatic islets	Promotes breakdown of glycogen, elevates blood sugar level
Estrogens (estradiol, estrone, and the like)	Ovaries	Promote female secondary sexual characteristics, development of genital tract
Progesterone	Ovaries, corpus luteum	Promotes and maintains pregnancy
Androgens (testosterone, androsterone, and the like)	Testes	Promote male secondary sexual characteristics, sexual drive
Gastrin	Gastric mucosa	Promotes flow of gastric juice
Enterogasterone	Duodenal mucosa	Inhibits flow of gastric juice
Secretin	Duodenal mucosa	Excites flow of watery component of pancreatic juice
Pancreozymin	Duodenal mucosa	Stimulates secretion of pancreatic enzymes
Cholecystokinin	Duodenal mucosa	Stimulates contraction of gall bladder and release of bile
Angiotensin	Blood factor activated by renin from kidney	Elevates peripheral blood pressure, stimulates adrenal cortex to secrete aldosterone
Thymic hormone	Thymus	Promotes immunological competence of lymphoid tissues
Chorionic gonadotrophin	Placenta	Maintains pregnancy by promoting growth of corpus luteum of pregnancy
Relaxin	Placenta	Relaxes pubic symphysis before partuition

ous system is the major control system, integrating stimuli from both environments with appropriate responses. Yet the nervous system is not merely a connection between stimulus and response. The central nervous system (CNS), which includes the brain and spinal cord, does more than simply relay messages. The CNS acts as a "clearing house" for stimuli, like a programmed computer selecting appropriate responses. The brain of man programs itself, sometimes creating new responses to imagined experiences in ingenuity or insanity.

It seems that man and many animals are more than simple machines. Consciousness is a wondrous thing. The seeming independence of the complicated activities of the "higher centers" of advanced nervous systems led men to the devout belief that they could escape their bodies and still exist. This is neither scientifically provable nor refutable. It does not appear that the brain will ever turn out to be merely "head guts," but it is quite likely that there will be a unique mechanistic explanation of its function. Perhaps, then, we shall have prophetically fulfilled the ancient directive inscribed upon the temple of Apollo at Delphi: "Know thyself."

Now man uses extrasomatic, extracorporeal devices to secure the homeostasis of his technological life, brain-wrought, in a sense, unnaturally evolved. Certainly the Belgian physiologist, Leon Fredericq, had the higher nervous system in mind when he wrote in 1885:

The living being is an agency of such sort that each disturbing influence induces by itself the calling forth of compensatory activity to neutralize or repair the disturbance. The higher in the scale of living beings, the more perfect and the more complicated do these regulatory agencies become. They tend to free the organism completely from the unfavorable influences and changes occurring in the environment.[1]

THE NERVE CELL AND ITS FUNCTION

The structural unit of the nervous system is the nerve cell, or *neuron,* whose long cytoplasmic processes ("fibers") reflect its function of conduction (Fig. 3-6). The nerve impulse involves electrochemical events, but it is not an electric current. The rate of conduction is too slow, 1 meter per second (mps) in unsheathed nerve fibers and 100 mps in insulated fibers, as compared with the speed of electricity, which is hundreds of millions of meters per second. This is not to say that the nerve impulse is crude; its speed is more than adequate.

[1] Walter B. Cannon, *The Wisdom of the Body* (New York, N. Y.: W. W. Norton & Company, Inc., 1963), p. 21.

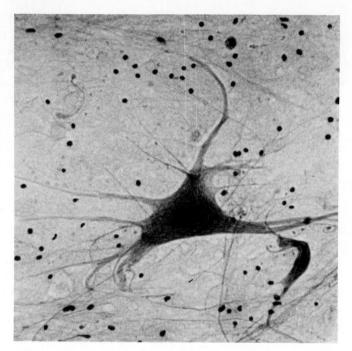

Fig. 3-6

**Nerve cell. Note the numerous cytoplasmic processes. These
are the conductile nerve "fibers."**
From M. J. Zbar and M. D. Nicklanovich, *Nervous System,* Saturn
Scientific, Inc., Fort Lauderdale, Fla., 1971.

There is a charge across the nerve cell membrane with a potential
difference between the outside and inside of about 70 millivolts (mv)
[$\frac{1}{1000}$ volt (v) equals 1 mv]. The outside is positive with regard to the inside (Fig. 3-7), due to a differential distribution of ions. The extracellular
fluid (ECF) is rich in sodium ions (Na^+) and poor in potassium (K^+). The
intracellular fluid (ICF) is poor in sodium, rich in potassium, and contains an abundance of negative ions, charged organic acids, proteins, etc.
(Fig. 3-8). At the physiological pH of 7.4, most proteins bear a negative
charge. They are nondiffusible. When the nerve cell dies, potassium ions
move out and sodium ions come in until their concentrations on both
sides of the membrane become nearly equal. The unequal distribution
found in life is maintained at the expense of energy, and it is believed
that "sodium pumps" and "potassium pumps" with carrier enzymes are
distributed all along the nerve cell membranes and actively transport
sodium outward and potassium inward. The pumps are "driven" by the
breakdown of ATP. So far, we have described the nerve cell at "rest,"
ready for conduction.

The nerve impulse is somehow generated when stimuli, such as pressure, heat, or certain chemicals, change the properties of the membrane at a site. Sodium rushes in, the *resting membrane potential* of 70 mv drops first to zero, then overshoots so that the charge is reversed. The membrane becomes positive on the inside and negative on the outside. The site is said to be *depolarized* at zero and reversely polarized when the charge is switched (Fig. 3-9). After sodium has influxed, potassium effluxes. Before another impulse can be generated, the original distribution of these ions must be reestablished. It is here that the work of the nerve is done. The energy is not expended in conduction but in preparation for it. Electrical stimuli will discharge a nerve. When reverse polarity is established at a discharge site, the switched potential causes the conduction of miniature electric currents through the adjacent, normally charged areas of the membrane, which, thus stimulated, fire (Fig. 3-10). Whenever one part of a membrane changes its potential (whether it reaches zero or not), there is current flow, and adjacent areas of the membrane begin to be affected. The impulse spreads over the cell membrane of the cell body of a neuron and along the membranes enclosing cylinders of cytoplasm in the processes (fibers). A circular wave of depolarization moves along the fiber's membrane, and concentric circles

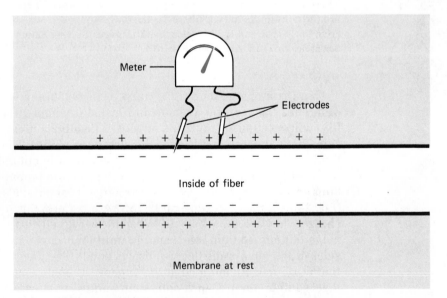

Fig. 3-7   At rest, the membrane of a nerve fiber is electrically polarized. The inside is negative in relation to the outside. A millivoltmeter will register a potential difference of 70 to 100 mv across the membrane.

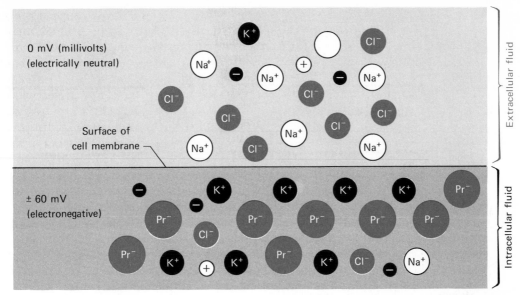

Fig. 3-8   **Distribution of ions in the intracellular fluid (ICF) and the extracellular fluid (ECF). Note the large number of nondiffusible cytoplasmic proteins (Pr⁻). The ECF is rich in sodium (Na⁺) and poor in potassium (K⁺). Their concentrations are just reversed within the cell. These concentration differences (a diffusion gradient) are maintained by the cell's active transport pumps, which bail out sodium and concentrate potassium. Because of the efficiency of the sodium pump, the resting membrane can be said to be impermeable to sodium for all practical purposes, even though a small amount of sodium continually diffuses inward. The negative ions of the ICF outnumber those of the ECF.**

of depolarization spread out from the original site of excitation, as ripples fan out beyond a rock's splash in water. The insulation of sheathed nerve fibers is interrupted at places (the nodes), and only these sites depolarize. The impulse jumps from node to node, and the electrical stimuli of reverse polarity are conducted through the electrolyte (current-conducting) salt solution of the ECF. This saltatory (jumping) conduction is the quickest, since whole sections of fiber are skipped (Fig. 3-11). The reestablishment of the "resting" membrane potential is called *repolarization*. Waves of repolarization follow on the heels of depolarization. So far, nothing has been said about the direction of impulse conduction.

Nerve cells come in a variety of shapes and sizes. The "typical" neuron usually described is a motor (command) neuron with short, stubby processes (dendrites) projecting from one side of the cell body and a

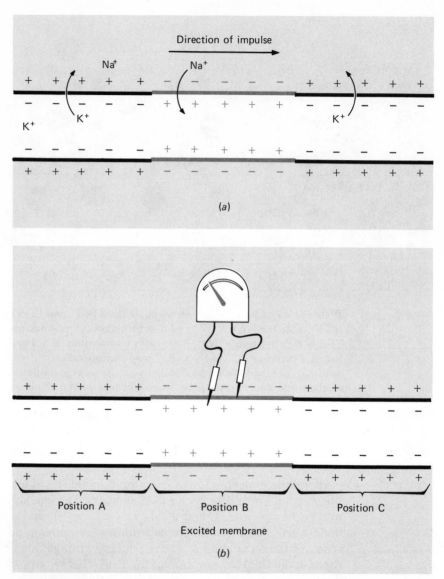

Fig. 3-9   (a) The polarity of the membrane is reversed during discharge, because the excited part of the membrane becomes more permeable to sodium. (b) As the nerve impulse passes along the fiber, the accompanying wave of reversed polarity registers a reverse potential at the discharging site. The resting potential is restored in the wake of the impulse.

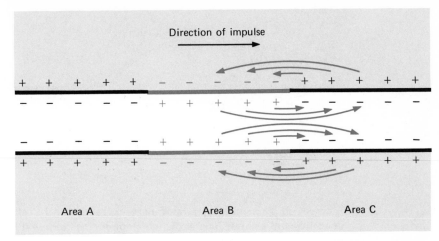

Fig. 3-10   **The excited part of a nerve cell membrane electrically stimulates adjacent areas. It is as though miniature currents flowed from discharging to resting sites.**

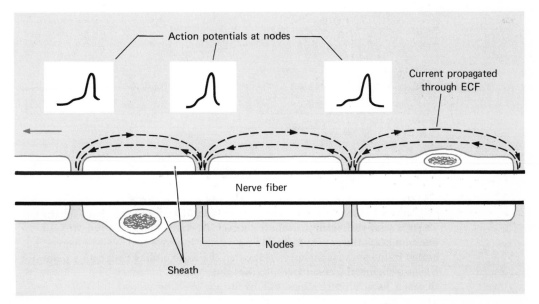

Fig. 3-11   **Saltatory or jumping conduction from node to node along an insulated fiber.**

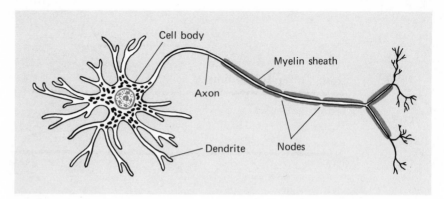

Fig. 3-12   A nerve cell. This multipolar motor nerve cell consists of a cell body with short
dendrites and a single long axon, which undergoes terminal branching before
the nerve endings. Axons can be 1 m long.

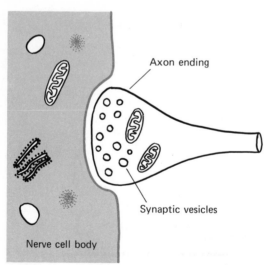

Fig. 3-13   The synapse, where one nerve fiber ends upon the dendrite or
cell body of the next nerve cell in the chain of conduction.
There is a gap rather than intimate contact. The nerve impulse
does not cross the gap, but rather a new nerve impulse is generated in the successive nerve cell when the axon ending releases a chemical transmitter into that gap. Some substances
released here inhibit rather than excite the next nerve cell.
The synaptic vesicles are believed to store the transmitter. An
electron microscope would be required to reveal the detail
shown here.

long, thin process (the axon) extending from the other side (Fig. 3-12). The dendrites are said to conduct toward the cell body, and the axon carries impulses away from it. The "dendrons" of the sensory neurons are quite like axons structurally, but they conduct in the opposite direction. Actually, the impulse can spread in either direction but seldom does, since (with perhaps one exception) no nerve cell acts alone and the communication between neurons is polarized or unidirectional.

Most neurons act in chains, yet never meet physically. When the impulse of one nerve cell reaches the end of the link, it dies, but a new impulse is generated in the next cell in consequence of it. The place where the fiber ending of one nerve cell comes close to the cell body of the next is called a *synapse,* and a chemical "transmitter" substance is released here by the nerve ending (Fig. 3-13). This compound does not really carry the same impulse across the synapse. The transmitter alters the properties of the cell membrane of the next nerve cell in the sequence so that it is excited and discharges, generating a new impulse, etc. (Acetylcholine is a common synaptic transmitter substance in the peripheral nervous system, but the central nervous system transmitter is unknown.) The nerve endings of motor neurons release acetylcholine upon the muscle cell membrane, which is also polarized and excitable (Fig. 3-14). The events are much the same as in nerves, except that contraction follows conduction and excitation. If acetylcholine remained at

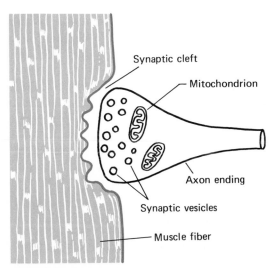

Synaptic cleft

Mitochondrion

Axon ending

Synaptic vesicles

Muscle fiber

Fig. 3-14   **The neuromuscular junction is very similar to the nerve-to-nerve synapse. Acetylcholine is believed to be stored in these synaptic vesicles.**

the synapse (or the neuromuscular junction), the nerve cell (or muscle cell) membrane could not repolarize and conduct again. An enzyme acetylcholinesterase breaks it down. New acetylcholine is synthesized and stored in the nerve endings. The breakdown of acetylcholine is extremely important, for the strength of a stimulus is reflected in the number of impulses conducted per unit of time. Some nerve terminals secrete an inhibitory transmitter instead of an excitatory transmitter. Gamma aminobutyric acid (GABA) has been nominated as the inhibitory transmitter within the CNS. What do you think would be the mechanism of inhibition? See Fig. 3-15.

THE REFLEX

It used to be said that the functional unit of the nervous system was the *reflex arc,* the circular pathway between stimulus and response. Some investigators believed that all behavior would eventually be explained as a series of reflexes. Although this is far from the modern view, the reflex was the place to begin. A pure reflex is a simple method of response, involuntary, unmodified by the will, automatic, completely programmed, rapid, unimpeded by deliberation. Yet, even conscious behavior has its beginnings in sensations, the registration and recognition of stimuli, and ends in response, frequently slow, to be sure, and far from automatic. Although the pathway between stimulus and response is not fixed and will intrudes, nevertheless, an arc of neurons is involved. The number of nerve cells employed in voluntary behavior is greater than that found in reflex activity, and the route of the neuron chain is long and twisted, unlike the few nerve cells of the simple reflex arc, which follow a simple loop. The nervous reflex is a perfect example of a control system. Let us analyze its components.

The *receptor end organs* are usually sensory nerve endings associated with connective tissues (Fig. 3-16). There are receptor end organs just beneath the surface of the skin that are sensitive to changes in pressure (light touch and heavy pressure), temperature (heat and cold), and pain (injurious distortion produced by several of the preceding stimuli). Nerve wrappings of tendinous and muscular fibers are receptors sensitive to stretching. There are internal organ receptors sensitive to changes in blood pressure and the chemical composition of the body fluids. These are the detectors or "sensors" of the control system (reflex). They translate mechanical, thermal, and chemical stimuli into nerve impulses bound for the CNS by way of *afferent* (toward) *fibers.* These particular

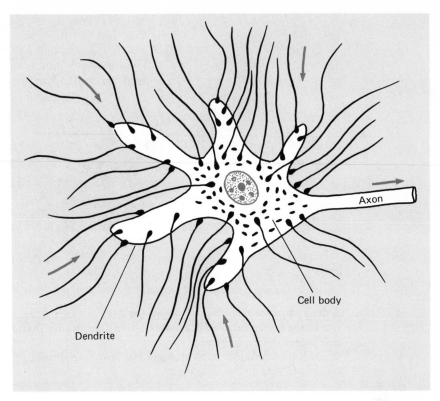

Fig. 3-15   **Hundreds of nerve fiber terminals synapse with the dendrites and cell body of a motor neuron. Whether it fires depends upon the algebraic sum of excitatory and inhibitory discharges upon the membrane.**

fibers that lead from the receptor end organs to the cell bodies of the sensory neurons are known as the *peripheral processes*. The cell bodies of the sensory neurons are found in collections called ganglia alongside the spinal cord or the brain. (A *ganglion* is a collection of nerve cell bodies outside the CNS.) A *nucleus* is an aggregation of nerve cell bodies within the CNS. Ganglia are associated with both cranial and spinal nerves. From the sensory nerve cell bodies *central processes* arise and conduct impulses into the cord or the brain.

In the simple spinal reflex arc, the central processes usually synapse with the cell bodies of *associational neurons* in the cord (Fig. 3-17). In cross section, the cord contains a butterfly of grey matter surrounded by white matter. The grey matter consists of nerve cell bodies and un-sheathed fibers, and the white matter is made up of the insulated fibers

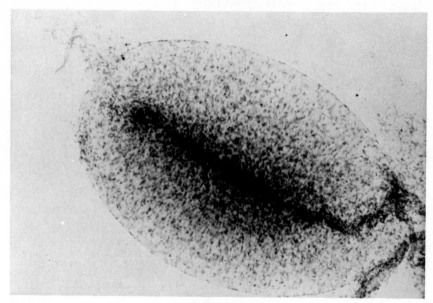

Fig. 3-16   **A deep pressure receptor end organ, the Pacinian corpuscle. A bulb of connective tissue layers surrounds a central nerve fiber ending. Distortion of the tissue discharges the nerve fiber.**
From M. J. Zbar and M. D. Nicklanovich, *Nervous System,* Saturn Scientific, Inc., Fort Lauderdale, Fla., 1971.

of the *ascending* and *descending tracts* coming and going from the brain. (A *tract* is a bundle of nerve fibers in the CNS, and a *nerve* is a bundle of nerve fibers outside the CNS.) The grey matter of the cord has a dorsal (backside) "horn" and a ventral (bellyside) "horn." The cell bodies of the associational neurons, which connect sensory and motor nerve cells, are located in the dorsal horn of grey matter, and their processes (axons) sweep around to the ventral horn to synapse with the cell bodies of the *motor neurons*. Sometimes there is no intervening associational neuron, and the central processes of the sensory neurons synapse directly with the cell bodies of the motor nerve cells. The axons of the motor cells leave the cord by way of ventral "roots," which join the dorsal (sensory) roots to form the spinal nerve. The spinal nerve then branches dorsally and ventrally, sending branches to the dorsal body wall and musculature and the ventral body wall and limbs. The axons of the motor neurons therein are bound toward the *effector end organs,* primarily muscles. Motor fibers are said to be efferent, since they conduct impulses away from the CNS. The endings of motor nerve fibers upon muscle fibers are called *motor end organs* (Fig. 3-18). Upon receiving "commands" from the motor neurons, the effectors react, producing the response to the stimulus. A branch of a spinal nerve is "mixed," carrying both sensory and motor fibers.

Some cranial nerves are purely sensory or solely motor, but many are mixed. In a simple cranial reflex, sensory neurons eventually connect with motor nuclei in the brain base. Although the cranial reflex is not anatomically identical to the spinal reflex, it is functionally parallel.

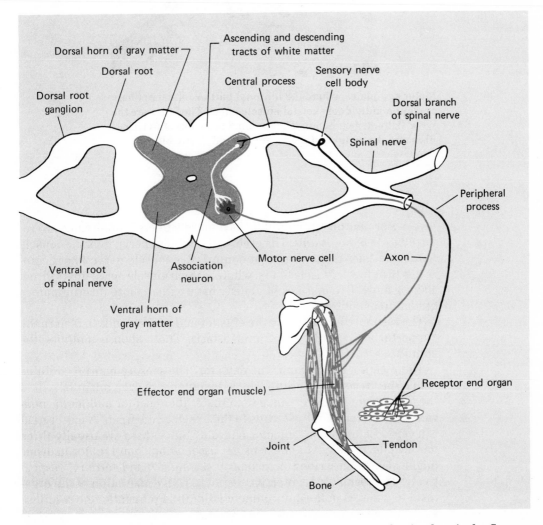

Fig. 3-17   **Cross section of the spinal cord and the elements of a simple spinal reflex arc.**

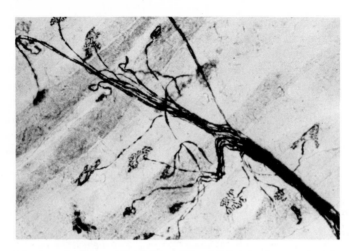

Fig. 3-18   **Motor end plates, where the terminal buttons of nerve fibers end upon individual skeletal muscle fibers. The latter are the wide, light-striped bands. The terminal branches of the nerve fiber and the motor end plates are stained nearly black.**
Courtesy of Carolina Biological Supply Company, Burlington, N. C.

A motor neuron and the 10 to 1000 muscle fibers it innervates are referred to as a *motor unit*. The greater the ratio of nerve fibers to muscle fibers, the finer is the degree of control. Eye muscle nerve fibers command 10 or fewer muscle fibers, whereas thigh muscle nerve fibers direct 1000 or more. Eye muscles obviously have a finer, more discriminatory gradation of control.

The response produced by the effector end organs tends to restore the "unperturbed" steady state, homeostasis. The response mollifies the stimulus.

The receptor end organ is the detector. The sensory neuron could be called the "receptor neuron." The associational neuron might be called the "connector neuron," since it connects the receptor and motor neurons. The motor nerve cell leads to the effector end organ. A motor neuron could be called an "effector neuron." Since there are usually three nerve cells in the chain, the sensory, associational, and motor neurons might also be described as primary, secondary, and tertiary, respectively. The result of the effective response is the adaptation of the organism to changes in the environment (stimuli). Frequently, the response shuts off (negative feedback) the stimulus.

The familiar knee-jerk reflex provides an excellent example. Although the patellar (kneecap) ligament ties the kneecap to the shin bone, the band is really the tendon of insertion of the massive, four-headed, front thigh muscle, the quadriceps femoris, which acts across the knee joint to extend the lower leg. (A *tendon* is a band of dense connective tissue that ties muscle to bone. *Ligaments* tie bone to bone.) The kneecap is simply a bone that develops within a tendon. When one stands stiff-legged with locked knees, the heads of the quadriceps are easily felt as bulges on the front and sides of the thigh, and the patellar tendon stands out tautly between the kneecap and shin bone. The lower leg is passively flexed (bent) and the great extensor relaxed, when one sits on the examination table with the leg dangling over the edge. The blow below the knee stretches the tendon and its muscle above, stimulating receptors (musculo-tendinous spindles) (Fig. 3-19) that relay impulses via the sensory neurons and their processes to the CNS (cord), where associational neurons synapse with both sensory and motor neurons. Impulses are conveyed back to the muscle fibers of the quadriceps by way of motor nerve fibers. The muscle fibers contract in response, the muscle shortens, and the stretch stimulation of the receptors is relieved (Fig. 3-20a).

The withdrawal reflexes perhaps better illustrate the rapidity and automatic nature of the reflex. The hand is withdrawn from the flame long before pain is consciously perceived. The pathway arcs through the cord; no higher levels are required to effect the response (Fig. 3-20b).

If one blade of a scissors is placed in the corners of a frog's mouth and

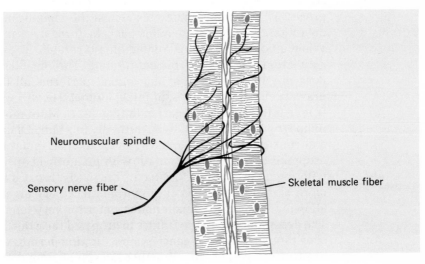

Neuromuscular spindle

Sensory nerve fiber

Skeletal muscle fiber

Fig. 3-19   **Neuromuscular spindle, a receptor end organ.**

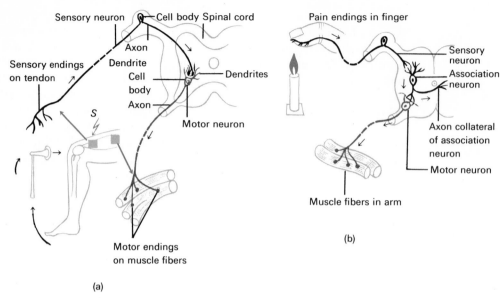

Sensory neuron — Cell body Spinal cord

Axon

Dendrite

Sensory endings
on tendon

Cell
body

Axon

Dendrites

Motor neuron

S

Pain endings in finger

Sensory
neuron

Association
neuron

Axon collateral
of association
neuron

Motor neuron

Muscle fibers in arm

(b)

Motor endings
on muscle fibers

(a)

Fig. 3-20   **Spinal reflexes. (a) The knee-jerk response, and (b) withdrawal. The hand is withdrawn before pain is consciously perceived.**

the other blade put across the top of the head just behind the eyes, it is possible to prepare a *"spinal" animal* or decerebrate animal with one stroke. (The forebrain, which includes the cerebrum, is removed in the process.) This animal maintains almost the same posture as before the operation. If he is placed on his back, he turns over and rights himself. When a normal animal is rotated on a turntable, he points his head in a direction opposite the route of rotation. The "headless" spinal animal does this likewise. The spinal animal performs all the typical withdrawal reflexes as well as an intact animal. He will withdraw his foot from a mildly acid (vinegar) bath (Fig. 3-21). More remarkably, he will jump when prodded and swim normally in a tank. Of course, he is blind and runs into objects ordinarily avoided. The intact frog breathes by dropping the floor of his mouth with his nostrils open, then raising it with them closed, thus forcing air down his throat into his lungs. The moist skin of the frog is an important accessory organ of respiration, allowing the frog to remain under water for very long periods of time. The decapitate animal continues to drop and raise the floor of his mouth even though the roof is gone! Skin respiration permits the spinal animal to survive several hours. Whether the spinal animal is dead or alive is a question of semantics. If a dissecting needle is now run down the spinal cord, injury potentials are generated in all the nerves and all the muscles contract tetanically. Then, the body goes limp and can no longer perform any of the previously described reflexes. Now the animal is absolutely dead, unquestionably. This "back pithing" destroys not only the cord but also the remaining hindbrain and brain stem. A number of integration centers for reflexes concerned with blood pressure, heart beat, and breathing are located in the base of the brain.

Decerebrate cats have been prepared for laboratory studies. Their behavior is similar to that of the decapitated frog, except that the experimental cats are exceedingly rigid. Spinal animals react to stimuli, but otherwise remain passive, not originating activity.

War wounds have occasionally produced cases of paralysis in humans that show many similarities to the spinal animals just discussed. The severance of the spinal cord in midback sometimes produces subjects retaining complete control of their upper body with immobility and insensitivity of the lower extremities, which will after a while respond reflexly to stimuli. Such men have been able to flex their legs uncon-

Fig. 3-21

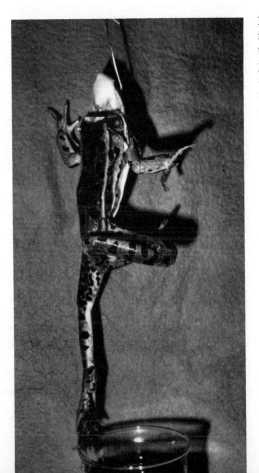

Decerebrate frog withdrawing its foot from vinegar (acid). The top of the head has been cut off just behind the eyes, yet this animal is still capable of performing a meaningful act unconsciously, automatically.

Courtesy of Saturn Scientific Inc., Fort Lauderdale, Fla.

sciously following the pinching of their toes. The fact that some time is required before the reflexes appear full blown, and then are relatively crude at that, suggests that the involvement of higher centers in "spinal" reflexes is considerable in advanced vertebrates. However, it would be surgically possible to prepare a decerebrate man whose behavior would closely resemble that of cats so treated.

Thus far, we have discussed stimulatory impulses and only briefly mentioned inhibitory impulses. A man can hold his hand in the flame if he wills it. Obviously, the withdrawal reflex has been overcome, inhibited. Conscious concentration, awareness, can either retard or facilitate the knee-jerk reflex, strengthening or weakening the response. The higher centers can modify a reflex. Muscles work in antagonistic sets. To extend a limb, the flexor muscles must be relaxed as the extensors contract. To flex a limb, the extensors must be relaxed as the flexors contract. Thousands of nerve endings synapse with the cell body of a motor nerve cell in the cord. Whether or not it discharges is the result of the algebraic sum of excitatory (facilitatory) and inhibitory postsynaptic ("after the synapse") potentials generated. No reflex is actually as simple as the three-neuron unit described. Thousands of nerve cells are involved in most reflexes. Many of their chains follow parallel pathways, however. When a frog flexes one leg in withdrawal reflex, he often reflexly extends the other leg simultaneously. Although the stimulus originates on one side, it must be carried across the cord to bring about the contralateral (opposite side) response.

Reciprocity is a feature of nervous control. Stimulation of some elements results in the inhibition of others. In the "chemical reflexes" of hormonal control, stimulation is followed by inhibition (negative feedback). Push is followed by, or accompanied by, pull in control systems. Response is hardly ever distinct, except in the wrenching reflexes of emergencies. The smooth curves of hormones and the fluid flexion of voluntary limb movement are the product of antagonism, a playoff of opposites, in which one set of opponents temporarily dominates. Control is something like Newton's law of mechanics: "For every action, there is an equal and opposite reaction." Yet, if this were precisely so, organisms would be paralyzed. Animals move toward or away from certain external stimuli. The stems of plants grow toward the light, and their roots point to the gravitational center of the earth. Even a stationary animal, properly oriented in space, is continually falling slightly off balance and regaining it. Slight tremors pass through the postural muscles. Some fibers contract as others relax, first on one side and then on the other. The same stretch receptors encountered in the knee-jerk reflex also provide information about the position of the limbs in space. These proprioceptors ("one's own initiators") are sensitive to tensional stimuli.

There are also receptor end organs in the internal ear that give the animal information about the direction of gravity and detect rotational movements of the body. All these spatial receptors initiate reflexes for righting the organism, restoring equilibrium. They provide an unconscious "body awareness" and sensory information for balanced, coordinated motion as well as immobility. This system of balance is probably best developed in the cats, whose grace and uncanny equilibrium are legend. Homeostasis applies to motion as well as "rest." Imbalance, internal or external, is only temporary. As the physiologist Langley states, "If there is a deviation in one direction, there is a reaction in the opposite direction."

THE AUTONOMIC NERVOUS SYSTEM

The *autonomic nervous system* (ANS) can function independently of consciousness and the will. (Autonomy means independence.) Sometimes it is called the involuntary nervous system. Since the ANS is concerned primarily with regulating the internal environment, and since its effector end organs are most often the smooth muscles of internal organs or viscera, it is occasionally called the visceral nervous system. By strict and traditional definition, however, the ANS includes only the *visceral motor neurons* and *not* the visceral sensory neurons and their internal receptor end organs. Thus far, we have spoken mostly about external receptors, somatic (body) sensory nerve cells, and somatic motor neurons. Since the ANS operates reflexly, the internal receptors and the visceral sensory neurons are unavoidably included in the pathways of visceral reflex, despite the archaic definition.

The ANS is further divided into *sympathetic* and *parasympathetic* divisions. Many internal organs are dually innervated by these two divisions, the effects of which are often antagonistic. The normal function of such viscera is the result of a playoff between these opposing influences. Usually, one division or the other dominates.

Not all internal organs are dually innervated. Some are controlled exclusively by the sympathetic division. It is puzzling that dual innervation was not followed throughout the interior; it seems to be a superior control system compared with single innervation. Yet, blood vessels, innervated by sympathetic vasoconstrictor fibers alone, are quite adequately controlled. Their normal tone is one of partial constriction, maintained by the arrival of an average rate of sympathetic impulses. When the rate of sympathetic stimulation rises above average, the smooth muscle in the vessel walls contracts more strongly. The vessels dilate as their smooth muscle relaxes following a decrease in the number of sympathetic impulses arriving per unit of time. Single innervation seems to be a conservative design. Still, the heart, stomach, and intestines are dually innervated.

   The sympathetic nerves arise from the chest level and midback regions of the cord, whereas the parasympathetic nerves arise from the brain stem and cord of the lower back (Fig. 3-22). In contrast to the so-

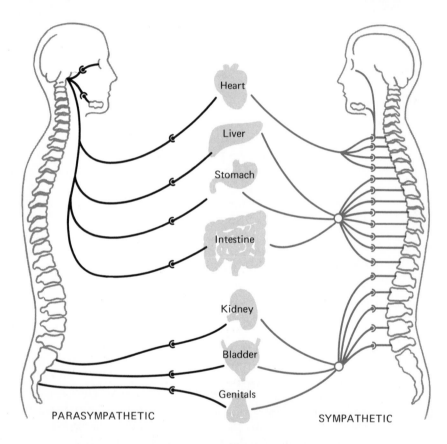

Fig. 3-22   **The autonomic nervous system (ANS) carries impulses to the internal organs and regulates their activities. Note that there is a chain of two nerve cells from the central nervous system to the organs innervated. Because of the regions of the cord from which its fibers exit, the sympathetic division (in solid color) of the autonomic nervous system is sometimes called the thoracolumbar outflow, whereas the parasympathetic division (black) is referred to as the craniosacral outflow for the same reason. Many parasympathetic centers are located in the brain stem.**

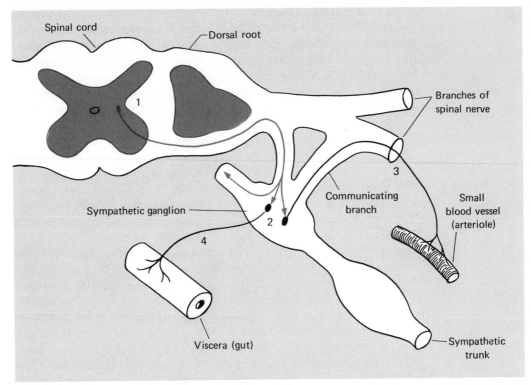

Fig. 3-23   **Sympathetic pathway of autonomic nervous system. At (1), the nerve cell body of the first neuron (of two) is located within the intermediate mass of grey matter. At (2), the fiber of the first neuron synapses with the cell body of the second neuron, whose fiber either (3) rejoins the spinal nerve and proceeds to end organs in the periphery or (4) leaves the chain by way of a visceral nerve bound for internal end organs. The presynaptic fiber can travel up or down the sympathetic trunk to synapse with a sympathetic ganglion cell at a lower or higher level.**

matic reflex pathways, which contain but a single motor neuron, the autonomic paths bear impulses over *two* visceral motor neurons to the effector end organs. The cell bodies of the first in the two-linked chain of command are located either in the grey matter of the cord between the dorsal and ventral horns (in the notch along the butterfly's wing), or in centers in the brain base. The cell bodies of the second visceral motor neurons are located in sympathetic ganglia just outside the cord or in parasympathetic terminal or collateral ganglia adjacent to or within the organ of innervation (Fig. 3-23). The endings of the fibers from the second neurons of sympathetic nerves release norepinephrine, or *norad-*

renaline, a compound very close in structure and action to adrenaline, or epinephrine. The interior of the adrenal gland (the medulla) is, in fact, a modified sympathetic ganglion whose cells lack processes and pour their neurohormone directly into the bloodstream, thereby producing widespread and generalized, rather than localized, responses. The parasympathetic nerve endings release acetylcholine. As the sympathetic nerves produce their effects by means of the chemical agent noradrenaline, they are said to be "adrenergic." The parasympathetic nerves are said to be "cholinergic," since they release acetylcholine.

The largest parasympathetic (and autonomic) nerves are the paired tenth cranial nerves, the *vagus* nerves. They descend through the neck to the chest, where branches are distributed to the heart and lungs. In the abdomen, branches of the vagus nerves are sent to the stomach and intestines. The long course of the tenth cranial nerve so impressed early anatomists that they called it the vagus, which is Latin for "wanderer." Although the vagus nerves contain 80% of the total parasympathetic fibers, only 20% of their fibers are parasympathetic. Many vagal fibers are visceral sensory and cannot be included in the ANS by definition. For instance, certain vagal sensory fibers carry impulses from stretch receptors in the lungs and form the sensory link in the reflex control of breathing. The vagal fibers to the heart and digestive tract are motor and therefore truly autonomic (parasympathetic).

The acetylcholine released in the heart by vagal nerve endings slows the cardiac rate. The heart's pacemaker determines an intrinsic rhythmicity: a frog's heart will beat even when all nerves to it are cut. The rate of a denervated heart is faster than an innervated one. The heart is normally under vagal inhibition. If acetylcholine solution is dropped on an exposed frog heart, it is brought to a near standstill. The sympathetic nerves to the heart seem to have little influence upon the heart rate except in emergencies, when their endings release noradrenaline in the heart and increase the rate and strength of contraction. Physicians sometimes inject adrenaline into the heart in cardiac emergencies.

The effects of the two divisions of the ANS on the digestive tract are just the reverse of heart control. The branches of the vagus nerves in the abdomen are the major motor nerves of the gut. They stimulate the movement of the bowel and the secretion of its glands. (A vagotomy is a section of the motor nerves of the stomach, occasionally performed in the treatment of gastric ulcer.) When acetylcholine is dropped upon an isolated length of rabbit intestine, the frequency and power of its contractions increase. Adrenaline has the opposite effect; sympathetic stimulation of the gut inhibits its motility.

In what has been called the "emergency reaction," sympathetic responses predominate. The blood vessels beneath the skin constrict, and the person blanches in "white rage" or "pale fear." The arteries to the skeletal muscles and the heart dilate, the greatest volume of blood becomes pooled in the internal veins, and blood is diverted from the gut and the kidneys in a virtual renal shutdown. The heart rate and strength of contraction increase. Blood pressure rises. Circulation becomes more rapid. The bronchioles (the smallest branches of the windpipe) dilate as the rate and volume of respiration increase. Adrenaline causes the breakdown of liver glycogen: the blood glucose level is elevated. The pupils dilate, permitting more light to enter the eye. Hair or feathers are raised as their associated smooth muscle contracts, creating "goose bumps." As a result, the contours of the body become vague, and the creature appears much larger, more imposing. Superficial wounds bleed slightly due to peripheral vasoconstriction. In this state, the animal is capable of amazing feats of strength as well as rapid escape. Perhaps that is why it is called the "fight or flight" syndrome (collection of symptoms). The cold "clammy sweat" of fear or anger appears in the palms. The organism is "energized to respond to danger before it strikes." The emergency reaction seems to be dramatically adaptive.

The earliest account of the evolutionary value of "emotion" (fear and rage) is to be found in *The Wisdom of the Body* by Walter Cannon, coiner of the word homeostasis and originator of the emergency theory. Although the emergency theory has been much criticized, no alternative theories have supplanted it or provided a more meaningful interpretation of the crisis response. In calmer states, the parasympathetic division frequently dominates autonomic tone. The use of the expression "pumping adrenaline" to describe an excited person has become proverbial.

The hypothalamus of the brain base, cited earlier as a hormonal control area, is also the principal center integrating autonomic functions. In view of its vital and varied roles, the physiologist, Terence Rogers,[2] was quite justified in stating, "The maintenance of homeostasis is absolutely dependent on the integrity of the hypothalamus since all homeostatic mechanisms are under some degree of autonomic control."

BRAINS

Those animals that remain attached or stationary throughout their lives have little in the way of heads with specialized sense organs and the associated, concentrated aggregates of nerve cells that are called

---

[2] Terence Rogers, *Elementary Human Physiology* (New York, N. Y.: John Wiley & Sons, Inc., 1961), p. 332.

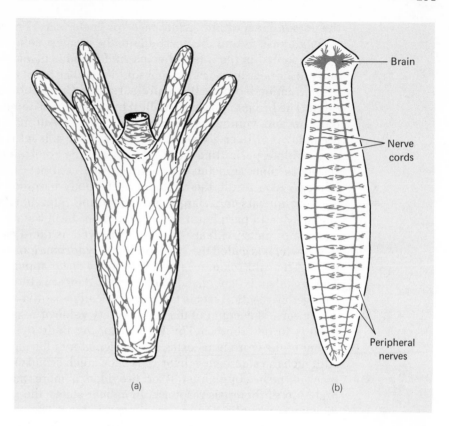

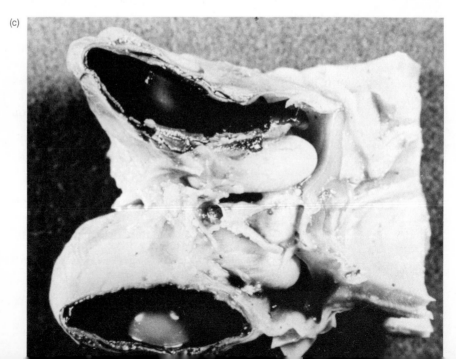

"brains." Even the jellyfish, which moves up and down randomly in the sea, has—like most attached organisms—the symmetry of a wheel, *radial symmetry*. (It can be sliced like a pie into equivalent halves along any number of planes.) The clam, which seldom moves and mostly filter-feeds, is like the legendary "headless horseman," although it lacks radial symmetry. Those creatures which move—and in particular hunt—such as the clam's relative, the squid, and the sluggish, lowly flatworm, planaria, have eyes and "brains" (Fig. 3-24). Specialized sense organs and expansions of the nervous system are located in the head region, which makes of that end a "head end." Such animals are said to be *cephalized*. The adjective "cephalic" is derived from the Greek word *kephale*, which means "head." Cephalized animals have back and belly sides in addition to "tail ends," and can be divided into equivalent halves (right and left) along one plane only. They are said to possess *bilateral symmetry*. The head end moves into new environments first, and, with its analytic computer brain, decides whether to proceed, after processing sensory information. If the stimuli are positive, the animal advances; if the stimuli are negative, the animal withdraws or turns. Human anatomists have often unjustly termed the brains of invertebrates "ganglia."

Like other organs, the brain evolved. There can be no question that the most advanced brain belongs to man (Fig. 3-25a). In the human brain, it seems, evolution overcame its major drawback, hindsightedness. Man "knows that he knows" and "has learned how to learn." His brain has turned upon itself, so to speak, for perhaps the final introspection.

The spinal cord leads up into the *brain stem*, which includes the *medulla oblongata* of the hindbrain as well as the hypothalamus of the lower forebrain (Fig. 3-25b). Above the hypothalamus lie the *thalami* (hypothalamus means "below the thalamus"), which are paired (thala-

---

Fig. 3-24   OPPOSITE PAGE: **Invertebrate nervous systems. (a) The primitive, uncentralized nerve net of hydra. (b) The centralized, ladder-type nervous system of planaria; the enlargements of the flatworm CNS are found in the head region in close association with the sense organs concentrated there: the eye spots and the projecting sensory "ears." (c) The squid eyes and brain; the tops of the major eye cavities have been removed, revealing the pearl-like lenses and the nervous film of the retina on the inside of the eyeball. The retina is surrounded by a pigmented layer, a "black box," and the overall design of the eye is the "camera type," similar to that of man. The squid brain lies to the right between the two dissected eyes. The most highly developed portions are the optic lobes, serving the great eyes.**

Photo from M. J. Zbar and M. D. Nicklanovich, *Nervous Systems*, Saturn Scientific, Inc., Fort Lauderdale, Fla., 1971.

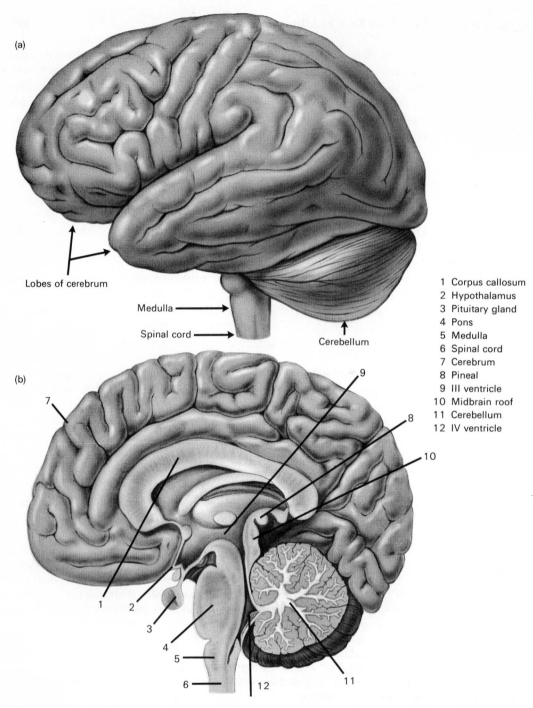

(a)

Lobes of cerebrum

Medulla

Spinal cord

Cerebellum

1 Corpus callosum
2 Hypothalamus
3 Pituitary gland
4 Pons
5 Medulla
6 Spinal cord
7 Cerebrum
8 Pineal
9 III ventricle
10 Midbrain roof
11 Cerebellum
12 IV ventricle

(b)

Fig. 3-25   **(a) Side view of the brain. Little can be seen other than the highly expanded cerebrum, which overlies the older parts of the brain. The cerebral cortex or surface is extensively folded, increasing the surface area and number of nerve cells located here. (b) Longitudinal section through the brain, exposing the inner surface of the right hemisphere and the buried structures of the brain stem.**

mus is singular) relay stations that project impulses to the "highest center," the cerebrum, the "seat of consciousness." The thalami are clearly seen in Fig. 3-26a. They are said to be the "seats of crude awareness." The thalami and the hypothalamus are clearly the "oldest" and lowest, most primitive parts of the forebrain. They sort out or filter sensory impulses from below, "deciding" which are to be sent to the highest centers, the cerebral hemispheres, the superior forebrain. The consciousness of man and other animals appears with the development of the cerebrum. The lower forebrain (the thalami and hypothalamus) is thought of as the highest level of the "old brain" which is capable of interpreting the external and internal environments sufficiently for primitive existence, without a great degree of discrimination in the recognition of stimuli or the selection of responses. The "elementary" sensations of hunger, thirst, fear, anger, and sex drive seem to originate in the layer of tissue around the thalami and their related structures. By inserting electrodes in this region, it is possible to call forth any of these "base" drives, such as "sham rage," by simply turning on the current. The cerebral hemispheres of man are the "newest" portions of his forebrain.

The youngest part of the hindbrain is the *cerebellum,* which overlies the older hindbrain, the medulla, just as the cerebrum overlies the lower forebrain (see Fig. 3-25a and b). The cerebellum is said to be concerned with the "smooth coordination of muscular activity." Centers in the medulla integrating balance and posture and motor areas of the higher brain (cerebrum) connect with and conduct impulses to cerebellar areas.

The arrangement of grey and white matter in most of the brain is the reverse of what is found in the cord: the grey matter surrounds the white. The white matter constitutes the bulk and consists of insulated fibers that connect areas of the brain (associational fibers), brain centers and the spinal cord (projection fibers), and the halves of the brain (commissural fibers). These are the bundles of cables of the "computer." Inside the brain masses lie small spaces filled with cerebrospinal fluid. Thus, the brain can be roughly compared with a house, having a roof, walls, a floor, and rooms. This state is the result of the fact that the brain develops from expansion of an embryonic neural tube (see Fig. 3-26b). Within the cerebral hemispheres lie the lateral ventricles (I and II). The lower forebrain contains the third ventricle, and the fourth lies above the medulla and below the cerebellum. A passageway runs through the midbrain, connecting III and IV, and two openings connect I and II with III. The fourth ventricle narrows to form the minute central canal of the spinal cord. The fluid surrounding the brain and the cord is derived from the interior.

The thalami may be viewed as the walls of the third ventricle, and the hypothalamus forms its floor. In the rear of the roof of the third ventricle

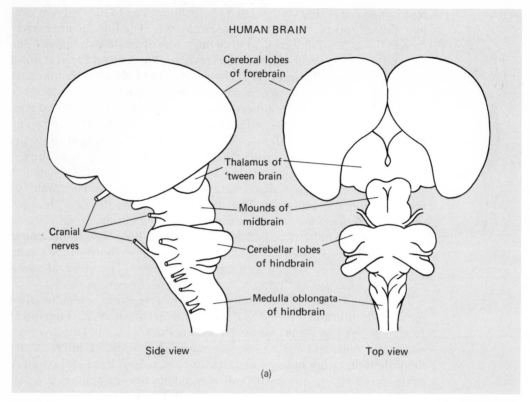

HUMAN BRAIN

Cerebral lobes
of forebrain

Thalamus of
'tween brain

Mounds of
midbrain

Cranial
nerves

Cerebellar lobes
of hindbrain

Medulla oblongata
of hindbrain

Side view   Top view

(a)

Fig. 3-26   **(a) Human brain at about three-month embryonic development. These views show the major divisions of the brain more clearly than can be seen in the adult brain, in which the great cerebral lobes overgrow the rest of the brain. At this stage, the cerebral surface is smooth, as in primitive mammals. (b)** OPPOSITE PAGE: **These earlier views of brain development show how the nervous system develops from a hollow tube (1). At the head end of the tube, three bulging structures become conspicuous and develop into the brain [(2) and (3)]. Roman numerals I through IV represent the ventricles, or hollow spaces, within the brain, which remain at maturity and reflect the tubular origin of the nervous system.**

is found the small, rounded *pineal gland* (Fig. 3-25b). In many lizards, the pineal body is a third eye. Even in frogs, where the pineal underlies the roof of the skull, photoreceptor cells similar to those found in the retina of the eye are retained. In man and the mammals, the pineal is overgrown by the cerebral hemispheres and the cerebellum, so that it is buried in the near center of the brain mass. Descartes proposed that here the body and soul interacted. Physicians have known for years that tumors of the pineal led to precocious puberty or the opposite, delayed

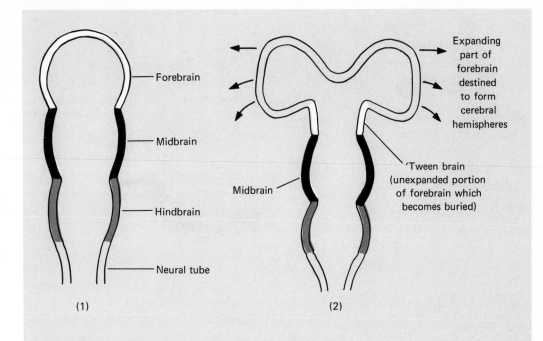

(1)   (2)

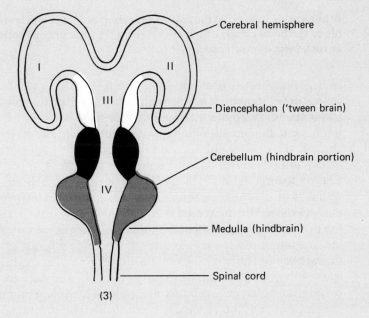

(3)

(b)

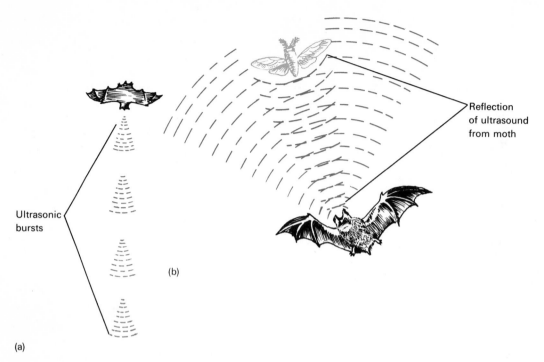

Reflection
of ultrasound
from moth

Ultrasonic
bursts

(b)

(a)

Fig. 3-27   Echolocation in bats: (a) the animal emits bursts of high-pitched sound (inaudi-
ble to the human ear), which are reflected (b) from prey or solid objects, such
as rock formations in caves.

sexual development. It has been shown that sensory nerves from the eye
connect with sympathetic ganglion cells in the neck, which, in turn, send
impulses to the pineal. When rats are kept in continuous light, the pineal
shrinks, the ovaries enlarge, and the estrous cycle speeds up. The cycle
of the animals slows down if female rats are kept in continuous dark.
The exposure of male hamsters to periods of 1 hour of light and
23 hours of darkness caused the shrinkage of their testicles. This effect
was prevented by the removal of the pineal. It was then discovered that
the exposure of experimental animals to long periods of light inhibited
the formation of an enzyme essential to the synthesis of *melatonin,* a
pineal hormone that inhibits sexual development and retards the sexual
cycle. Blinded rats do not respond to the light and dark treatments. The
correlation between reproductive cycles in birds and mammals and day
length (season of the year) may be controlled by the pineal gland. Some
investigators have gone so far as to suggest that the decreasing age for
puberty and menstrual onset in the civilized countries of the world is the
result of electrical lighting. It is not known whether the pineal hormone

exerts any influence on the hypothalamus, but radiolabeled hormone tends to be taken up by the gonads. Here is another example of a neurohormone.

In the roof of the midbrain are located four hills (Fig. 3-26a) of tissue that are reflex centers integrating body responses with light and sound stimuli. In the bat, which navigates by ultrasonic radar (Fig. 3-27) and has no time for conscious consideration of response to these high-pitched echoes, the lower hills of the midbrain are huge and allow automatic adjustment of flight in the dark. Again, we see the complementarity of structure and function.

The outer grey matter of the cerebral hemispheres, the *cerebral cortex,* is the seat of conscious thought, memory, and imagination. Yet, there are many areas of the cortex that are devoted to body functions and special senses (Fig. 3-28). Here are the body sensory and motor areas, as well

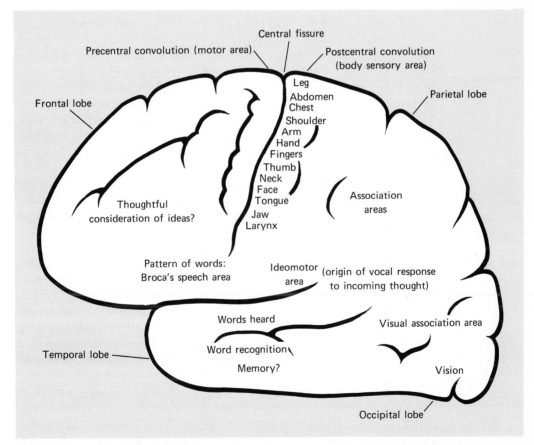

Fig. 3-28   **Tentative map of the cerebral cortex. Side view of cerebral hemisphere.**

as visual and auditory regions. Impulses from the toes, fingers, and tongue rise to the cortex where they are consciously perceived. Very large sensory areas are devoted to the lips and fingers. The wide visual sensory region occupies the rear surface, and the auditory centers lie on the sides. Corresponding motor areas lie in front of the great body sensory area. The regions described so far account for less than half the surface of the cerebral cortex, whose area is increased by extensive folding in man and a few other animals.

The remainder of the cortex was formerly classified as "silent areas," but this phrase has been abandoned in favor of *"association areas,"* where activities such as logic, memory, and planning probably take place. Many of the cerebral areas have been mapped by inserting microelectrodes into the brains of patients undergoing conscious neurosurgery. In addition, the localization of damage in impaired function has provided much information about the normal roles of brain areas. Many motor neurons of the cerebrum convey impulses to the motor neurons of the cord, so that the latter are designated the "final common pathway." Those impulses that arise in the cortex and inhibit simple spinal reflexes below are particularly striking. Other impulses take a less direct route, following side roads to other areas of the brain before descending. Motor impulses are shunted from the cerebral hemispheres to the cerebellum. The cerebrum thus plays an important role in fine coordination. The coordination of human eyesight and hand manipulation, accurately judged manual skill, is most highly developed in man. There are relatively large speech motor areas on the sides of the human cerebrum as well as auditory association areas that interpret heard sentences as conceptual thoughts. The speech center controls the muscular patterns for forming different sounds by the voicebox and the mouth. The cerebral cortex is only 3 to 5 mm thick, and nerve cells do not regenerate. Ridges on the skull interior often damage the speech area of boxers.

The sides of the cerebrum also bear *gnostic* ("knowing") or common integrative areas, where all sensory signals are integrated into coordinate thoughts. Closely associated with this region is the *ideomotor area,* which decides upon a course of action in response to coordinated information from the gnostic area. The ideomotor and gnostic areas are almost always located on the left side of the brain, and the corresponding areas on the opposite side are seemingly suppressed. Presumably, this prevents confusion. The left cerebral hemisphere is slightly larger than the right.

The frontal areas of the cerebrum (the prefrontal lobes) operate in conjunction with the ideomotor areas to allow judgment before action. The prefrontal area can suppress or excite the ideomotor region. The prefrontal areas are the location of abstract thought, future planning, mathematical operations, and anxiety. The memory banks are also here. Animals with destroyed prefrontal lobes rapidly lose information gained by experience. Sometimes prefrontal lobotomies (brain sections) are performed upon psychotic patients. These patients frequently exhibit little judgment postoperatively, and their emotional response varies from euphoria (unjustified glee) to rage. Their personalities have been described as those of "giddy adolescents." Their moods are variable; happiness and rage are quickly forgotten. Association is one of the great triumphs of the human cerebrum, but it seems that the memory of painful experience can subvert this advantage.

One of the crucial questions of biology and philosophy concerns the anatomy of memory, the "parchment of the soul on which are writ the commandments and the sins." The first theory of memory proposed that new circuits were established in the multitude of neurons and potential pathways. Recent evidence implicates protein synthesis in the process. The ultimate cellular code is the genetic one, which directs protein synthesis, but memory is not physically hereditary. Custom has conditioned us against the notion that the mind should have an anatomy, even a molecular one. The substantial or material explanation of the operation of the ill-defined mind will undoubtedly gravely threaten the traditional dual concept of body and mind, just as the discovery of psychosomatic illness has done. Although physiologists are far from understanding the simplest intellectual processes, it is quite possible that current biochemical investigations may eventually lead to a mechanical explanation of those precious processes of imagination and artistic creativity.

BRAIN EVOLUTION

The brain stem of lower vertebrates is essentially the same as that of higher vertebrates. However, the cerebral hemispheres of lower vertebrates are small, anterior lobes primarily serving the sense of smell, the old "nose brains." The chemical sense of smell is quite important in aquatic animals. Smell receptors are scattered all over the body skin of the catfish. Enlargement of the cerebral hemispheres and their devotion to additional functions began in the reptiles (Fig. 3-29). The cerebrum underwent enormous expansion in mammals (and man in particular), where the number of cells and the surface area are tremendously increased by the addition of cell layers and the wrinkling and infolding of the brain substance, until finally the brain stem is dwarfed and overlaid. As this neural evolution involved a multiplication of nerve cells, so it

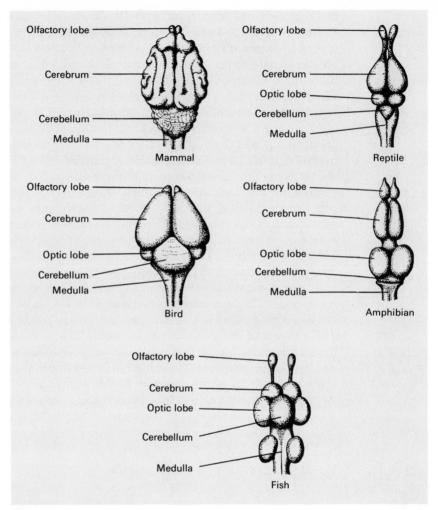

Fig. 3-29 **The comparative anatomy of vertebrate brains. They are drawn to the size they would be if they were all the same weight. Even so, note the difference in the size of the cerebrum. The cerebrum has undergone the most radical changes. Note that the cerebellum of modern fishes is better developed than it is in either amphibians or reptiles.**

increased the number of their possible interconnections. The expanded cerebral cortex and its underlying fibers make up the "new brain." The new has been grafted upon the old, as is so often the case in evolution. The spinal cord, brain stem, and cerebrum are levels of ascending hierarchy, and, in many cases, the new has come to dominate the old. Evolution is a conservative rather than a radical process, and none of the revolutions was total. Most external body reflexes are still controlled by the cord. The internal reflexes of circulation, excretion, and temperature control are yet directed by the autonomic or "vegetative" nervous system. A man remains nearly wholly unconscious of these activities. Even many "willed" (voluntary) actions, such as walking, involve a multitude of automatic operations as well as a myriad of nervous pathways.

The differences between the toad and man are differences in degree more than kind. There is excellent evidence that the lower levels of command in the brain are the most primitive or "older." The lower levels of command deal with automatic functions, such as heart rate, respiration, etc. Stimuli from the external environment can be dealt with at the level of the cord. The sensory impulses from the internal environment are integrated with reflex responses in the lower brain, the brain stem. Only in the highest centers of the highest vertebrates is the reflex escaped. Yet, even here there is an interaction with lower levels.

The nerve net of the jellyfish lacks centralization, and the animal's repertory of response is extremely limited and disproportionate to stimuli. Instead of a single tentacle being withdrawn, the whole body is retracted. With increasing size, complexity, and centralization of the nervous system, the stores of potential response became quite varied, and the degree of response became proportionate. The shackles of automatic response were finally broken. In a sense, security was cast to the winds. Evolution carried out a revolution within itself.

THE NERVOUS SYSTEM
AND THE COMPUTER

If any stimulus is strong enough to generate a nerve impulse at all, then the entire nerve cell and its fibers fire. A weak stimulus never causes a weak impulse. It is either strong enough to stimulate the nerve cell or not. This is called the all or none law. This feature of nervous activity is reminiscent of the binary system of notation used in programming computers. The circuit is either "on" or "off." It has been said that both nervous and electronic systems utilize electric pulses. Although miniature electric currents do occur along a nerve fiber, the nerve impulse is not an electric current.

Stimuli such as light, heat, pressure, and chemical reactions are converted into nerve impulses, and these, in turn, are reconverted into muscular activity. Energy translation or transformation occurs. The energy input (stimulus) is frequently much smaller than the output (response). This amplification is a characteristic of both living and electronic communication systems. Computers also convert mechanical, chemical, and magnetic changes into electric pulses, and sometimes back to magnetic or mechanical function (typing). Both the brain and the computer have "memory banks."

The brain is capable of providing memory storage that is a billion times greater than the memory bank of even the largest computer. This is attributed to the small size of neurons. The billions of neurons and their interconnections dwarf the few thousand tubes and transistors of a computer. Even the smallest transistor is still 10 times larger than the largest nerve cell body. A transistor requires $\frac{1}{10}$ watt (w) to operate; the energy required to operate a neuron is equivalent to $1/1,000,000$ w. A large computer weighs over 100 tons, whereas the brain weighs only 1 pound.

In the analogue computer, there is proportionality between input and output. Analogue computers can process a great range of quantities that are continuously variable and have linear properties. The digital computer merely identifies "on" or "off" conditions in the input as information for problem solving. Here, there is no proportionality between input and output. The brain demonstrates functional characteristics of both analogue and digital computers.

The differences between brains and computers are more impressive than the similarities. The human brain is more complex and flexible than any existing or planned computer. Complexity as well as flexibility seem to be the prerequisite of creativity. The computer is, after all, a "brain child." In minutes, computers can perform calculations that would require years of human labor.

The attempts to compare biological systems with mechanical models have met with a great deal of success. Our grasp of principles is much better for it. The mechanical viewpoint is indispensable for the research biologist. Man will probably be at the problem of his mind for centuries to come. Computers are not yet brains. "They" cannot take over any but the dull, routine tasks, abhorred by the mind, and which have wasted the time of men's lives long enough. It may be necessary for man to understand the workings of his own brain before he will be able to construct a computer like it. Prudence requires a mention of another John von Neumann admonition:

The problem of understanding the animal nervous action is far deeper than the problem of understanding the mechanism of the computing machine. Even plausible explanations of nervous reaction should be taken with a very large grain of salt.[3]

One day, man may not have to confess such ignorance of his "self." Perhaps then he will say, confidently, "The brain is a machine."

THE KIDNEYS

Until his death in 1962, Homer Smith, American scientist and writer, was the world's foremost expert on the kidney. He held this homeostatic organ in such high esteem that he entitled his work on the evolution of the vertebrate kidney *From Fish to Philosopher*. The following quotation from this famous book gives some insight into the meaning of its title:

The lungs serve to maintain the composition of the extracellular fluid with respect to oxygen and carbon dioxide, and with this their duty ends. The responsibility for maintaining the composition of this fluid in respect to other constituents devolves on the kidneys. It is no exaggeration to say that the composition of the body fluids is determined not by what the mouth takes in but by what the kidneys keep: they are the master chemists of our internal environment, which, so to speak, they manufacture in reverse by working it over completely some fifteen times a day. When, among other duties, they excrete the ashes of our body fires, or remove from the blood the infinite variety of foreign substances that are constantly being absorbed from our indiscriminate gastrointestinal tracts, these excretory operations are incidental to the major task of keeping our internal environment in an ideal, balanced state. Our bones, muscles, glands, even our brains, are called upon to perform an innumerable variety of operations. Bones can break, muscles can atrophy, glands can loaf, even the brain can go to sleep, and endanger our survival; but should the kidneys fail in their task neither bone, muscle, gland nor brain could carry on.

Recognizing that we have the kind of internal environment we have because we have the kind of kidneys that we have, we must acknowledge that our kidneys constitute the major foundation of our physio-

---

[3]Quotation of John von Neumann from V. Lawrence Parsegian, Paul R. Shilling, Floyd V. Monaghan, and Abraham S. Luchins, *Introduction to Natural Science* (New York: Academic Press, Inc., 1970), p. 437.

logical freedom. Only because they work the way they do has it become possible for us to have bones, muscles, glands, and brains. Superficially, it might be said that the function of the kidneys is to make urine; but in a more considered view one can say that the kidneys make the stuff of philosophy itself.[4]

The kidneys more than any other organs maintain "the constancy of the composition of the internal sea," the body fluids, which include the intracellular as well as extracellular fluids. The extracellular fluid (ECF) can be divided into the fluid portion of the blood, the plasma, and the tissue or interstitial fluid, which bathes the tissue cells and exchanges with the blood stream. The function of the kidneys is inseparable from circulation, since it is the blood that brings compounds to them from the cells and their tissue fluid baths. The blood is filtered and "cleansed" in the kidneys. With the exception of gas exchanges carried out in the respiratory system and nutrient absorption performed by the digestive system, it is the kidneys that maintain the normal composition of the plasma and through it the constitution of the cellular fluids and their milieu.

The "ashes of our body fires" are the nitrogenous wastes of protein respiration and nucleic acid breakdown. There is no storage form of protein in the body. Structural and functional body proteins are constantly being broken down into their constituent amino acids and either resynthesized or respired. Some amino acids do circulate in the blood. New amino acids are added to the pool from the diet; an excess of protein is usually ingested, and many of the amino acids resulting from digestion are respired rather than assimilated following absorption. The nitrogen containing group ($-NH_2$), the *amino group*, is cleaved from the carbon skeleton, which is then quite like an organic or fatty acid that can be respired for energy purposes. The ammonia (NH_3) so formed is very toxic, but it poses no elimination problem for many aquatic creatures that are surrounded by gallons and gallons of water. Land animals, such as reptiles, birds, and mammals, have a water conservation problem. Unlike aquatic organisms, they cannot continually urinate to wash their poisonous wastes away. Therefore, many land animals convert ammonia to the less toxic compound, *urea* ($(NH_2)_2CO$), by combining two molecules of ammonia with one of carbon dioxide.

$$2NH_3 + CO_2 \longrightarrow (NH_2)_2CO + H_2O$$

Actually, the process of urea formation involves a more complex biochemical cycle, but the preceding generalizations are diagramatically

---

[4]Homer W. Smith, *From Fish to Philosopher* (Garden City, N. Y.: Doubleday & Company, Inc., 1961), pp. 2–3.

correct. The urea cycle occurs primarily in the liver, where amino acids are deaminated and the resultant ammonia "detoxified." The loss of functional liver cells and their replacement with fibrous connective tissue resulting in the "hardening of the liver" (cirrhosis) leads to death by autointoxication. Birds, reptiles, and even some terrestrial invertebrates convert their protein nitrogen to *uric acid* rather than urea. The uric acid found in the urine of urea producers is the product of nucleic acid breakdown rather than protein degradation. Creatinine, a product of muscle metabolism, is also excreted in the urine. The clearance of nitrogenous waste products from the body fluids by the kidneys in the urine they prepare is a major part of *excretion*. Excretion is simply defined as "the discharge of the waste products of metabolism." Of course, CO_2 is an excretory product for animals, eliminated by the respiratory system, and wastes are excreted by the sweat glands; but, nevertheless, the kidneys are the primary excretory organs. The phrase "excretory system" refers to the urinary system, the kidneys and their associated structures (Fig. 3-30). The kidneys also regulate the water and salt composition of the ECF as well as the acidity (pH) and alkalinity of the body fluids.

The functions of the kidney units are threefold: ultrafiltration (very fine filtration), reabsorption, and secretion. The kidneys' fluid filtrate of blood carries urea, uric acid, and creatinine out in the urine. Valued substances are reclaimed from the filtrate, and some compounds are transported from the blood into the forming urine. The microscopic units of function in the kidneys are the *nephrons,* which are primarily tubules that begin with blood and end with urine (see Fig. 3-31).

The renal (kidney) arteries branch and rebranch until they are the smallest, microscopic arteries, the arterioles, in the outer portion of the kidney. There, they form complex capillary knots, the *glomeruli* (Fig. 3-32). Glomerular capillaries are unique in more than their knotted arrangements. The pressure of blood entering the glomerular capillaries is higher than that of blood flowing into the capillary beds of other sites in the body, and the walls of these kidney capillaries are also unusual in that they are perforated by minute pores (Fig. 3-33). In body capillary beds, the arterioles lead in and the venules (the smallest veins) lead out. An *afferent arteriole* leads into the glomerulus and an *efferent arteriole* leads out. Applied to the exterior of the porous glomerular capillary walls are an extracellular membrane and the foot processes of the inner cells of the *renal capsule.* The renal capsule is like a closed bag into which the glomerulus punches and from which a twisted and looped *renal tubule* commences. The renal capsule and glomerulus together make up the *renal corpuscle.* A renal corpuscle and its tubule make up a *nephron.* Each kidney contains a million nephrons in addition to their associated blood vessels.

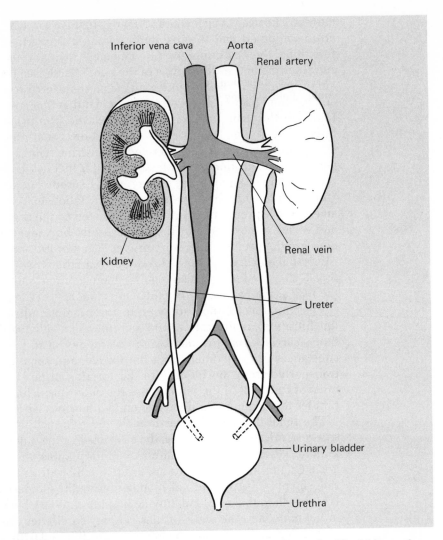

Fig. 3-30 **The human urinary system and associated blood vessels. The kidneys form urine by removing materials from the blood passing through them.**

The blood cells and plasma proteins do not pass through the glomerular pores, but the fluid portion of the blood minus the above components does filter through to form the glomerular or *nephric filtrate*, which first appears in the capsule and begins the journey down the renal tubule. This fine filtrate (ultrafiltrate) is far from urine. As noted, it is blood minus its cells and plasma proteins. No organism can afford to void it

Fig. 3-31   **General structure of kidney. (a) Cross section of rat kidney. The outer dark-staining portion of the kidney is the cortex, wherein lie the blood filtration units. The inner, light-staining zone is the medulla, into which kidney tubules loop and which is drawn out in the form of a fingerlike projection, the renal papilla. (b) The kidney consists of nearly one million single units, nephrons, one of which is shown in the kidney in the lower left. In the lower right, a nephron is enlarged and diagrammed to show how it begins in a capillary knot, where an initial fluid is formed by filtration. This fluid is then modified and reduced in volume as it passes down the tubule until it reaches the ureter as urine.**

(a) Courtesy of Saturn Scientific, Inc., Fort Lauderdale, Fla.

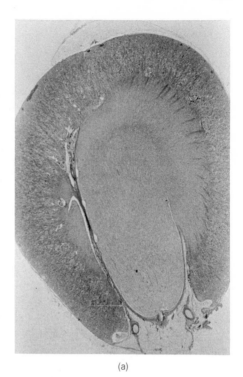

(a)

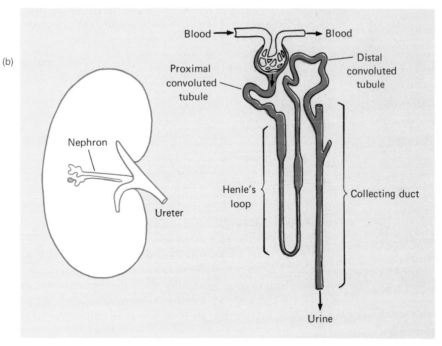

(b)

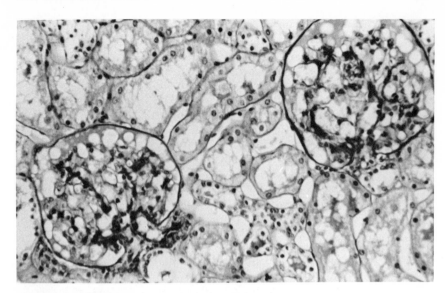

Fig. 3-32

Two renal corpuscles, consisting of capillary knots, glomeruli, and their capsules, Bowman's capsules, which empty into the first portion of the kidney tubule. In this preparation, the white spaces within the knots are capillary channels, and the stain distinctly outlines the outer wall of the capsule. The kidney corpuscle is the unit of blood filtration. The glomerular filtrate first forms within the capsule. Sections of the surrounding twisted tubules are scattered between the two kidney corpuscles. The corpuscles and the proximal and distal convoluted tubules (pct and dct) are limited primarily to the cortical region of the kidney.

From M. J. Zbar and M. D. Nicklanovich, *Urinary Function and Malfunction,* Saturn Scientific, Inc., Fort Lauderdale, Fla., 1971.

without reclaiming from it the nutrients, glucose, amino acids, salts, and—in land vertebrates—even water itself. In 24 hours, about 180 liters (l) of filtrate are formed, 4.5 times the entire body fluid volume of 40 l. Only 2 l or less normally become urine. Pounds of glucose, amino acids, and salts are reabsorbed (Fig. 3-34), in addition to the 40 gallons (gal) of fluid (178 l). The whole body fluid is thus processed and reprocessed several times per day by the kidneys. The blood that leaves the glomerulus via the efferent arteriole then flows around the tubule in a *peritubular capillary network* (pcn). This mesh surrounds the twisted and looped portions of the renal tubule (Fig. 3-35) before emptying into the veins that drain the kidney. It is the blood within this network that regains the valuable substances lost in filtration.

In the twisted first part of the renal tubule, the filtrate, which is now

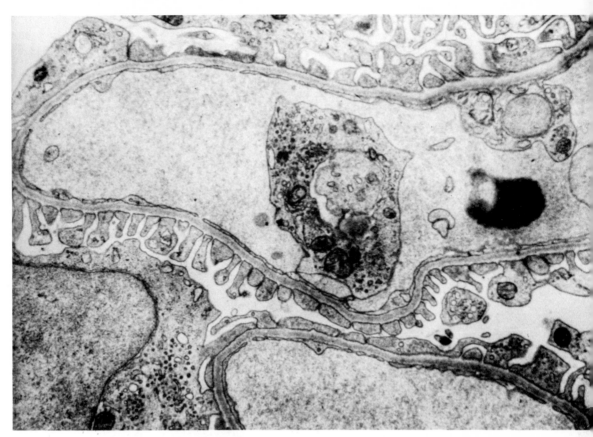

Fig. 3-33   Electron micrograph of cross section of glomerular capillaries. The white space
is the urinary space of Bowman's capsule, in which the glomerular filtrate first
appears. It is an ultrafiltrate of blood, free of blood cells and the plasma proteins. Therefore, the urinary space here is white rather than grey, as in the
glomerular capillaries. Portions of two glomerular capillaries occupy most of
this field. The fine, grey granular material filling their channels is the protein of
blood, the plasma protein molecules. The black spot just to the right of center
is the cut edge of a red blood cell. The fringe surrounding the capillary walls
consists of the foot processes of the podocytes ("foot cells"), one of which
can be seen in the lower left. Between the fringe of foot processes and the
thin wall of the capillaries, a narrow band of basement membrane courses.
The glomerular capillaries differ from most capillaries in that their walls are
perforated by numerous pores. The plasma proteins are restrained by the
filtration barrier, which includes the porous capillary wall cells, the macromolecular meshwork of the basement membrane, and the clefts between
foot processes.

Reprinted with permission from K. R. Porter and M. A. Bonneville, *Fine Structure of Cells
and Tissues*, Lea & Febiger, Philadelphia, 1964.

called the "tubular fluid," has reabsorbed from it all glucose, amino acids, and most of its water and salts. The osmotic pressure of the blood in the pcn is higher than that of the tubular fluid, and water reabsorption follows an osmotic gradient. The cells of the twisted portion of the tube are the specialized transport cells (Fig. 3-36) that were described in Chapter 2, and the nutrients and salts do more than simply diffuse; they are actively transported back into the blood.

The tubular fluid then descends one limb of a loop which reaches down into the inner portion of the kidney, the *medulla,* and next rises back up to the cortex by means of the ascending limb of the loop (Fig. 3-37). Glomeruli are found only in the cortex: they are absent from the

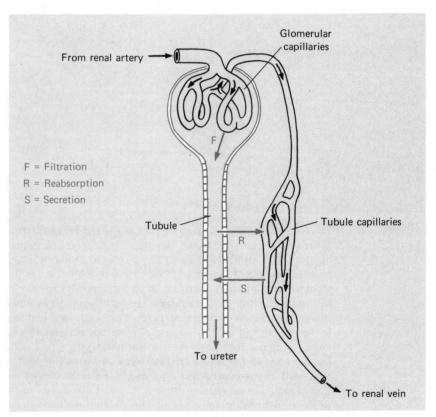

Fig. 3-34   **In the formation of urine, water and small molecules are first filtered through the glomerulus into the tubules; then the composition of the filtered fluid is changed by tubular *reabsorption* and *secretion.***

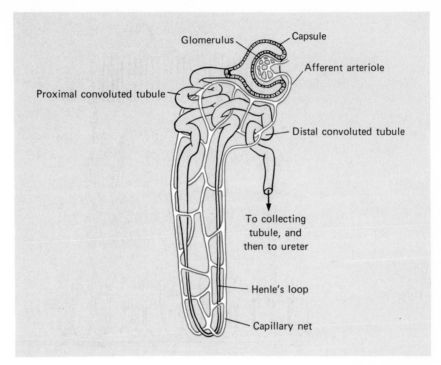

Fig. 3-35   **This drawing of a nephron shows the close relationship established between the twisted and looped portions of the kidney tubule and the peritubular capillary network.**

medulla. The loop was a puzzling pathway until it was learned that it was a means of increasing the salt concentration in the kidney's interior for the purpose of producing a final concentration of the fluid, which is at this point in its course a watery urine. Successive slices of the kidney proceeding from the cortex to the medulla show progressively higher freezing point depressions. (Recall that salt is sprinkled on winter walks to retard the formation of ice.) The salt concentration of the tissue fluid bathing the looped tubules and their embracing capillaries in the medulla is much greater than it is in the cortex; in fact, the concentration of salt in this medullary interstitial fluid is three times as great as it is in the tissue fluids of any other sites of the body. The cells lining the ascending limb of the loop actively secrete sodium ions (Na^+) from the tubular fluid into the medullary tissue fluid surrounding the tubules, and chloride ions (Cl^-) follow the sodium, so that the sodium chloride

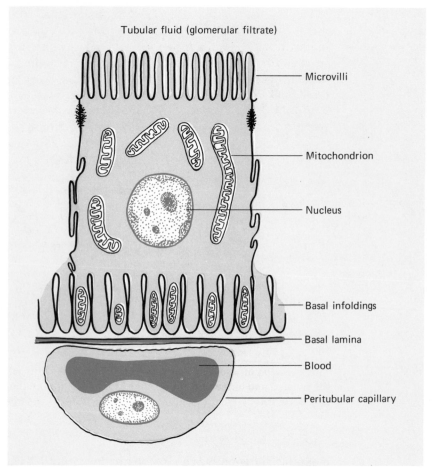

Tubular fluid (glomerular filtrate)

— Microvilli

— Mitochondrion

— Nucleus

— Basal infoldings

— Basal lamina

— Blood

— Peritubular capillary

Fig. 3-36   **Fine structure of proximal convoluted tubular (pct) epithelial cells. The cells making up the walls of the pct are transport cells, probably the most active in the body. They bear numerous microvilli on the free surface (upper right) and are extensively infolded at their bases, adjacent to the thin-walled capillaries of the peritubular network. Note the numerous mitochondria. What is their function here?**

(Na$^+$Cl$^-$) or salt concentration rises. The tubular fluid descending the loop passively absorbs salt as it drops down into the salt-enriched medulla (Fig. 3-38).

The peritubular capillary blood (which has received the products of reabsorption) does not empty into the veins in the medulla, but rather rises to veins in the cortex. Thus, the peritubular capillaries form a kind of loop in the medulla also (like the renal tubule's loop, but more complex). As the blood flows downward in the loop, salt diffuses into the blood until the concentration in the blood is almost as high as that of the surrounding extracellular fluid at the bottom of the loop. As the blood flows upward in the "loop" and out of the medulla, the gained salt diffuses back out of the blood into the tissue fluid (Fig. 3-39). In this way, the *countercurrent flow* of blood in the loop prevents the loss of excessive salt from the medullary interstitial fluid. If the capillary blood had followed a straight course through the medulla, the accumulated salt would have been removed, but, by means of this countercurrent mechanism of the kidney, the high salt concentration of the deep tissue fluids

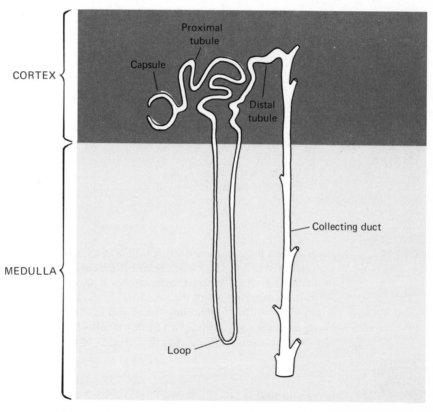

Fig. 3-37   The distribution of the parts of the nephron within the kidney regions. Note that the medulla contains only loops and collecting ducts. The glomeruli, their capsules, and the proximal and distal convoluted tubules are found primarily in the cortex. This anatomical arrangement in the mammalian kidney has great functional significance in the formation of concentrated urine.

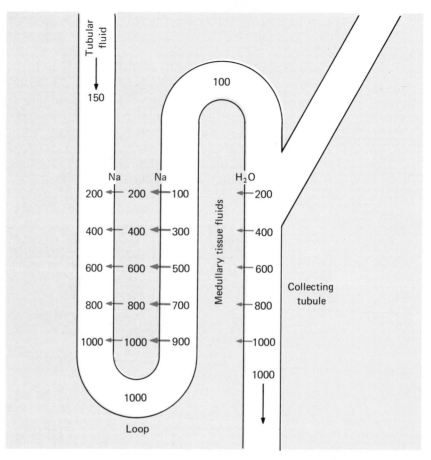

Fig. 3-38   **Diagram of a loop and a collecting tubule of a nephron during the formation of concentrated urine. The figures refer to sodium concentrations. Heavy arrows indicate an active transport, light arrows a passive diffusion. Sodium (Na) is actively transported out of the ascending limb of the loop, causing an accumulation of Na in the surrounding tissue and in the fluid of the descending loop. Water is osmotically withdrawn from the collecting tubule because of the high salt concentration in the surrounding tissues of the medulla.**

is maintained with little loss. Thick-walled veins do plunge down from the cortex through the medulla, carrying cleansed blood from the kidney, but their walls are much thicker than those of capillaries and so retard the diffusion of salt into the venous blood to any great extent.

The countercurrent mechanism is employed in other situations. Warm-blooded whales often swim in the cold depths or polar seas. Al-

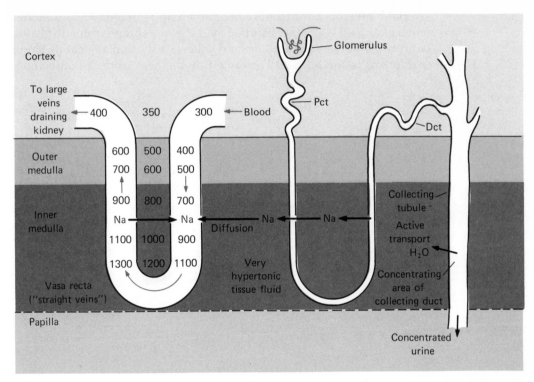

Fig. 3-39   **Countercurrent mechanism by which the kidneys concentrate the urine. The blood (left) does not drain straightaway from the kidney but rather follows a looped pathway similar to the tubular loop. Special vessels, the vasa recta, originate in the outer medulla, pass deep into the inner medulla, form a second peritubular capillary network, and then turn back out of the medulla. As the blood in them descends, it picks up sodium by diffusion from the concentrated tissue fluids of the medulla. As it rises in the loop, the blood loses sodium (Na) to the surrounding tissues and leaves the medulla hardly any richer than when it entered. If the blood were to drain straightaway from the kidney, the high concentration of sodium in the medulla would be decreased. The salty fluids of the medulla are required to pull water osmotically from the collecting tubule. Water, reclaimed from the urine, is carried from the medulla by the blood.**

though their bodies are well insulated with a heavy layer of blubber, their fins are not so covered, and, therefore, constitute a potentially dangerous avenue of heat loss from limb circulation. The arteries that proceed into the limb and the veins that drain it have come to lie alongside each other. As warm blood from the body interior enters the limb, its heat is transferred to the cooled, returning, venous blood. By the time the arterial blood reaches capillary beds at the tip of the limb it is cool, but much of its heat has been absorbed by the venous blood instead of being entirely lost to the surrounding cold water. As the cooled venous blood is returned to the body, it is warmed by the heat from the outcoming

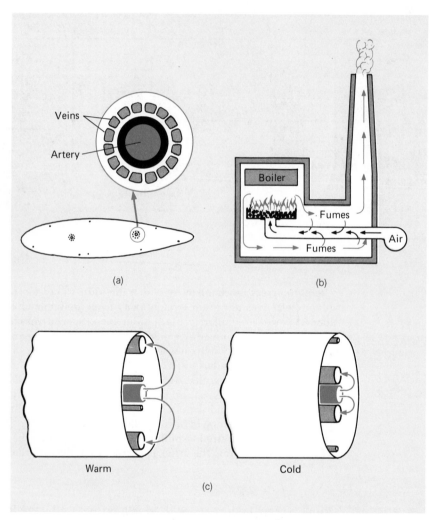

Fig. 3-40   **(a) Blood vessels in a porpoise fluke. (b) A countercurrent heat exchanger. The cold air is returned to the furnace by way of a conduit through the exhaust, and picks up part of the heat otherwise lost with the exhaust fumes. (c) Diagram representing the shift in venous blood flow through the extremities from the superficial blood vessels in a warm environment to deep vessels in the cold.**

blood of the neighboring arteries (Fig. 3-40). This countercurrent circulation helps to hold the temperature of the whale constant, just as the kidney countercurrent maintains the high medullary salt level.

Why is the high medullary salt concentration desirable? Let us return to the watery tubular urine that we left, returned to the cortex. This dilute urine passes through a second twisted conduit and then drains into straight collecting tubules and ducts, which pass down into the salty medulla again. Here in the medulla the watery urine is finally concentrated (Fig. 3-41). Water moves easily from the dilute urine into the salty tissue fluid of the medulla *provided* the walls of the collecting tubule are permeable to water, allowing it to "pass," following the osmotic gradient. When the body fluids are dilute, the cells making up the collecting tubule are impermeable to water, and a dilute (watery) urine is passed. If excessive quantities of water are drunk in a short period of time, an increased flow of urine (diuresis) follows. When the body fluids are concentrated (relatively dehydrated), the collecting tubules become quite permeable to water, and the urine voided is concentrated.

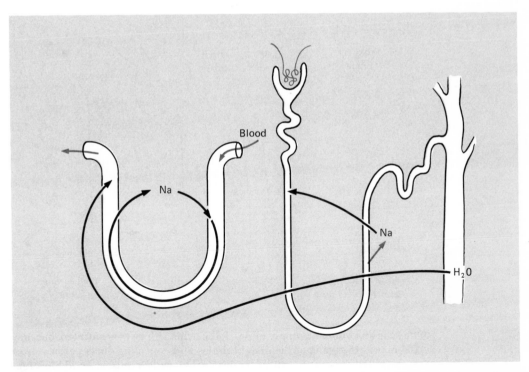

Fig. 3-41   **Summary of salt movement and water conservation in the mammalian kidney.**

In osmosis, water diffuses from a high to a low concentration of water, *but* from a solution with low osmotic pressure to one of high osmotic pressure. When the osmotic pressure of the blood is high (hemoconcentration or low blood water content), certain nerve cells in the hypothalamus, the *osmoreceptors,* are stimulated. The nerve fibers of the osmoreceptors project down in the pituitary stalk to the rear lobe of the pituitary, where they end in the near vicinity of capillaries. The permeability of the collecting tubules to water is controlled by the *antidiuretic hormone* (ADH), which is released into the bloodstream by osmoreceptors' fiber endings in the posterior pituitary. (ADH is a neurohormone, secreted by nerve cells.) Antidiuretic hormone increases the permeability of the collecting tubules to water, and the urine is thus concentrated (Fig. 3-42). (The osmotic pressure of dilute urine is less than that of the medullary tissue fluids, and water flows from the tubules into the ECF.) The water so retained lowers the osmotic pressure of the blood (hemodilution) and other body fluids. Stimulation of the osmoreceptors ceases, and the secretion of ADH is brought to a halt as the osmotic pressure of

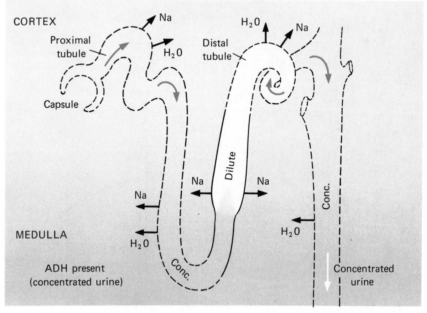

Fig. 3-42   **If there is an abundant supply of ADH, the urine will be concentrated, because ADH makes the walls of the distal tubules and collecting ducts permeable to water. Dotted lines represent water permeable walls of tubule; solid lines represent waterproof walls.**

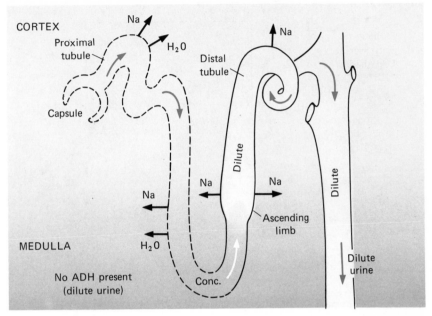

Fig. 3-43   **If ADH secretion is low, the urine will be dilute because the walls of the distal tubule and the collecting duct are impermeable to water. Dotted lines represent water permeable walls of tubule; solid lines represent water impermeable walls.**

the body fluids undershoots and the blood is hemodiluted (too much blood water). In the absence of ADH, the collecting tubules become relatively impermeable to water (Fig. 3-43) and a dilute urine is passed, until the osmotic pressure of the body fluids rises too high and the osmoreceptors are stimulated to release ADH once again.

High osmotic pressure of the body fluids is a stimulus (a change in the internal environment) that activates the osmoreceptors and leads to the release of ADH. The collecting tubules are the targets of this neurohormone, and the response is the absorption of water from the urine, thereby lowering the osmotic pressure of the blood and thus the other body fluid compartments with which the blood exchanges. The response abolishes the stimulus. Low osmotic pressure of the body fluids deactivates the osmoreceptors and inhibits ADH secretion. The water conservation system is shut off, a negative feedback (Fig. 3-44).

There is an ADH-deficiency disease called diabetes inspidus ("the tasteless urine disease"), in which 15 l of watery urine are voided each day in contrast to the normal 1 or 2 l. The severance of the pituitary stalk

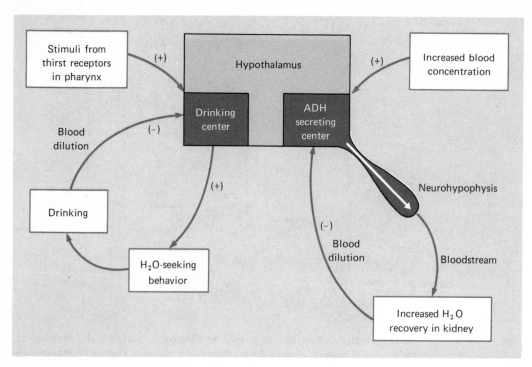

Fig. 3-44   **Hypothalamic control of blood volume.**

can also produce the disease. The volume of urine formed can be varied from 0.35 ml/min to 15 ml/min. The nervous shutdown of blood flow to the kidneys can play a significant role in the decreased urinary output of shock. Antidiuretic hormone can control the rate of urine formation over at least a fifteenfold range.

We have seen that the ability to take water from the urine depends upon the capacity to hold salt. "If you can't hold salt, you can't hold water," no matter how much ADH is secreted. The hormone *aldosterone* from the adrenal cortex controls the rate of sodium excretion by the kidneys. When the sodium levels of the body fluids are too low, aldosterone secretion increases, causing the kidneys to retain sodium until the sodium levels of the tissue fluids are raised (Fig. 3-45). When sodium levels in the ECF are too high, aldosterone secretion diminishes and the kidneys allow sodium to remain in the urine in increasing quantities.

The salt concentration of the ECF determines its osmotic pressure to a great degree. The osmotic pressure of the body fluids is largely a function of the salt-water ratio. Water excretion via the kidneys is a function

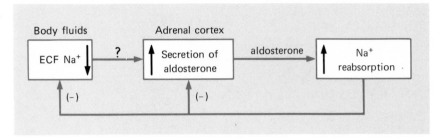

Fig. 3-45   **The maintenance of the constancy of sodium in the internal sea.**

of sodium excretion; conversely, water retention by the kidneys is a function of sodium retention. Body fluid volume is an important factor in blood pressure. It has been hypothesized that the kidneys respond to lowered blood pressure by releasing a substance called renin into the blood. Renin combines with a plasma protein to form a protein that stimulates the secretion of aldosterone by the adrenal glands (see Fig. 3-46).

Addison's disease involves a destruction of the aldosterone-producing adrenal cortex, and its cardinal symptoms include low blood pressure (low blood volume) and frequent urination. Why?

At the same time that aldosterone stimulates the kidneys to retain sodium, it promotes the excretion of potassium. Any one of three different stimuli increase aldosterone secretion: low ECF sodium, high ECF potassium, and low ECF volume. The mechanism by which these stimuli call out aldosterone is incompletely understood. Although electrolyte (salt) balance and osmoregulation are intertwined, the concentrations of the various salts in the body fluids have consequences other than merely osmotic ones.

The excitability of nerves, muscle, and heart depends upon the charge across their cell membranes, which is due to a differential distribution of sodium and potassium ions in particular. Calcium ions have a profound effect on membrane permeability and are intimately involved in the mechanics of muscular contraction. High ECF potassium levels decrease membrane potentials, thus depressing nerve conduction and muscle excitability. Potassium becomes elevated in kidney shutdown. The heart relaxes and slows. Low ECF potassium raises the membrane potentials so high that the nerves fail, no longer transmitting impulses to the muscles, and muscular paralysis results. Low ECF sodium also decreases the strength of heart contraction. A low level of calcium ions in the ECF will cause the heart to dilate and stop in a relaxed state. In addition, low calcium causes hyperexcitability of the nerves and tetanic

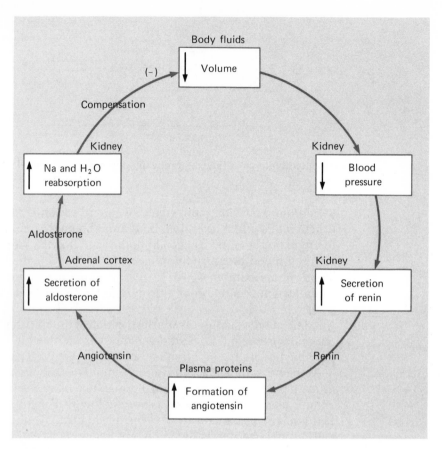

Fig. 3-46   **If the body fluid volume is low, the secretion of aldosterone rises. Aldosterone stimulates an increase in sodium (and water) retention by the kidney. This response compensates for the decrease in body fluid volume. The feedback loop shown in the diagram may be involved in stabilizing the volume of body fluids and the related blood pressure. (A crucial factor in blood pressure is blood volume.)**

muscular contraction. The latter effect is known as low calcium tetany. High ECF calcium depresses the neurons of the CNS and promotes overcontraction of the heart. Calcium ions normally decrease the permeability of cell membranes to sodium, thereby decreasing the excitability of irritable tissues. The kidneys play an important role in regulating blood and tissue fluid calcium ion levels.

The parathyroid glands secrete a hormone in response to low ECF calcium levels that tends to elevate the ion by mobilizing it from bone stores and increasing calcium reabsorption by the renal tubules. Para-thyroid hormone also reduces kidney tubular reabsorption of phosphate ions, causing large quantities of phosphate to be excreted in the urine. This effect tends to increase the dissolution of calcium phosphate bone mineral (Fig. 3-47). High ECF calcium is rare, but it would produce the opposite chain of events.

The kidneys also regulate the acidity of the body fluids. The pH is a mathematical expression for the concentration of hydrogen or acid ions (H^+). Actually, pH is the logarithm of the reciprocal of the hydrogen ion concentration [H^+]. The higher the [H^+], the lower is the pH; the lower the pH, the higher is the [H^+]. A pH of 10 is in the alkaline or basic range (the hydrogen ion concentration is low); a pH of 3 is in the acid range ([H^+] is high); a pH of 7 is the point of neutrality, neither acidic nor basic.

The normal pH of the body fluids ranges between 7.2 and 7.4, slightly alkaline, but near neutrality. A person passes into a coma when the pH of his ECF falls below 6.9, and the lower limit at which a person can live more than a few minutes is about 7.0. The upper limit is approximately 7.8. Above even 7.4, a person is afflicted with a mild *alkalosis,* and in severe alkalosis the nerves and muscles become hyperexcitable. Tetanic muscular contractions culminate in convulsions and death. In *acidosis* (excessive acidity of the body fluids), the resulting coma is the result of the depression of nervous activity. From these figures, it is obvious how important to life are the adjustment and control of the pH of the body fluids.

An acid is a substance that releases hydrogen ions (H^+); a base is a compound that combines with hydrogen ions. Some chemicals are both acids and bases, for, depending upon circumstances, they can liberate or bind hydrogen ions. These substances are known as *buffers,* since they tend to resist changes in pH by absorbing or freeing acid ions (H^+). Proteins are the most important buffers in the body, for their constituent amino acids have both acidic and basic groups. Large protein molecules bear many such groups, and much of the body construction is pro-teinaceous. Salts of weak acids (poor H^+ liberators) also serve as buffers. The biocarbonate and phosphate salts of the body fluids are important buffers, although less so than proteins.

Carbon dioxide dissolves in the body fluids and forms weak carbonic acid, which slowly forms hydrogen and bicarbonate ions:

$$CO_2 + H_2O \rightleftharpoons \underset{\text{(carbonic acid)}}{H_2CO_3} \rightleftharpoons H^+ + \underset{\text{(bicarbonate ion)}}{HCO_3^-}$$

Bicarbonate functions primarily as an alkaline buffer in the body, pick-ing up hydrogen ions, but the carbonic acid so formed can liberate them as well. In conjunction with the buffers, the respiratory and renal sys-

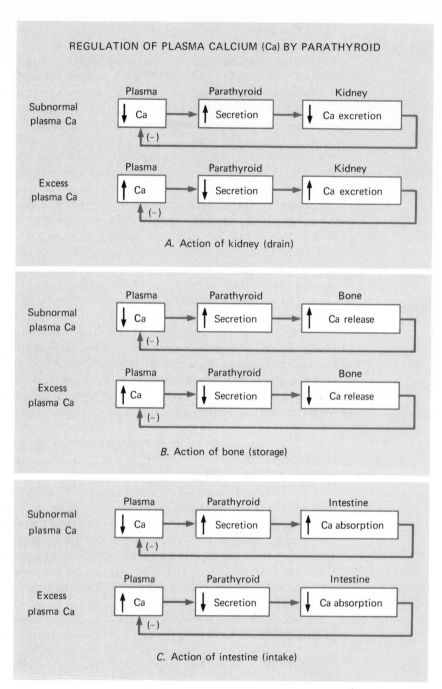

REGULATION OF PLASMA CALCIUM (Ca) BY PARATHYROID

A. Action of kidney (drain)

B. Action of bone (storage)

C. Action of intestine (intake)

Fig. 3-47   The body tends to maintain itself in a steady state. An excellent example of this continual striving for stability is the mechanism for the maintenance of the constancy of the blood calcium level. Here again, the kidneys play a key role.

tems play important parts in regulating the acid-base balance of the body.

Respiratory adjustment of body pH is rapid. When the body fluids become too acidic, reflex hyperventilation (rapid and deep respiration) ensues, and carbon dioxide is blown off in large quantities. The removal of CO_2 draws the whole chain of the preceding reversible reactions to the left, thus decreasing the hydrogen ions (acid) on the right, adjusting the acidity back toward normal. When the body fluids become too alkaline, the respiratory rate and depth decreases; carbon dioxide builds up in the body fluids, and the equilibria are shifted to the right, increasing the hydrogen ion concentration and pulling the pH down toward neutrality. A person can induce alkalosis and a giddiness near fainting merely by hyperventilating for a few minutes. Alone, the respiratory system is capable of only partial compensation (50 to 75%) of acidosis or alkalosis. Although the kidneys' acid-base regulation requires longer, it can be complete. Even if the acidotic patient breathes poorly, his pH will be returned to normal in 12 to 24 hours. The pH of a hyperventilating, alkalotic patient will also be adjusted in the same period of time.

When the body fluids are acidic, the kidneys make an acid urine. When the ECF is too alkaline, the "masters of the internal sea" prepare a basic urine. Normally, when the rate of hydrogen ion secretion into the kidney tubules equals the rate of bicarbonate filtration, nothing happens to the hydrogen concentration of the ECF. In acidosis, H^+ secretion exceeds bicarbonate filtration; no bicarbonate is lost in the urine, but hydrogen ions are excreted as the pH of the body fluids is shifted toward the alkaline side. The kidneys can prepare a urine with a pH as low as 4.5 in acidosis. In alkalosis, bicarbonate filtration exceeds H^+ secretion, and large quantities of bicarbonate are not reabsorbed by the kidney tubules. Thus, a major alkaline buffer (bicarbonate) is lost in the urine and decreases in the ECF, drawing the preceding reaction sequence to the right, increasing the H^+ concentration and dropping the pH toward the acid range. Urinary pH can be as high as 8.0 in alkalosis. More acid than alkali is formed in the body each day. Therefore, the normal pH of urine is slightly acidic (6.0).

Let us summarize the functions of the human kidney.

(1) The kidneys hold constant the osmotic pressure of the body fluids. This function is called *osmoregulation*.
(2) The kidneys eliminate *nitrogenous wastes*.
(3) The kidneys regulate the salt, or *electrolyte balance* of the body fluids.
(4) The kidneys control the *acid-base balance*.

The kidneys maintain the normal composition of the blood and through it that of the other body fluids (including the intracellular fluids) with which the blood exchanges. In addition, there is good experimental evidence that the kidneys produce two compounds, renin and erythropoetin, that stimulate the elevation of the blood pressure and the production of red blood cells, respectively. It is not surprising that these multifunctional organs inspired Homer Smith. There is wonder in their work, yet there is mechanism.

The symptoms of kidney shutdown are negative reflections of normal renal function. High and finally fatal concentrations of urea and other nitrogenous wastes result from the failure to excrete these metabolic end products. The tissues become "puffy," or edematous. Edema is the accumulation of excessive tissue water, and, in renal shutdown, it results from water retention, an osmoregulatory failure. Potassium levels in the ECF rise to high concentrations, as potassium excretion halts. Electrolyte balance is upset. Last, but not least, acidosis results from the inability of the kidneys to rid the body of the normal acidic end products of metabolism. This drastic fall in pH culminates in coma and death.

It is no small tribute to the ingenuity of man that an artificial kidney has been designed. The artificial kidney pumps arterial blood from an arm or leg vessel through a cleansing bath and back into an arm or leg vein. The blood is separated from the bath fluid by semipermeable cellophane or plastic sheets with pores large enough to permit small molecules, such as urea, glucose, and the various salt ions, to diffuse through them, yet small enough to prevent the escape of plasma proteins and, of course, blood cells. The separation of small molecules from large by means of a semipermeable membrane with selective pore size is called *dialysis,* and the artificial kidney is really a dialyzer (Fig. 3-48). The bath solution is a dialysis fluid, which contains the same concentration of all the ions and small molecules normally found in the blood plasma, except that urea, uric acid, creatinine, and phosphate are omitted. These substances therefore diffuse from the blood into the dialysis fluid. Dialysis fluid contains a slightly higher concentration of glucose than is normally found in the blood. Used dialysis fluid is discarded and fresh fluid continually led into the bath during the 6 to 12-hour treatment, which is required biweekly or triweekly. It is astounding that this simple procedure replaces the master chemists of the internal environment. Of course, artificial kidney patients are kept on special diets, restricting protein and fluid consumption as well as salt intake, particularly potassium, in which certain foods such as bananas and avocados are rich. The treatment is expensive and inconvenient, but lifesaving. Many patients receive kidney transplants from related donors after a few years of such treatment, yet, tens of thousands of Americans die of kidney failure each year.

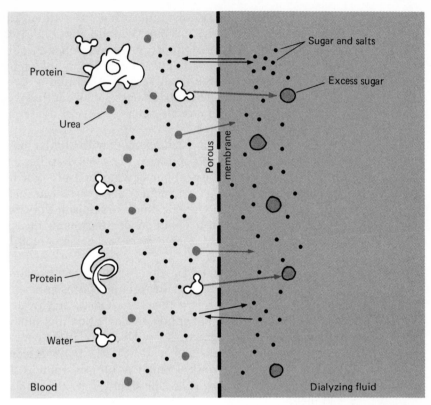

Fig. 3-48   **Diagram of dialysis, the principle upon which the artificial kidney operates. A porous membrane separates blood (left) from dialyzing fluid (right). Proteins are too large to pass through the pores. Sugars and salts (black dots) required by the body are duplicated in the dialyzing fluid, so that they will diffuse both ways, preserving their blood levels. Wastes, such as urea, are absent from the dialysis bath, of course, and diffuse more to the right than back. Some water is attracted by the excess sugar molecules in the dialyzing fluid.**
After Longmore, *Spare-Part Surgery*, Doubleday & Company, Inc., New York, 1968.

THE EVOLUTION
OF THE VERTEBRATE KIDNEY

The kidney that we have described so far is the mammalian kidney, and its anatomy and physiology show wide variation, depending upon the environment to which the given mammal is adapted. Desert ground squirrels have elongate medullary regions of their kidneys, and thus longer tubular loops (Fig. 3-49). What do you think is the consequence of this anatomy? There is hardly any evolutionary problem that has only one solution. Many small desert rodents have anatomically typical kidneys, but secrete large quantities of ADH and so pass an extremely concentrated urine. The medullary regions and loops of several freshwater, semiaquatic mammals, such as beavers, otters, and muskrats, are shortened.

The cortico-medullary arrangement with tubular loops is not found in the kidneys of lower vertebrates (Fig. 3-50). Reptiles and their bird offshoots have solved the problem of water conservation by excreting their nitrogenous wastes as uric acid. These vertebrates lack urinary bladders, and the uric acid is deposited as a semisolid, crystalline paste along with the feces from a common compartment, the cloaca ("sewer"), whose walls reabsorb water from the urine. Amphibians have little water problem.

Kidneys of freshwater fish do not eliminate nitrogenous wastes to any great extent. Most of their nitrogenous wastes are excreted in the form of simple ammonia across the gills. When a freshwater fish is inserted into a tightly fitting rubber hole in a partition that subdivides the aquarium, almost all nitrogen waste will be found in the water on the head and gills side of the tank. The water on the tail side contains negligible ammonia, probably only as much as diffuses from the skin, even though the "urinary" opening is on this side! What then is the function of this kidney?

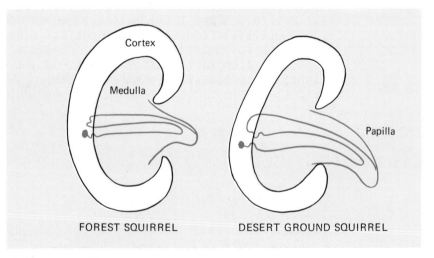

FOREST SQUIRREL   DESERT GROUND SQUIRREL

Fig. 3-49   **Compare the lengths of loops, ducts, and papillae of these two forms. What is the significance of the breadth of the medullary region and its components?**

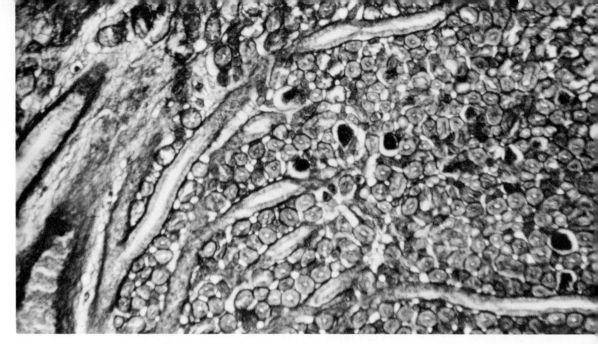

Fig. 3-50   **Reptilian kidney. This photograph of a section of turtle kidney was taken at a region that would correspond to the medulla of higher forms, but, as you can see, glomeruli (dark) are scattered throughout the field among the sections of tubules. Thus, it resembles the cortex of the mammalian kidney. There is no cortical-medullary arrangement. Similar kidneys are found in birds, amphibians, and most fish.**
From M. J. Zbar and M. D. Nicklanovich, *Urinary Function and Malfunction,* Saturn Scientific, Inc., Fort Lauderdale, Fla., 1971.

Fresh water is hypotonic to the body fluids of fish. It contains more water and less salt than fish body fluid. Water, therefore, diffuses into the body fluids, whereas salts tend to diffuse out into the fresh water. Certain gill cells are known to actively transport salts from the fresh water into the fish blood (Fig. 3-51). Presumably, the primary function of the glomerular, filtering kidney of freshwater fish is to eliminate the excess water that continually seeps into the body fluids from the hypotonic environment. Freshwater fish urine is dilute and voluminous. The microscopic architecture and function of the freshwater fish kidney provide one of the strongest arguments for the theory of the freshwater origin of vertebrates.

Most marine fish have glomerular kidneys, yet they live in an environment that is hypertonic to their body fluids. Sea water contains more salt and less water than marine fish body fluid. Water tends to diffuse out of their bodies and salt in. Ocean fish drink sea water to compensate for water loss, and their gills actively excrete the excess salt taken up (Fig. 3-52). Disappointingly, the marine fish kidney is unable to produce a concentrated urine to help conserve water and eliminate salt. It would

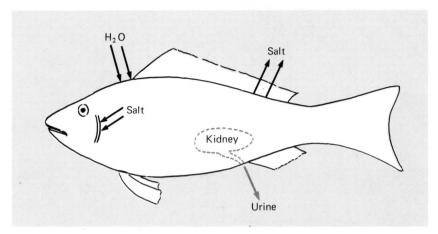

Fig. 3-51   The freshwater fish has a higher salt concentration than the water, and therefore loses salt while gaining water. Water is eliminated as urine, and the salt
loss is compensated for by its active absorption by the gills.

seem that a kidney with such powers would be ideally suited for the
marine environment. In fact, one would predict such a kidney, considering the osmotic problems of marine fish. Nevertheless, other solutions
have been found: the ingestion of sea water and the gill excretion of salt.
Why do the kidneys of marine fish filter their blood at all?

Several *aglomerular* (without glomeruli) marine fish are known: the

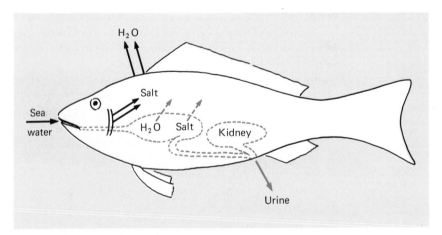

Fig. 3-52   A marine fish steadily loses water because the salt concentration of sea water
is higher than that of the fish. To compensate for the water loss, the fish drinks
sea water. Excess salt taken in is excreted by the gills.

anglerfish, the toadfish, the sea horse, and the pipefish (Fig. 3-53). These forms demonstrate that glomeruli are not at all essential to urine formation. Rather than filtering and reabsorbing, these fish rely upon renal tubular secretion for excretion. Aglomerular kidneys do excrete uric acid, creatinine, and trimethylamine oxide (a substance excreted only by fish). Presumably, their glomeruli have degenerated, been lost in the course of evolution. In this way, the excretion of these fish is more "perfect" than that of the other marine fish, in which filtration seems to be a wasteful process. In many marine fish, the flying fishes, cod, anchovy, puffer, needlefish, sergeant major, billfish, and the blue-striped grunt, the glomeruli are small and poorly vascularized, and their functional activity is probably greatly reduced.

Sharks and their relatives have solved the osmotic problems of salt water life in another way. The waste product urea, normally excreted by the kidney in other animals, is retained by sharks. Urea and trimethylamine oxide are built up in the blood and body fluids of sharks until the osmotic pressure is so high that, not only is body water not lost to the sea, but water is actually taken from the brine. A noted physiologist has said that they are the only vertebrates to have achieved "osmotic dominance" over the oceans. Imagine, a waste product has become a means of survival! Evolution seems to have been a perpetual scavenger hunt among the living, dying. Apparently, the sharks and their relatives were the first vertebrates to invade the sea. The bony

Fig. 3-53   **Toadfish kidney. A few marine fish have entirely lost glomeruli in the course of their osmoregulatory evoltion. Excretion in the aglomerular kidney is wholly by tubular secretion. As you can see in this photomicrograph of sectioned toadfish kidney, only tubules are present; glomeruli are absent.**
From M. J. Zbar and M. D. Nicklanovich, *Urinary Function and Malfunction,* Saturn Scientific, Inc., Fort Lauderdale, Fla., 1971.

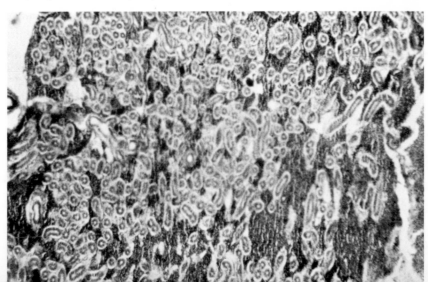

fishes entered the oceans at a later date and found a different answer to the problems of salt water life. The paradoxical, radical solution of the sharks reflects what Julian Huxley has called the "opportunism" of evolution. By and large, evolution has been hindsighted rather than foresighted, "making do with what is at hand," instead of inventing the entirely new, bound no place in particular, locked in the ever-present struggle for existence with the materials of the past. A few aglomerular fish have returned to the brackish or fresh water. None have regained glomeruli. As Homer Smith said: "It is not in the nature of evolution that the clock can be turned back." A few sharks have returned to the fresh water; the heart of the freshwater shark will not beat in the absence of urea.

This information has been presented from the viewpoint that the first vertebrates evolved in fresh water, which is not to say that lower life did not arise in the sea. The relative proportions of the various ions in vertebrate body fluid are approximately those of the ions of sea water, but the salt of vertebrate blood is less than 1%, whereas that of sea water is almost 3%. The proponents of the marine theory of vertebrate origin explained this discrepancy by maintaining that the ancient seas were more dilute than the present-day oceans. (The salinity of the seas has been built up with salts washed from the land.) The latest evidence indicates, however, that the primeval sea was at least twice as concentrated as modern vertebrate blood, even in the Ordivician period of 450 million years ago, when the first fish appeared.

When the higher vertebrates went again to sea "permanently," they took with them land kidneys, just as ancient fish presumably brought freshwater kidneys along on their oceanic invasions. Whales, dolphins, and seals can no more drink salt water and live than can man. Their predicament is like that of Coleridge's Ancient Mariner: "Water, water, everywhere, but not a drop to drink." They exist entirely upon the body water of the fish they consume and their metabolic water, byproduct of the respiration of food, just as many desert animals do. The land kidney is more fit for life in the open sea than the freshwater kidney, but the kidneys of sea mammals have evolved no astounding concentrating power. The most concentrated urine that the seal can pass is 5.6 times as concentrated as its plasma, greater than the dog (4.5) and man (4.2), but considerably less than that of the kangaroo rat (17 or higher). Nevertheless, marine mammals can pass a urine which is more concentrated than sea water. This ability is not newly evolved, however; it is merely the heritage of their land nephrons that possess great powers of water reabsorption. The blood of sea mammals is more concentrated than that of land mammals, but not enough more to hold water against the higher osmotic concentration of the salt water. Nursing is a potential avenue of water loss for marine mammals; whale milk is 10 times as concentrated as cow milk.

Marine birds, such as the albatross, shearwater, and petrel, spend weeks or months—even years—at sea, returning to the land only to lay and hatch their eggs. Coastal birds, gulls, cormorants, guillemots, and auks also appear to be independent of fresh water. At first it was thought that they had solved their problem the way that seagoing mammals had: depending upon food and metabolic water. Then, in 1956, Knut Schmidt-Nielsen and his group gave salt water to a cormorant and found to their surprise that a concentrated salt solution was exuded from the nostrils. Experiments with the albatross and other seabirds yielded the same results. Further investigations revealed a specialized, salt-secreting gland in the nasal region (Fig. 3-54). The *nasal glands* of marine birds are now also called "salt glands." From their position inside the eye cavity, ducts lead into the nasal passages. The ability of the nasal glands to actively excrete salt allows sea birds to drink salt water. The bird kidney has glomeruli and loops and therefore some ability to concentrate urine, but all cold-blooded vertebrates, including the reptilian ancestors of birds and mammals, lack the loop structure and are unable to form a urine more concentrated than their blood, not to mention sea water.

The nasal glands of marine reptiles, sea turtles, saltwater crocodiles, sea snakes, and marine iguanas have the same function as those of sea birds. Reptilian nasal glands open into their eyes, so that they literally "weep salt tears." Nasal glands are also found in terrestrial birds and reptiles, but these are small, less glandular, and lack a rich blood supply. Their function is not clear. These glands were taken by marine birds and reptiles and developed into large, glandular organs with rich blood supplies; they came to serve as safety devices, protecting against excessive intake of salt or loss of water. Salt water is ingested accidentally or only in the case of an emergency. Like the whales and many desert animals, sea birds and reptiles do depend upon their diets for osmotically palatable water and the "water of biological oxidation" (respiration). The nasal gland is absent in marine mammals. It would be interesting to learn if this organ were present in the marine fish-lizards, the ichthyosaurs, plesiosaurs, and the mososaurs of the Age of Reptiles.

Dollo's "law" states that evolution is irreversible. The land vertebrates that invaded the sea did not modify their land kidneys. They found other solutions for life in the oceans, which, for vertebrates, were "osmotic deserts." Life was born in the sea. The concentration of salts in the body fluids of invertebrates is the same as that of sea water. The vertebrates most likely were born in the fresh water. Their kidneys, which were bailers of excess water, became organs of survival, maintaining the constancy of the internal sea on the deserts of the land.

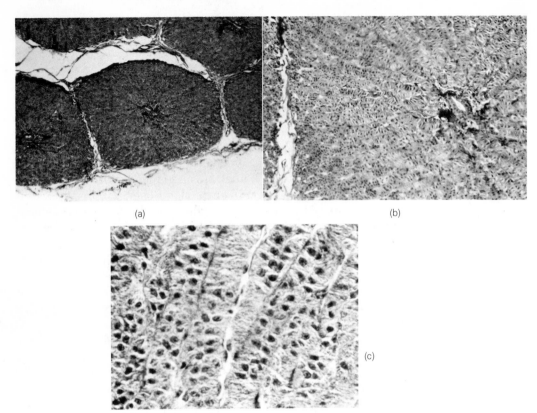

(a)   (b)

(c)

Fig. 3-54   **Salt glands of a gull. Sea birds are able to drink salt water, because specialized salt glands eliminate the salt that their kidneys cannot handle. (a) Under low power the lobules of the gull's salt gland exhibit straight tubules radiating about the hub of the central canal. Capillaries between tubules carry blood in the direction opposite to the flow of secretion. (b) A portion of a lobule at intermediate magnification. (c) At high power the cytoplasm of the tubule cells seems fragmented. It is actually highly involuted, increasing the surface area for transport.**
From M. J. Zbar and M. D. Nicklanovich, *Urinary Function and Malfunction*, Saturn Scientific, Inc., Fort Lauderdale, Fla., 1972.

Perhaps the concept of homeostasis need not be limited to physiology. Nearly half a century ago, Walter Cannon suggested,

In our study of the effects on the organism of a controlled stability of the fluid matrix we noted that just insofar as the stability is preserved the organism is released from the limitations imposed by internal and external disturbances. Is it not probable that similar results will flow from

*control and stabilization of the fluid matrix of the social organism?
. . . The main service of social homeostasis would be to support bodily
homeostasis. It would therefore help to release the highest activities of
the nervous system for adventure and achievement. With essential
needs assured, the priceless unessentials could be freely sought.*[5]

DISCUSSION QUESTIONS

1. *How does the external environment of the organism influence the internal
environment?*

2. *Why has the designation of the pituitary as the "master gland" been criticized in recent years?*

3. *How does the behavior of the spinal animal differ from that of the intact organism?*

4. *How is energy expended by the nerve cell?*

5. *How does the artificial kidney work? How does its function differ from that
of the natural organ?*

6. *What arguments for the fresh water origin of vertebrates have been derived
from the study of the comparative anatomy and physiology of vertebrate kidneys?*

REFERENCES

BENZIGER, T. H., "The Human Thermostat," *Sci. Am.*, Jan., 1961.

ECCLES, J. C., "The Physiology of Imagination," *Sci. Am.*, Sept., 1958.

GUYTON, A. C., *Function of the Human Body*, W. B. Saunders Co., Philadelphia,
1968.

LANGLEY, L. I., *Homeostasis*, Reinhold Publishing Corp., New York, 1965.

MAISEL, A. Q., *The Hormone Quest*, Random House, New York, 1965.

ROGERS, T. A., *Elementary Human Physiology*, John Wiley & Sons, Inc., New
York, 1961.

SCHMIDT-NIELSEN, KNUT, *Animal Physiology*, Prentice-Hall, Inc., Englewood
Cliffs, N. J., 1964.

SMITH, HOMER W., *From Fish to Philosopher*, Doubleday & Company, Inc., New
York, 1961.

WIENER, NORBERT, *Cybernetics*, Technology Press, Cambridge, Mass., 1948.

---

[5] Walter B. Cannon, *The Wisdom of the Body* (New York, N. Y.: W. W. Norton & Company,
Inc., 1963), pp. 322–323.

OBJECTIVES

1. List and describe the phases of the menstrual cycle.

2. Compare and contrast the estrous and menstrual cycles.

3. Correlate the events of the ovarian and uterine cycles.

4. Enumerate the hormones involved in the reproductive cycles and give their sources, their targets, their effects, their interactions, and their relative levels at various times in the cycles.

5. Describe the makeup and mechanism of action of the birth control pills.

6. Distinguish between the processes and the results of mitosis and meiosis.

7. Outline and contrast the typical plant and animal life cycles.

8. Contrast the methods and results of asexual and sexual reproduction.

9. Explain the adaptive significance of sexual reproduction.

10. Describe the photoperiodic regulation of animal and plant reproduction.

11. Label diagrams of the male and female reproductive organ systems.

12. Describe three types of sexual determination.

Reproduction

REPRODUCTION IS PERHAPS THE PRIME CRITERION OF LIFE, for, no matter how diligently a single organism strives to maintain the steady state, it nevertheless declines and death follows. In reproduction, life renews itself. In the Old Testament's *Ecclesiastes*, it is written,

To every thing there is a season, and a time to every purpose under the heaven: A time to be born, and a time to die; a time to plant, and a time to pluck up that which is planted. . . .[1]

Most higher organisms reproduce seasonally. In the late summer and early fall, following the breeding season, the primary reproductive organs, the *gonads*, the testes and ovaries of many birds, mammals, and other animals shrink into inactive vestiges. They no longer produce sex cells, the *gametes*, or sex hormones. Sexual behavior is absent from the winter flocks of temperate zone residents. The energies of the species are diverted and consumed in the "struggle for existence" in the "hardest time of the year." Then, in the spring, with the lengthening of the day, the gonads are stirred to activity. They swell and resume their production of sex cells and hormones. The gonads make a comeback, a recrudescence, and mating and brooding follow in the wake of their return.

In the 1920's, it was shown that increasing the hours of light by artificial illumination—even for birds kept outside in cold weather—would

---

[1] III, 1–8.

result in an increase in the size of the gonads and a tendency to migrate north. The breeding gonads of the Pacific Coast white-crowned sparrow are a hundred times as large as the nonbreeding organs.

The maturation and release of ripe eggs (ova) from the ovaries, *ovulation,* occurs once a year in annual breeders. Deer, antelope, kangaroo rats, amphibians, and many freshwater fish are such seasonal ovulators. In several female mammals, such as the raccoon, the wild rabbit, and the ferret, the increasing ratio of light to darkness as spring approaches sets into action a chain of events that ends with ovulation. In many species, the time of ovulation coincides with the period of maximum sexual behavior, "passion." This period of sexual excitement has been called *estrus* (from the Greek *oistros,* meaning "gadfly," "sting," or "frenzy") or animal "heat." It is the only time during which the female is receptive to the male's advances. Males become sexually aggressive at about this time also.

The period of male sexual excitement, corresponding to female estrus, is known as the *rut* (from the Latin *rugites,* meaning the "roaring" or "bellowing," as of male animals in rut). Rut and male fighting for territories and harems often occur some time prior to female estrus. The sex hormones from the recrudescing gonads produce many changes in secondary sexual features. The neck of the rutting stag enlarges as his musk glands become very active. The odor of the male deer is even detectable by the poor sense of human smell at many yards distance. The actual mating of the deer and fertilization occurs in the early fall, and the fawns are born early in the following spring, in April and May, and mature sufficiently by the following autumn to withstand the rigors of winter. The pregnancy or *gestation period* of the deer lasts 8 months.

The synchrony of the reproductive process is often amazing. The testicles of the kangaroo rat are held inactive within the abdomen until the breeding season, when they descend into the scrotum and resume the production of sperms and male sex hormone. The majority of the females in a population of kangaroo rats sampled in March is in the same period of their 20 to 30-day pregnancies.

The entire female reproductive process of most animals has been termed the *estrous cycle.* Those species with but one "heat" per year are called monestrous; a species with several is polyestrous. "Domestic" rats and mice, the "hangers-on" of our civilization, come into heat once every 5 days and breed the year round. Many domesticated animals whose relatives are monestrous in the state of nature increase their breeding times throughout the year in the security of man's care. The great apes and men (like rats) also breed throughout the year. Dogs and cats have an estrous period once every 4 to 6 months, or two or three times a year. The lining of the uterus, the *endometrium,* expands and shrinks, waxes and wanes in activity, in correlation with the rise and fall of ovarian hormones. The growth of the endometrium reaches its peak after ovulation at about the time the developing embryo would implant.

If fertilization and implantation fail in man and his relatives, most of the endometrial tissues and some blood are shed in *menstruation* (from the Latin *mensis,* meaning "month"). The bleeding of dogs and cats is not menstruation, but rather hemorrhagic accidents during the period of uterine vascular engorgement. Except for the bleeding and behavior, the menstrual and estrous cycles are quite similar. It seems that the menstrual cycle was derived from the estrous cycle in the course of evolution, but the reasons for it remain obscure.

ESTROUS CYCLE

The examination of the ovaries of a female rat just after heat reveals several yellow bodies, the *corpora lutea* (singular—corpus luteum), in each. If the animal is not impregnated, the yellow bodies degenerate and are replaced by scar tissue. The removal of the ovaries or the excision of their yellow bodies early in pregnancy results in the abortion of the embryos. The yellow bodies of the ovary apparently produce a substance essential to the maintenance of pregnancy.

Microscopic studies have shown that the mammalian eggs (ova) develop within fluid-filled spaces, the *follicles* of the mammalian ovary (see Fig. 4-1). Following ovulation (the rupture of the follicles and the release of their eggs), the follicles are transformed into yellow bodies, which persist, provided that fertilization of the eggs and the implantation of embryos occur. Follicle development, ovulation, yellow body formation (luteinization), and degeneration constitute the major events of the *ovarian cycle* of estrus.

The tissue lining of the uterus, the *endometrium,* undergoes periodic changes correlated with the ovarian cycle. The alterations of the endometrium during the estrous cycle can be referred to as the *uterine cycle* (see Fig. 4-2). Shortly after estrus, when the ruptured follicles of ovulation have transformed into yellow bodies, the endometrium reaches its greatest size and activity. The expanded *uterine glands* of the endometrium produce glycogen-rich secretions. The blood supply of the endometrium is great, the number of capillaries being the highest of the cycle. An elaborate preparation has been made to receive the implanting embryos. If fertilization does not take place, the endometrium shrinks at the same time that the ovarian yellow bodies regress; the uterine glands halt

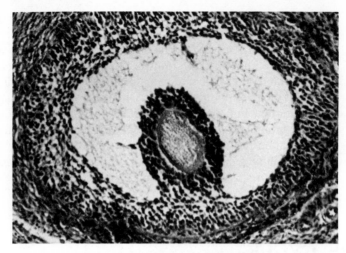

Fig. 4-1

Ovarian follicle. In this dog ovary, we can see a semimature follicle. The egg and surrounding cells project into the fluid-filled cavity. At ovulation, the egg and its neighboring cells are released. The ruptured follicle is transformed into a yellow body.

From M. J. Zbar and M. D. Nicklanovich, *Biology of Reproduction,* Saturn Scientific, Inc., Fort Lauderdale, Fla., 1971.

Fig. 4-2

Uterine cycle of lower mammal. (a) "Out of season," the anestrous condition. Note the few simple tubular glands in the uterine wall. (b) "In season," or estrus, heat. Note the numerous, well-developed, twisted glands in the uterine lining (endometrium).

From M. J. Zbar and M. D. Nicklanovich, *Biology of Reproduction,* Saturn Scientific, Inc., Fort Lauderdale, Fla., 1971.

(a)   (b)

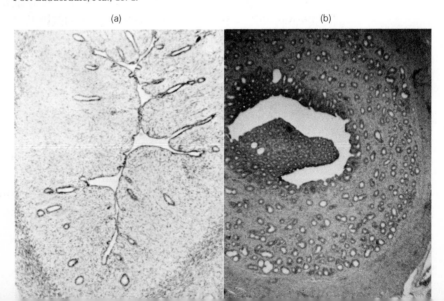

secretion and decrease in size, and the endometrial blood supply is reduced. As a new crop of ovarian follicles matures, the endometrium increases its thickness again. Maximal thickness and activity are attained following ovulation and the luteinization of the follicles. In summary, the endometrium swells along with follicular development and shrinks as the corpora lutea decline.

The events of "heat" (estrus), ovulation, and maximal uterine development almost coincide. This is no accident; there is "design" in it. The eggs are matured and released at the time that the female becomes sexually excited and receptive to the male. The uterine lining is prepared at the proper time to receive and nourish embryos. If pregnancy occurs, the cycles are suppressed, ovulation ceases, the corpora lutea persist, and uterine development is maintained. The cycle repeats again and again in the rat if pregnancy does not take place.

THE MENSTRUAL CYCLE

The ovarian events of estrus and the menstrual cycle are identical, but the uterine cycle of menstruation ends with the destruction of all the endometrium except its deepest layers. By definition, the menstrual cycle begins on the first day of flow and ends the day before the next period of bleeding. In the human female, the average menstrual cycle is 28 days in length (see Fig. 4-3). The cycle of the chacma baboon is 32 days and amazingly regular in contrast to the wide individual and monthly variability of the human cycle. Menstruations are 5 weeks apart in the chimpanzee. The average primate cycle occupies about one lunar month. Primates (apes and men) do not have a single breeding season, but are fertile throughout the year. The human female reproductive system is diagrammed in Fig. 4-4.

The lining of the uterus at the end of menstruation has been compared with a "raw wound surface." Epithelial cells from the remaining deep portions of the uterine glands leave their depressions and glide out over the denuded surface. This phase of "healing" begins even before bleeding stops. Days 4-6 through are termed the *reparative phase*. By this time, a new crop of follicles has begun to mature in the ovaries.

Their growth continues during the next stage, the *proliferative phase*, which ends with ovulation on the fourteenth day. The week of proliferation (days 7 through 15) sees a doubling of the thickness of the endometrium, including an increase in the number of uterine glands as well as their elongation. The connective tissue cells surrounding the glands

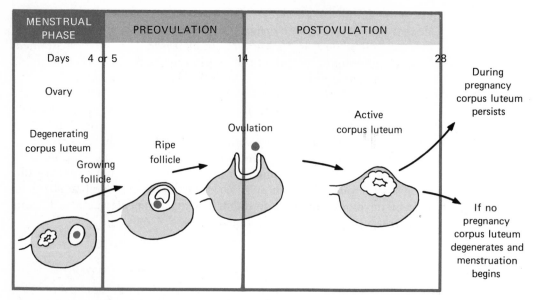

MENSTRUAL PHASE	PREOVULATION	POSTOVULATION

Days   4 or 5   14   28

Ovary

Degenerating corpus luteum

Growing follicle

Ripe follicle

Ovulation

Active corpus luteum

During pregnancy corpus luteum persists

If no pregnancy corpus luteum degenerates and menstruation begins

Fig. 4-3   **Ovarian cycle and the days of the menstrual cycle.**

multiply and produce quantities of tissue polysaccharides and protein fibers. The proliferative period of the uterine cycle is sometimes called the follicular phase, since it coincides with the stage of the ovarian cycle in which the follicles ripen and rupture (see Fig. 4-5a). Although many follicles begin to develop in a human cycle, usually only one reaches full maturity; the rest degenerate.

The *secretory phase* (days 16 through 28) of the uterine cycle is also termed the progestational, or progravid ("before pregnancy") stage as well as the luteal phase, for it occurs simultaneously with the period of the ovarian cycle dominated by the corpus luteum. During the secretory stage, the uterine glands further elongate and finally become twisted, and saclike expansions appear along their length. They secrete a thick, mucoid fluid, rich in glycogen. The development of the surrounding tissues and blood vessels (begun in the proliferative period) continues in the secretory phase until the endometrium has again doubled its thickness (see Fig. 4-5b). The state of the progestational endometrium has been described as "succulent," having juicy tissues. The uterine lining is thus well prepared to receive and nourish the embryo by the time it descends from the oviducts to the uterus. The egg is usually fertilized in the upper portion of the oviduct and its downward migration requires 3 to 5 days. The implantation day is anywhere from days 18 through 22.

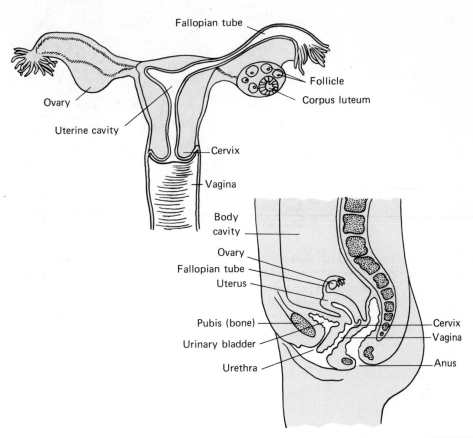

Fig. 4-4   **Human female reproductive organs. (a) Removed and (b) as revealed in midline hemisection of the torso. Note the relationship to other organs in (b).**
Redrawn from *Scientific American*.

Fertilization is most often close to ovulation (see Fig. 4-6). Since the unfertilized egg lives only 24 hours and sperm survive no more than 48 hours in the female tract, the time of maximal fertility is just a few days, 2 days before and 1 day after ovulation.

If fertilization does not occur, the corpus luteum begins to degenerate even before menstruation commences. At the end of the twenty-eighth day, the *menstrual phase* begins (see Fig. 4-5c). The sloughed endometrial tissue debris and blood make up the menstrual flow or menses. The twenty-ninth day is the first day of the next cycle, and the blood flow lasts until the fifth day. Initiating this destruction of tissue is a constriction of uterine arterioles. More than three-quarters of the endo-

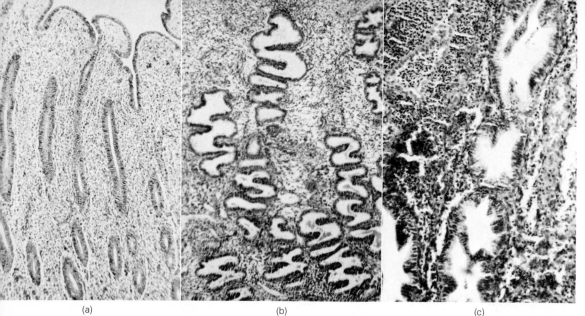

(a)   (b)   (c)

Fig. 4-5   **Phases of the uterine or endometrial cycle in the human female. (a) Proliferative**
phase, with straight tubular, nonsecretory glands. (b) Secretory phase, with
twisted, expanded active glands. (c) Menstrual phase, with bleeding into the
tissue spaces and glandular breakdown. See text for further description.
From M. J. Zbar and M. D. Nicklanovich, *Biology of Reproduction,* Saturn Scientific, Inc.,
Fort Lauderdale, Fla., 1971.

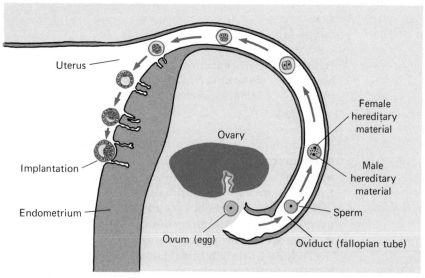

Uterus

Female
hereditary
material

Ovary

Male
hereditary
material

Implantation

Endometrium

Sperm

Ovum (egg)

Oviduct (fallopian tube)

Fig. 4-6   **Fertilization usually occurs high in the oviduct. The fertilized egg undergoes**
several cell divisions in its descent to the uterus, before the early embryo
implants in the endometrium.

metrium is lost. The famed embryologist, L. B. Arey, described menstruation as "a violent demolition of the premenstrual edifice some days after the expected tenant fails to arrive." Bradley Patten portrayed menstruation as the "protest of a disappointed uterus, that all its elaborate preparations for embedding and nourishing a fertilized ovum have gone for naught."

THE HORMONES OF REPRODUCTION

The anterior pituitary produces several gonad-stimulating hormones. The *follicle stimulating hormone* (FSH) governs the growth and ripening of ovarian follicles. Along with FSH, *luteinizing hormone* (LH) is one of the "hormones of ovulation," and, as its name implies, it plays an important role in transforming the ruptured follicles into yellow bodies, or corpora lutea. The *luteotrophic hormone* (LTH) stimulates secretion by the yellow bodies but is also known as the lactogenic hormone, since it stimulates milk production in the breast ("lactogenic" means "milk-producing").

The developing ovarian follicles secrete *estradiol,* the most potent of the *estrogens* ("heat generators"), which are in charge of the reparative and proliferative phases of the uterine cycle. The corpus luteum produces *progesterone,* the hormone of the preparation for pregnancy, which directs the secretory or progestational period of the uterine cycle. The corpus luteum also continues to produce estradiol.

Ovarian sex hormones feed back negatively against the pituitary gonadotropins, FSH and LH. Estradiol actually stimulates LH secretion before ovulation, but, afterward, when progesterone and estrogen levels are high, FSH and LH levels are low. Perhaps the ovarian hormones feed back against FSH- and LH-releasing factors from the hypothalamus. At any rate, at the end of the menstrual cycle the levels of progesterone and estradiol drop precipitously. Thereafter, FSH is secreted again, and a new cycle begins. Study the graph of hormone levels during the cycle in Fig. 4-7.

During pregnancy, when the levels of estrogen and progesterone remain high, follicular development and ovulation are held in abeyance. The implanting embryo secretes a hormone, human *chorionic gonadotrophin* (CG), which supports the corpus luteum, stimulating its secretion and prolonging its life. In the human female, the corpus luteum degenerates in about the fifth month of pregnancy, when the placenta assumes its role, producing progesterone and estrogen. After birth, the

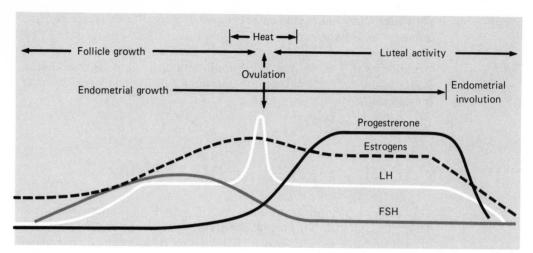

Fig. 4-7   **Hormone levels and the ovarian and uterine events in a reproductive cycle. This is an estrous cycle. Hormone levels are essentially the same in the menstrual cycle.**

From N. M. Jessop, *Biosphere*, Prentice-Hall, Inc., Englewood Cliffs, N. J., 1970.

endometrium of the pregnant uterus is sloughed, cast off. Thus, the gestation period is really like a prolonged secretory phase of the menstrual cycle.

The first birth control pill, Enovid®, contained norethynodrel, a synthetic progesterone without some of the latter's undesirable side effects but at least 10 times as effective in suppressing ovulation, and an artificial estrogen, mestranol. These pill steroid hormones, like normal estradiol and progesterone, inhibit FSH and LH secretion. Thus, follicles do not develop, and, of course, no ovulation occurs. In this way, a woman under birth control therapy is like a pregnant female. To outward appearances, the cycle is normal: bleeding occurs on time. The pills contain graded dosages. Interestingly, the "birth control" pill has been used in the treatment of cases of female sterility. Apparently, use of the pill repatterns the normal, internal hormonal cycles. When all criticisms, scientific and ethic (by the way, who is to say where one ends and the other begins?), have been met, perhaps the Nobel prize in medicine should be awarded to its developers, since the health problems of man stem increasingly from overpopulation. There are many competent scholars who would say that the number one health problem of modern man is overpopulation. Other contraceptive devices are shown in Fig. 4-8; the effectiveness, advantages, and disadvantages of 11 contraceptive methods are summarized in Fig. 4-9.

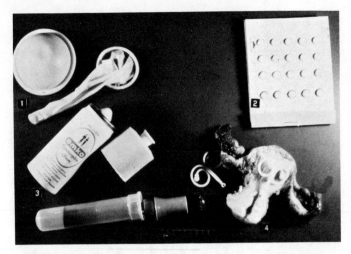

Fig. 4-8 **Contraceptive devices. (1) The female rubber contraceptive device, the diaphragm (upper left), which fits over the cervix. The male rubber contraceptive, the condom. (2) The birth control pill, which consists of female sex hormones. (3) Vaginal spermicidal foam and applicator. (4) A Saf-T-Coil intrauterine device, in and out of place in a dissected uterus.**
From M. J. Zbar and M. D. Nicklanovich, *Biology of Reproduction,* Saturn Scientific, Inc., Fort Lauderdale, Fla., 1971.

REPRODUCTIVE LIFE

Gonadal growth, commencement of function, and the development of secondary sexual characters are all parts of *puberty,* the attainment of sexual maturity. The onset of a girl's "menstrual career" is called *menarche,* and the failure of the cycle occurring in middle age (45 to 50 years) is the *menopause.* Puberty comes about the age of 12 to 14 years in girls and slightly later in boys. Puberty literally means "becoming covered with hair." The development of pubic hair in the armpits and the genital region is a feature of puberty in both sexes. In males, pubic hair extends upward from the groin to the navel and even onto the chest, whereas, in the female, torso pubic hair is restricted to a sharp triangle in the genital region. The distribution of pubic hair is an example of what are known as the *secondary sexual characteristics.*

Tissues other than those of the primary and accessory reproductive organs become the targets of the sex hormones. The male sex hormone

Method	Effectiveness	Advantages	Disadvantages	Medical Risk
Diaphragm	High, particularly with spermicide (about 17.5% failures)	No coital involvement	Initial medical instruction required; cannot be used by women with certain uterine variations; precoital interference	None
Oral contraceptive (progestin-estrogen combination)	Extremely high (under 1% failures)	No equipment or coital involvement; improved menses	Requires motivation and physical checkups; many side effects—major and minor	Reports of statistical association with thromboembolic disorders; metabolic and liver changes; possible relation to cancer; side effects include depression, weight gain, nausea, vomiting, headaches
IUD	Extremely high (2.7% failures)	No coital involvement; no continued motivation required	Requires physician for insertion and checkups; sometimes expelled	Pelvic inflammatory disease, perforation of uterus, bleeding, cramps, backache
Vasectomy	100%	No equipment or coital involvement; no continued motivation required; permanent	Physician visit required; not always reversible	None (occasional psychological reaction)
Tubal ligation	100%	No equipment or coital involvement; no continued motivation required; permanent	Hospital stay required; usually irreversible	Small surgical risk (occasional psychological reaction)
Postcoital douche	Very low (over 30% failures)	No coital involvement	Requires bathroom facilities and equipment	None
Intercourse during lactation	Very low; ovulation unpredictable	No equipment or effort needed	Unreliable	None
Coitus interruptus	High if practiced correctly (over 18% failures)	No mechanical or chemical equipment required	Timing essential for effectiveness; often psychologically unacceptable	None (occasional psychological reactions)
Spermicidal vaginal foam	Low (over 28% failures)	No prescription required	Messy; precoital involvement	None
Calendar rhythm	Fair if practiced correctly (over 24% failures)	Sanctioned by Roman Catholic Church; no prescription or equipment required	Requires motivation: intelligence, and long periods of abstinence	None
Condom	High when used correctly (16% failures)	No prescription required; male's responsibility	Requires strong motivation; occasional breakage or leakage of semen; some loss of sensation; precoital involvement	None

testosterone promotes muscular development, while estradiol brings about the deposition of fat in the connective tissues beneath the skin, particularly in the buttocks and breasts. The smooth contours of the female body are due to the lack of muscular definition and the smoothing influence of "sexual fat." Estradiol stimulates the development of the duct system of the breast, and progesterone promotes the development of the breast's secretory units. All these effects are secondary sexual characteristics.

In general, estradiol and testosterone are metabolic antagonists. Injections of estradiol in males will retard the development of masculine traits, and testosterone injections into females will block the development of feminine characteristics. In the "lower animals," we have seen that estradiol and testosterone markedly influence behavior, producing "heat" and "rut," respectively. A bull is a massive, powerful, and aggressive animal with tough meat. A castrated male calf develops into a steer, smaller, weaker, more easily handled, and with tender meat compared with that of the bull. An ox is a bull castrated at maturity, so that full size and strength are attained. An ox is therefore strong, but unaggressive, an ideal draft animal or "beast of burden." The subterranean clover of Australia produces estrogenic substances and has driven breeders to the point of insanity, since the eating of this plant results in nymphomania in cows and docility and sexual disinterestedness in bulls.

Until the eighteenth century, Italian choir boys were castrated to preserve their high-pitched voices. There was a whole class of men called the *castrati*. The deep male voice is a secondary sexual trait. Castration after sexual maturation will not raise the pitch of the voice. Castrated males, eunuchs, were used as domestic servants in the Middle East, particularly as harem guards.

Sperm formation is continuous, not cyclic. Unfortunately, the gonad-stimulating hormones have been femininely named. Follicle-stimulating hormone in males is more appropriately called the gamete-stimulating hormone, since it promotes sperm formation here. In terms of its male function, luteinizing hormone (LH) is known as the interstitial-cell-stimulating hormone (ICSH), since it stimulates the interstitial cells of the testicles to manufacture and release testosterone (see Fig. 4-10). The role of LTH (lactogenic hormone) in man is unclear, although male breasts can and do develop into full-fledged, femalelike mammary glands in males undergoing female sex hormone therapy, as for prostate cancer. Gynecomastia (female breasts in males) also sometimes result

---

Fig. 4-9   OPPOSITE PAGE: **A guide to contraception.**
Courtesy of Dr. Celso-Ramon Garcia.

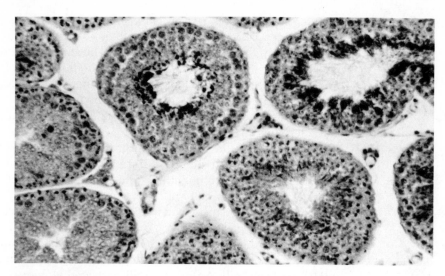

Fig. 4-10   **Testis. This section of the male gonad shows the seminiferous tubules in which**
sperm formation occurs. Hormone-secreting interstitial cells occupy the spaces
between tubules.
From M. J. Zbar and M. D. Nicklanovich, *Biology of Reproduction,* Saturn Scientific, Inc.,
Fort Lauderdale, Fla., 1971.

from digitalis heart therapy. Digitalis is an estrogenic substance produced by the foxglove plant and used in the treatment of certain cardiac
disorders.

The reasons for the onset of puberty are poorly understood at present.
Undoubtedly, the hypothalamus secretes releasing factors at that time
that stimulate the pituitary to release gonadotropins. The pineal gland,
discussed in Chapter 3, may play some role in this biologic clock mechanism. Puberty is part of the larger plan of body growth and development.
Reproductive powers are held in check until the rest of the body is mature enough to bear the physical responsibilities of parenthood. Puberty
is probably the greatest physiologic revolution following birth. In the
human female, at least, the decline of sexual fertility (menopause) is no
less dramatic.

In a woman's late forties and early fifties, ovulation and menstruation
halt, and the female secondary sexual characteristics fade as the levels
of the sex hormones drop. Many of the symptoms of the menopause are
defeminizing, and it has been rather appropriately called the "change in
life." The replacement of the fallen internal sex hormone with estradiol
in particular prevents the development of most of the undesirable physical and mental consequences of the menopause. In the male, the decline

of sperm formation and the fall of testicular hormone are more gradual than the corresponding events in the female, but sometimes the male sex career ends abruptly, quite like the female menopause.

THE SEX ACT

In many aquatic animals, the sex act is undramatic, involving a simple shedding of sex cells into the water. Of course, there is careful timing. One of the most effective stimuli for egg release by the American oyster is sperm water, and, conversely, egg water is an efficient stimulus for ejaculation. *Spawning* refers to the release of eggs and sperms in the water. In several aquatic species, pheromones (external hormonal secretions) are released into the water as chemical messengers from one sex to another in an effort to synchronize spawning. Not all aquatic animals practice *external fertilization*. Many carry out *internal fertilization,* and there are also intermediate processes.

In the crayfish and the lobster, the male grasps and inverts a female, stands over her, seizes all her walking legs with his pincers, bends his abdomen tightly over hers, then presses the tips of two modified abdominal appendages against the openings of the female's sperm receptacles and passes mucous packages of sperm into her receptacles. Days or even weeks later, the female alone cleans her abdomen, extrudes the stored sperm and several hundred eggs along with a slimy secretion that attaches the eggs to her abdomen. Although copulatory organs are involved and sperm are transported into the female body, fertilization is external since the eggs are fertilized at the moment of egg-laying.

In most fish fertilization is external, and no copulatory act occurs; however, in sharks, their relatives, and a few bony fish, intromittent organs have evolved and fertilization is internal. In sharks the inside portion of each male pelvic fin is modified to form a copulatory organ, the clasper (see Fig. 4-11). This elongation of the fin is provided with a groove, the edges of which overlap so that the groove is quite like a tube. During copulation, only one clasper is inserted into the cloaca of the female, and a siphon forces sea water down the clasper tube, ejecting spermatozoa into the female reproductive tract. In several teleosts (bony fish), the anal fin of the male is elongated to form an intromittent organ, the gonopodium. The common tropical aquarium fish, the "guppy," possesses a modified copulatory fin of this type.

Although fertilization in frogs and toads is external, male and female nonetheless participate in a "false copulation," or amplexus (see Fig.

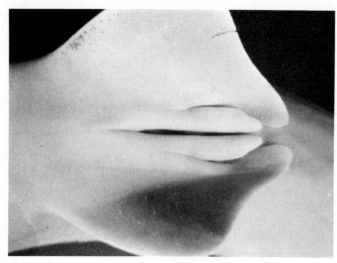

Fig. 4-11   **Pelvic fins of male shark. The modified inside portions of each fin are copulatory devices, the claspers, which are inserted into the female reproductive tract and bear grooves for seminal transport.**
From M. J. Zbar and M. D. Nicklanovich, *Biology of Reproduction,* Saturn Scientific, Inc., Fort Lauderdale, Fla., 1971.

4-12). The male mounts the female's back and clasps her firmly with his forelegs just behind her forelimbs. Swollen thumb pads apparently aid his grasp. It is extremely difficult to dislodge a male during amplexus. The compression of the body of the female by the powerful grasp of the male aids in the extrusion of the eggs. Sperm are shed over the eggs as they pass out the opening of the female cloaca. The male possesses no copulatory organ; his semen is shed directly from the vent opening of his cloaca, which is structurally the same as that of the female.

Internal fertilization occurs in salamanders, even though no copulatory organs are present. Males deposit cloudy packets of sperm on the pond or stream floor, and the females pick them up with the muscular movements of the lips of the cloaca. A pocket in the wall of the cloaca serves as a sperm storehouse until the eggs pass down the oviducts.

In most terrestrial forms, copulatory organs are present in the male. Birds are notable exceptions. Most birds copulate by cloacal apposition, simply by putting the lips of the vent together. Ducks, geese, swans, and ostriches do possess a penis that is quite similar to that of some reptiles; a clitoris is present in the females of these species. In many birds, a rudiment of the penis can be found. It is extremely puzzling that evolution should have taken this course.

The penis of turtles and crocodilians is erectile and grooved for the transport of semen. The penis of mammals contains a closed tube, the urethra, for the transport of urine and semen. The cavernous or erectile tissue of the mammalian penis is quite like that of reptiles. When blood flow into the organ increases during sexual excitement, spongy sinuses in the interior become swollen and compress the surrounding veins retarding drainage (see Fig. 4-13). The erect organ is then stiff enough to be inserted into the female vagina where seminal fluid is deposited by ejaculation. The *prostate* and the *seminal vesicles* are accessory glands of male reproduction and produce the seminal fluids (see Fig. 4-14). In male *orgasm*, the smooth muscle in the walls of the sperm ducts, prostate, and seminal vesicles contracts rapidly and rhythmically. The muscles in the perineal region (between the anus and the penis) also contract as semen is spewed from the external urethral opening at the head of the penis. Orgasm is a complicated reflex.

In many mammals, rodents, carnivores (such as dogs and cats), bats, whales, and some lower primates (members of man's own order), a *penis*

Fig. 4-12   **Frogs in amplexus. Although the male lacks a copulatory organ and fertilization is external, this sex act, in which the male mounts the female and grasps her abdomen, insures synchronous shedding of the sex cells in close proximity.**
Photo courtesy of Professor Lewis D. Ober.

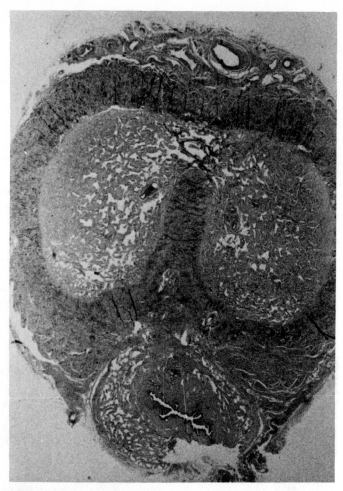

Fig. 4-13   **Human penis in cross section. Note the three cylinders of
                   spongy, erectile tissue. Blood fills its spaces to bring about
                   erection. The tube piercing the lower cylinder is the urethra,
                   ordinarily the urinary passageway, but also a passage for
                   the reproductive fluid in the sex act.**
                   From M. J. Zbar and M. D. Nicklanovich, *Biology of Reproduction,*
                   Saturn Scientific, Inc., Fort Lauderdale, Fla., 1971.

bone develops within the cavernous tissue of the penis and helps to increase the rigidity of the organ (see Fig. 4-15). The penis bone of some
species of whales reaches a length of 6 feet.

    The head of the penis, or the *glans,* of the cat and other species bears
a number of horny spines that function as sexual stimulants during

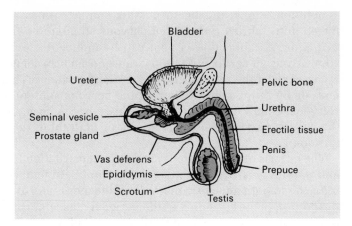

Fig. 4-14   **Human male reproductive organs (side view). All structures seen here with the exception of the penis and the prostate are repeated on the opposite side.**
From N. M. Jessop, *Biosphere,* Prentice-Hall, Inc., Englewood Cliffs, N. J., 1970.

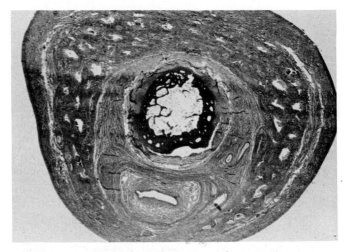

Fig. 4-15   **Cross section of dog penis. The dark-staining structure in the center of the organ is the penis bone, which develops within the surrounding erectile tissue of the penis of the dog and other carnivores. Walruses and whales also possess penis bones.**
From M. J. Zbar and M. D. Nicklanovich, *Biology of Reproduction,* Saturn Scientific. Inc., Fort Lauderdale, Fla., 1971.

copulation. Lions and tigers possess similar spines. The rabbit, cat, ferret, mink, shrew require the nervous stimulation of copulation for ovulation. A rabbit doe in heat can be induced to ovulate simply by stimulating her cervix with a glass rod, and such an act will interrupt the estrus cycle of the rat, causing a "pseudopregnancy." This is a reflection once again of the coordination of the nervous and endocrine systems. Nervous stimuli culminate in the release of the hormone of ovulation. (Recall the photoperiodic sensitivity of the rabbit, the ferret, and the raccoon.)

The clitoris of the female is similar in construction to the male penis and is quite sensitive to tactile (touch) stimuli. In female orgasm, or *climax,* the smooth muscle in the walls of the uterus and oviducts contracts in addition to the spasming of the perineal musculature. These contractions must play some role in fertilization. Following climax in the cow, the sperm travel the length of the oviduct in 5 minutes, whereas the journey of the same distance would require 50 minutes if the sperm traveled under their own power alone. Perhaps the contraction smears the semen over the folds in the walls of the female reproductive tract. As in male orgasm, the sensations that arise during female climax pass into the brain and somehow satisfy the sex drive.

MITOSIS AND MEIOSIS

Mitosis is a process of duplication. Daughter cells are identical and exactly like the parent cell. Mitosis is a sexless, or *asexual reproduction:* no sex cells are involved. In the body cells of all animal organisms are two sets of chromosomes, a maternal and a paternal set. The two sets are said to be *homologous,* since for each chromosome in one set there is a structural mate in the other set. Chromosomes carry genes along their length. Although homologous chromosomes are structurally alike, they seldom carry all the same genes. A cell or an organism with cells containing two sets of chromosomes is said to be *diploid.* A sex cell carries but a single chromosome set and is said to be *monoploid.*

In *fertilization,* the paternal set of chromosomes carried by the sperm unites with the maternal set borne within the egg. In *sexual reproduction,* the offspring are never identical to either parent. The formation of a variety of sex cells lies at the heart of the process of sexual reproduction.

After the chromosomes were studied in mitosis and the fusion of sex cells in fertilization became known, the German biologist August

Weissman predicted that the chromosome number of sex cells must be half that of body cells. Otherwise, chromosome numbers would be doubled from generation to generation ad infinitum. It turned out to be as he had predicted. *Meiosis* is the process by which the diploid chromosome number of sex cell precursors is reduced to the monoploid number of sex cells. Sex cells are also called *gametes* or *germ cells*. Meiosis is the central feature of gametogenesis, the formation of sex cells.

In mitosis, one cell division and one chromosome duplication occur. A diploid cell becomes two diploid cells. There are *two* cell divisions in meiosis, but only one chromosome duplication. When the chromosomes first appear in either mitosis or the first meiotic division, they are already doubled, so that the cell on the verge of division is in reality tetraploid, with four sets. In mitosis, this temporarily tetraploid cell is divided once into two diploid cells. In meiosis, the tetraploid cell is divided twice to form four monoploid cells, the gametes. In the formation of eggs, only one egg is usually formed, as the cytoplasm with its stored food materials is conserved in one cell, but the polar bodies are extruded nuclei, which, taken together with the egg, equal the number of sperm (four) formed from a single precursor cell. After the two meiotic divisions, there is finally only one set of chromosomes in each sex cell. Meiosis also differs from mitosis in that its products, the sex cells, are not identical to each other, but widely variable.

In mitosis, the doubled chromosomes of the two sets appear, line up at random along the cell equator, and split, traveling to the opposite poles (review Fig. 1-2). In the first meiotic division, the *homologous chromosomes*, the similar members of opposite sets, *synapse* or fuse end-to-end and length-for-length. During this fusion, they seem to exchange corresponding portions frequently, thus mixing genes, breaking up wholly maternal and entirely paternal chromosomes, and making new chromosomes that are both partly grandmaternal and grandpaternal. This process of the exchange of similar parts of homologous chromosomes, *crossing over,* has no counterpart in mitosis, in which the homologues remain apart (see Fig. 4-16). Even if crossing over did not occur and the chromosomes remained wholly maternal and paternal—unhybridized—there is no means of dictating that a sex cell shall receive the entire maternal set or the complete paternal set. The sets selected are almost without exception mixtures of maternal and paternal chromosomes, as well as hybrids resulting from the crossing over of homologues. Man has 46 chromosomes, two sets of 23. The odds against a sex cell carrying only maternal or paternal chromosomes are more than 8 million to 1; crossing over makes it even less likely. Without crossing over (which is unavoidable), the total possible types of offspring any

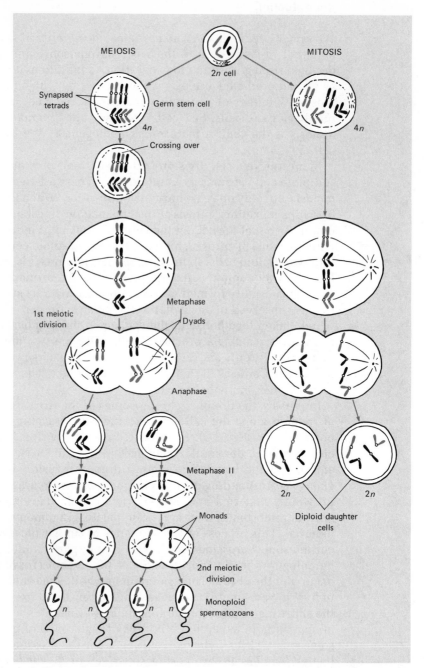

Fig. 4-16   **Mitosis and meiosis compared in a hypothetical organism with a chromosome number of 4. Maternal chromosomes are in red, paternal in blue. Note the increased variety afforded by crossing over.**

human couple can produce is over 64 trillion. With crossing over, the potential variety is even greater. Such is the source of variety provided by sexual reproduction.

EVOLUTION OF SEXUAL REPRODUCTION

Today, almost all organisms—from the lowest to the highest—carry out sexual reproduction at least sometime in their life cycles. Individuals of the unicellular animal (protozoan) *Paramecium,* which usually reproduce asexually by mitosis, occasionally conjugate and exchange micronuclei (see Fig. 4-17). Even the primitive bacteria, which are monoploid throughout most of their lives and reproduce primarily by mitosis, sometimes form conjugation tubes through which the donor cell passes part or all of its chromosome to the recipient. Viruses do not reproduce sexually.

In plants, sex cells are formed by mitosis, *not* meiosis. Meiosis precedes fertilization in animals but follows it in plants. The unicellular algal plant (protophytan) *Chlamydomonas* spends most of its generations as monoploid, asexually reproducing (mitosing) cells, but under unfavorable conditions the mitotic descendants of a cell become sex

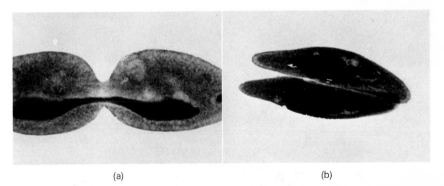

(a)   (b)

Fig. 4-17   **Reproduction in a unicellular animal, protozoon *Paramecium*. (a) *Paramecium* in fission. In this type of asexual reproduction, binary fission, the microorganism splits in two. The large nucleus is being divided amitotically (without mitosis) here; the micronuclei of this form do divide mitotically. (b) Sex in *Paramecium*. The two individuals are conjugated and will exchange micronuclei.**
From M. J. Zbar and M. D. Nicklanovich, *Biology of Reproduction,* Saturn Scientific, Inc., Fort Lauderdale, Fla., 1971.

cells that unite to form a diploid *zygote.* The latter forms a thick wall about itself and survives drought and cold. When conditions are favorable again, the zygote divides meiotically and produces four new monoploid cells. Interestingly, the organism resorts to sexual reproduction under changing environmental circumstances (see Fig. 4-18).

In animals only the short-lived gametes are monoploid. In most primitive plants, the monoploid phase of the life cycle dominates the diploid stage (see Fig. 4-19). In many plants there is an entire, multicellular, substantial plant body, the *gametophyte,* whose cells are monoploid and whose sex organs produce sex cells by mitosis. The zygote formed by the union of gametes develops into the next stage of the life cycle, the diploid *sporophyte,* which produces *spores* by meiosis. Spores germinate to form gametophytes, and the cycle repeats. In the higher plants, the gametophyte has been reduced to a microscopic form, dependent upon the sporophyte, which is the conspicuous plant body in all higher plants. This is just the reverse of the situation in lower plants. (More will be said about plant reproduction in Chapter 9, on evolution.)

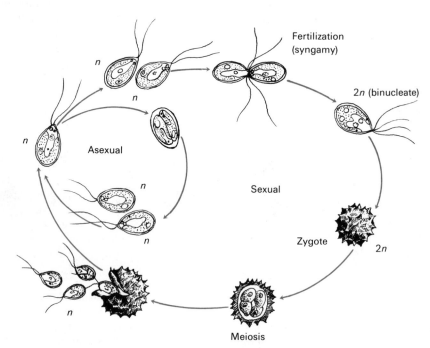

Fig. 4-18   **Life cycle of the one-celled plant (protophytan)** *Chlamydomonas.* **(See text for further description.)**

From N. M. Jessop, *Biosphere,* Prentice-Hall, Inc., Englewood Cliffs, N. J., 1970.

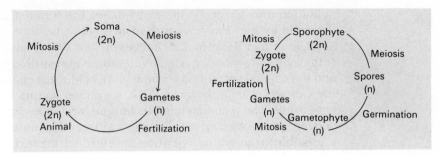

Fig. 4-19   **Comparison of the average plant and typical animal life cycle. Note the timing of meiosis and chromosome number in particular.**

Budding and fragmentation are types of asexual reproduction other than mitosis, but they both involve mitoses and its consequences (see Fig. 4-20). The major disadvantage of asexual reproduction by such mechanisms as mitosis is the lack of variety produced. All offspring are the same; the status quo is maintained. Sexual reproduction is a revolution, its primary advantage being the lack of duplicity in its offspring, the abundance of variety afforded. Duplicity with whom? Variety is more than the spice of life; it is the raw material of evolution. Without sexual reproduction, change in life is severely limited. It is not difficult to see why nature ruled in favor of sexual reproduction. Change is essential, not only for the survival of the individual but also for the perpetuation of the species.

SEX DETERMINATION

Of the 23 pairs of chromosomes in man, 22 match, but the twenty-third pair does not match. These unmatched chromosomes are called the *sex chromosomes*. The human female does carry a matched pair of sex chromosomes, designated XX; the unmatched pair of human male chromosomes is designated XY (see Fig. 4-21). Sex cells normally carry but one sex chromosome. All eggs carry an X chromosome; sperm may carry an X or a Y chromosome.

The male-determining sperm (Y bearing) is smaller and less hardy than the female-determiner (X carrying). Dr. Landrum Shettles recently suggested that it might be possible to substantially increase the chances of having a child of a desired sex. Intercourse by rear penetration, ejaculation on the cervix, following a bicarbonate douche, near the time of

ovulation are said to be the elements of the formula for boys. Male-
determining sperm survive well in an alkaline medium, and, since they
are lighter than X sperm, the chances of them first reaching the egg (high
in the oviduct) are better. After ovulation, as the egg descends the oviduct
and the vaginal secretions become more acidic, the chances of conceiving a male are poor, because the Y sperm perish quickly under acidic
conditions. The recommended technique for conceiving a girl involves
conventional intercourse following a vinegar douche after ovulation. It
is possible to determine ovulation time through the taking of daily rectal
temperatures, since ovulation is accompanied by a significant rise in
body temperature. The acidity of the female tract, as noted, also changes
during the cycle. Litmus paper (an acid-base indicator) has been used to
follow the cycle. Individual variations in vaginal pH might have something to do with the tendency of some females to have all boys or all

Fig. 4-20   **Hydra budding, a type of asexual reproduction.**
From M. J. Zbar and M. D. Nicklanovich, *Biology of Reproduction,*
Saturn Scientific, Inc., Fort Lauderdale, Fla., 1971.

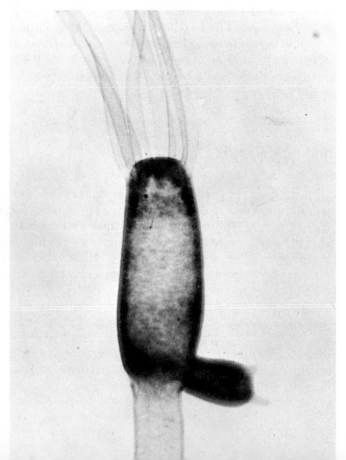

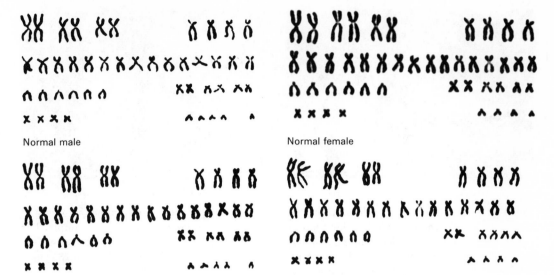

Normal male

Normal female

Klinefelter syndrome

Turner syndrome

Fig. 4-21   Human chromosomes and sex. The normal human chromosome number is 46, two sets of 23, one maternal and one paternal set. Every chromosome in one set has a mate in the opposite set, except in males, where the small Y sex chromosome does not match the X chromosome. A human with Turner's syndrome (XO) has only one sex chromosome and becomes a sterile female (see lower right). A person with Klinefelter's syndrome (XXY) has an extra sex chromosome and develops into a sterile male, sometimes with female breasts (see lower left). In the normal male and in Klinefelter's syndrome, the Y chromosome is found in the lower right of the Karyotype. The X chromosomes are located in the second row from the top in each Karyotype, and it is here that the difference in chromosomes is evident.

From Victor A. McKusick, *Human Genetics*, second edition, Prentice-Hall, Inc., Englewood Cliffs, N. J., 1969.

girls. In this way, the woman may well have something to do with determining sex. Some sperm samples contain nothing but male-determining sperm (androsperms; see Fig. 4-22). Although the sex of the firstborn will become increasingly less important to educated people in the future, the control of the sex of the second, most likely the legal last of the future, will be increasingly important.

The much-studied fruit fly has the same sex chromosome arrangement as man. In accidents of meiosis, the failure of chromosomal separation, *nondisjunction,* can result in the formation of sex cells either lacking sex chromosomes entirely or with two instead of the normal one. Fertilization of normal sex cells by such abnormal gametes can produce individ-

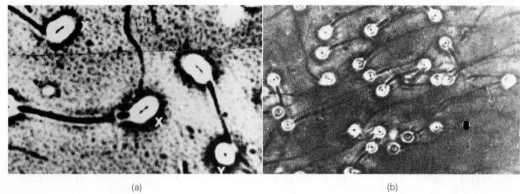

(a)   (b)

Fig. 4-22   **(a) The male-determining, Y-bearing sperms, androsperms, appear smaller and roundheaded in this phase contrast photomicrograph of a sperm sample. The female-determining, X-bearing sperms, the gynosperms, have larger, elongate, oval heads. (b) The donor of this sample comes from a family that for 256 years has produced almost nothing but boys. All sperms are "roundheads."**
Courtesy of Dr. Landrum B. Shettles.

uals with such abnormal sex chromosomal compositions as XO, XXY, XXX, YYY, etc. XO fruit flies are normal but sterile males. XO humans (Turner's syndrome) are defective females. At their puberty, menstruation, breast, and pubic hair development all fail; ovaries are virtually absent. An XXY fruit fly becomes a fertile female, but XXY humans (Klinefelter's syndrome) are normal but sterile males. Most "superfemales" (XXX) are infertile, even though the X chromosome seems to have a female-determining function.

It is possible to determine the sex of cells by examining the chromosomes of dividing cells. Recently, it has been discovered that it is even possible to "sex" the nuclei of undividing cells. A stain of cells scraped from the inside of the female cheek reveals a dark-staining body just beneath the nuclear membrane. This *sex chromatin,* or Barr body (see Fig. 4-23), is absent from the male cell nucleus. XO females have no Barr bodies; the cells of XXX females have two; XXY males' cells have one.

In amphibians, birds, butterflies, and moths, it is the female who bears two different sex chromosomes and so determines offspring sex. The sex chromosomes of such species are designated Z and W. A female is ZW; the male is ZZ.

In ants, bees, wasps, and a few other members of their order, fertilized eggs develop into queens (fertile females) and workers (sterile females), while unfertilized eggs develop *parthenogenetically* ("virgin born") into the male drones. The latter are monoploid.

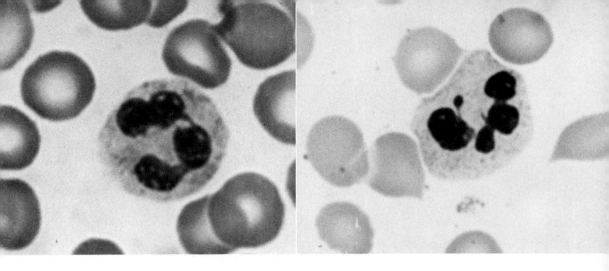

Fig. 4-23   **Male white blood cell (left) and female white blood cell (right). Note the "drum-stick" projecting from the female nucleus. Female cell nuclei of other body cells contain a dark-staining Barr body of sex chromatin just beneath the nuclear membrane.**
From M. J. Zbar and M. D. Nicklanovich, *Biology of Reproduction*, Saturn Scientific, Inc., Fort Lauderdale, Fla., 1971.

In the marine worm *Bonellia*, there seems to be an environmental de-termination of sex. The female is long with a slender proboscis. The male is minute and lives as a parasite in the female's kidney (Fig. 4-24). If the free-swimming larvae that develop from the fertilized eggs settle on the sea bottom, they grow into females; if they settle on the female proboscis, they become males. The hostess female secretes a hormone that determines male development. It has been experimentally shown that the larvae which settle on the female have not been committed to male development.

Many lower animals are true *hermaphrodites*, carrying a complete set of both male and female reproductive organs. (In Greek mythology, Hermaphroditus, the son of Hermes and Aphrodite, became united in a single body with the nymph Salmacis, while bathing.) Scallops, flat-worms, and earthworms are all hermaphroditic (see Fig. 4-25). Self-fertilization hardly ever occurs in hermaphrodites. Why not? Oysters and clams spend the first part of their lives as males, but later switch to the opposite sex. This condition is known as *protandry* ("males first"); some forms are females first and males later.

Male toads (ZZ) possess a vestigial ovary, Bidder's organ (see Fig. 4-26). Following castration, they transform into functional females, even though the sex chromosomal composition remains male (ZZ).

True hermaphrodites have never been found in mammals. Persons exhibiting abnormal development of the genitals usually have gonads of one sex or another. It is impossible to tell the sex of a developing human embryo by examination until the tenth to twelfth week. Prior to that time, the external genitalia of both sexes are identical. This early period of sexual development is aptly called the "indifferent stage." Both sexes possess a phallus. In the male, it comes to enclose the urethra and becomes the penis. In the female, the phallus of the indifferent stage becomes the clitoris, which does not contain the female urethra. The folds

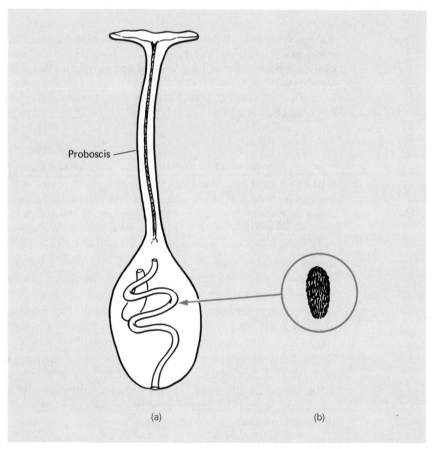

(a)   (b)

Fig. 4-24   **Sex determination in the marine worm *Bonellia*. If a larva settles on the bottom adult female worms, it develops into a female (a). If the ciliated larva comes close to a female, it develops into a nearly microscopic male (b), living in the female urinary tract. [The male in (b) is enlarged 10 times.] A hormone from the female determines the course of larval sexual development.**

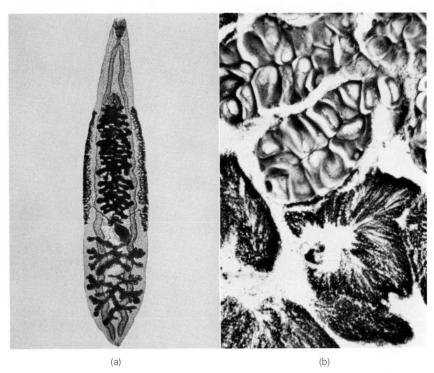

(a)   (b)

Fig. 4-25   **Hermaphrodites. (a) Human liver fluke, an hermaphroditic flatworm. The black coiled structure is a uterus filled with fertilized eggs. The branching structures below it are testes. This animal has both a penis and a vagina. (b) In this section of a scallop gonad ovarian follicles with developing large-celled eggs are shown (above) and spermaries or testicular follicles with maturing sperm (below) in the same organ.**
(a) Courtesy of CCM: General Biological, Inc., Chicago. (b) From M. J. Zbar and M. D. Nicklanovich, *Biology of Reproduction,* Saturn Scientific, Inc., Fort Lauderdale, Fla., 1971.

that fuse to form the scrotum of the male remain separate in the female to become the labia majora (see Fig. 4-27). Obviously, such a common pattern of development leads frequently to many errors in the development of external genitalia, and physicians occasionally mistake a newborn of one sex for the other. Rarely, surgeons have the task of "correcting" the external genitals to fit the sex of the internal gonads. The most famous cases of surgical "sex transformation" have involved transvestites, whose reproductive organs are clearly male or female, but who are psychotically obsessed with the desire to switch sexes.

The testes of the male fetus develop quite early and produce significant quantities of male sex hormone. Testosterone seems to direct the

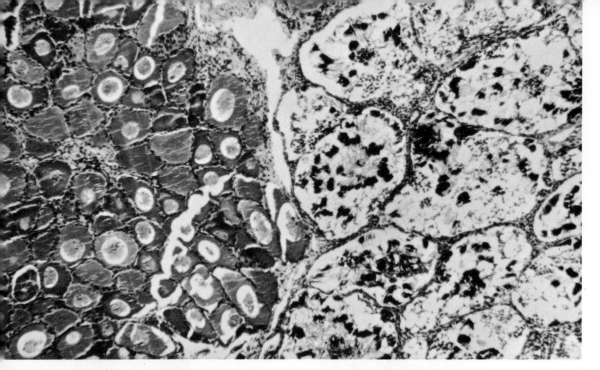

Fig. 4-26   **Bidder's organ from the toad. Note the immature egg nests of the vestigial ovary (left) and the dark-staining sperm heads (right) of the active testis. If this male toad were castrated, Bidder's organ (left) would become a functional ovary in less than 2 years.**
From M. J. Zbar and M. D. Nicklanovich, *Biology of Reproduction,* Saturn Scientific, Inc., Fort Lauderdale, Fla., 1971.

development of male structure. Female structures seem to develop independently.

Many plants bear flowers with both male and female organs on the same plant body, descended from a single seed (see Fig. 4-28). Such plants are said to be *monoecious* ("one house"), whereas those species which bear male or female flowers on separate plants are said to be *dioecious* ("two houses"). Sex chromosomes have been described in several plants. What would you predict about the sex chromosomes of monoecious or hermaphroditic plants and animals?

The geneticist Richard Goldschmidt prepared *intersexes,* individuals that are neither completely male nor completely female, but phenotypically intermediate. By crossing geographic races of the gypsy moth (*Lymantria*), he was able to produce phenotypic males with female sex chromosomes (ZW) and female phenotypes with male sex chromosomes (ZZ), as well as all grades of intermediate phenotypes. Goldschmidt attempted to explain these results by suggesting that the genes for determining maleness, carried on the Z chromosome, varied in strength from race to race, and that similarly variable female determinants lay in the

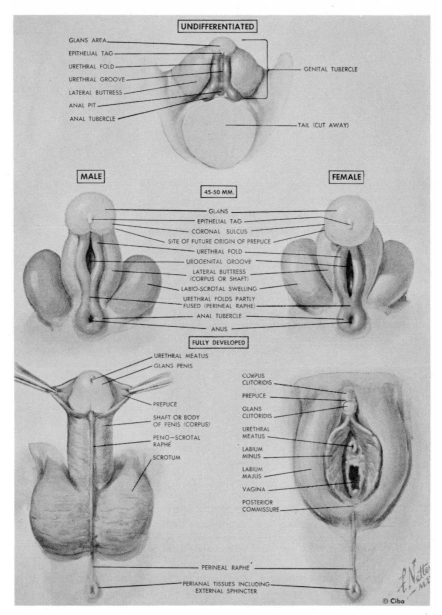

Fig. 4-27 **Drawings of the stages in external genital development.**
From *The Ciba Collection of Medical Illustrations*, Vol. II, *Reproductive Systems*. Drawing by Dr. Frank H. Netter.

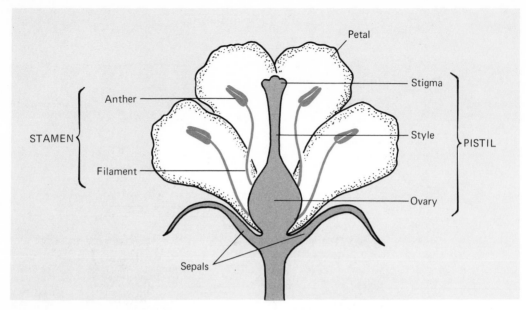

Fig. 4-28   **Diagram of a "perfect" flower with both male and female organs. Pollen grains (in whose germination sperm nuclei appear) develop in the anthers, and sperm nuclei fertilize egg nuclei, which develop in the ovaries. The ovary lasts longer than the other parts and the fruit is derived from it.**

cytoplasm. Careful studies of fruit fly intersexes have revealed that genes or chromosomes other than the sex chromosomes have a male-determining role, whereas the X chromosome is female determining.

The determination of sex is various. No general rule without considerable exception exists. If nature were simple, her study would not have fascinated man so long nor would it continue to do so, and in all likelihood man would not have evolved.

The chemical composition of the cell changes from moment to moment; the individuals making up a population change from generation to generation. Of course, reproduction is essential to the steady state of the species. Reproduction must accomplish two seemingly contradictory tasks if life is to remain and evolution is to occur: continuity must be maintained at the same time that variety is introduced. We have seen how sexual reproduction admirably achieves both ends. Weissman and others proposed that the germ plasm (sex cells) was immortal, whereas the body cells were mortal. The genes and DNA make up the germ plasm. They do continue via the sex cells all the way back to the Beginning, but hardly unchanged. The major mechanism of evolution, natural selection, has been equated with differential reproduction. Those that survive reproduce; those that fail do not.

DISCUSSION QUESTIONS

1. *What is the period of maximal fertility during the menstrual cycle? of minimal fertility?*
2. *How does the Prophase I of meiosis differ from a mitotic prophase?*
3. *What are two ways in which variety arises in sexual reproduction?*
4. *Why is human sexual development often confusing?*
5. *How is sex determined?*
6. *How does the "pill" work?*

REFERENCES

AREY, L. B. *Developmental Anatomy*, W. B. Saunders Co., Philadelphia, 1955.

PATTEN, B. M., *Human Embryology*, McGraw-Hill Book Company, New York, 1953.

PINCUS, G., *The Control of Fertility*, Academic Press Inc., New York, 1965.

YOUNG, W. C., *Sex and Internal Secretions*, The Williams & Wilkins Co., Baltimore, 1961.

OBJECTIVES

1. List and define three of Mendel's major laws of inheritance.

2. Use the Punnett square and algebraic methods to determine the genotypes and probabilities of offspring in monohybrid and dihybrid crosses.

3. Contrast linkage and independent assortment.

4. Distinguish between: dominant inheritance and intermediate inheritance (incomplete dominance); multiple gene and multiple allelic heredity; sex-linked and sex-influenced traits. Give examples of each.

5. Give the F_2 phenotypic and genotypic ratios characteristic of monohybrid and dihybrid crosses and their test crosses.

6. Describe the genetic code.

7. Explain how genetic information is translated.

8. List three classical definitions of the gene.

9. Name and define three terms frequently used by biochemical geneticists in place of "gene."

10. Compare the normal and the mutant gene.

11. List three characteristics of mutations.

12. List several causes of mutation.

13. Outline the experiments of Mendel, Beadle and Tatum, and Nirenberg and Matthei.

Heirs
and Sires

MENDEL

TODAY WE CALL THE SCIENCE OF HEREDITY *genetics* and honor the Austrian Augustinian monk Gregor Mendel as its founder, although the value of his work was not appreciated until 1900, 20 years after his death. In many ways, Mendel was an avant-garde scientist, and the birth of his science was incredibly delayed. These tragic elements of the story lend themselves well to dramatization. Yet Mendel did not die a disappointed man, and much of modern genetics is not simply Mendelian. The real tragedy is that Mendel did not live long enough to see how significant his experiments and conclusions were. It is safe to say that his discoveries were the key that unlocked the puzzle of heredity. Still, it must be confessed that the inheritance of traits was not considered an unexplainable phenomenon, a mystery, in Mendel's time and before.

The ancient Greeks believed that the embryo was conceived by the coagulation of menstrual blood by semen, and that heredity resulted from the "seeds," or particles that came from all parts of the parent bodies and were concentrated in the reproductive fluids. Thus, the ancients felt that inherited traits were the result of the "blending of bloods." With few exceptions, the scholars of the Middle Ages and the Renaissance held the Greek view. Even Darwin's theory of heredity followed the Greek idea.

What size children would you expect to be born to the marriages of tall men with tall ancestry to short women of short ancestry? Mostly average height, of course, with a few shorter than average and a few taller than average. Experience and common sense dictate this answer. When the noted German botanist Kolreuter crossed tobacco plants with long, narrow flowers with tobacco plants with short, wide flowers, all offspring were intermediate in character. When he interbred these intermediate hybrids, their offspring varied from one extreme to the other with most falling in between, like their parents. Quite often, when two extremes are crossed, the offspring are intermediate, and their descendants vary from one extreme to the other with most being average, however, like their hybrid parents. A graph of the second generation follows a bell-shaped curve, the normal distribution curve (Fig. 5-1). This is a continuous distribution: all gradations between the extremes can be found. One would think that common sense was correct: a blending had occurred. Darwin did not bother to count the types of offspring in a second generation:

The offspring from the first cross between two pure breeds is tolerably and sometimes (as I have found with pigeons) quite uniform in character, and everything seems simple enough; but when these mongrels are crossed one with another for several generations hardly two of them are alike, and then the difficulty of the task becomes manifest. . . .[1]

When Mendel crossed pure-breeding tall pea plants with short pea plants (also true breeding), all the offspring were as tall as their tallest parent; none were in between. A cross between members of the first generation of intermediates yields a second generation of three-fourths tall and one-fourth short, a ratio of 3:1. None of these plants is intermediate: all are either as tall as their tallest ancestors or as short as their shortest grandparents. This is a discontinuous distribution. Mendel found that only one-third of the tall plants of the second generation was pure breeding.

Other plant breeders had noted sharply contrasting characters and distinct ratios in peas before, but none had proposed a hypothesis of the mechanism of inheritance. It is unlikely that Mendel knew of the work of his pea-breeding predecessors. Perhaps, the discrepancy in the height of pea plants had come to his attention in the monastery vegetable gardens. Bees were also kept by the monks. Johann Dzierzon was a contemporary beekeeper and neighbor of Mendel. He found that when he crossed German with Italian bees, half the following generation's drones (males) were German and half Italian. This was a 1:1 ratio and a discontinuous distribution of types. In the words of Dzierzon, ". . . as if it were

---

[1] Charles Darwin, *The Origin of Species* (New York, N. Y.: Crowell-Collier Publishing Company, 1962), p. 40.

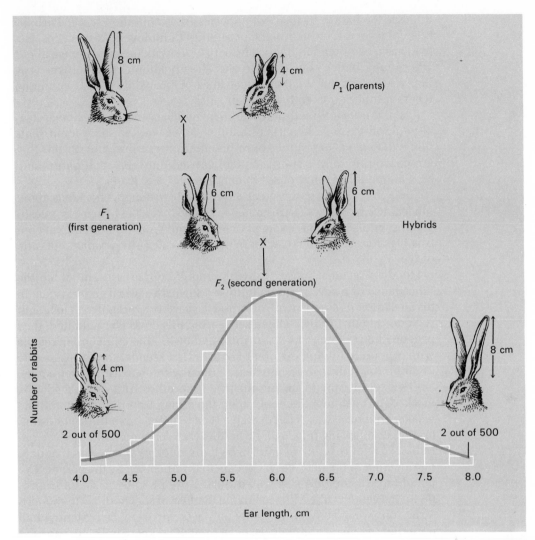

Fig. 5-1   A common pattern of inheritance. When two parents of opposite extremes are
                crossed, the hybrid offspring of the first generation (F₁) are average. The off-
                spring resulting from a mating of average hybrids show continuous variation
                from one extreme to the other with most again falling in between. The distribu-
                tion of types describes the bell-shaped curve of normal distribution. Many
                traits, such as size, intelligence, and skin pigmentation, seem to be inherited in
                a similar fashion.
                From N. M. Jessop, *Biosphere,* Prentice-Hall, Inc., Englewood Cliffs, N. J., 1970.

difficult for nature to fuse both species into a middle race."[2] Perhaps Mendel was thus alerted to the possibility of finding definite ratios. He, too, was a bee-breeder, but he chose to work with the garden pea, about whose inheritance considerable was already known. Mendel was very impressed with the mathematical laws of probability and wondered whether inheritance followed their rules.

Pea plants usually self-pollinate. The pollen grains of one flower usually fall on the female parts of another flower on the same plant. This inbreeding tends to produce pure-breeding or true-breeding strains that are constantly tall or short. Mendel carried out cross pollination and studied the inheritance of seven characteristics or traits.

For eight years he gathered numerical data to support his ideas about inheritance in plants. There can be little doubt that he began his investigations with certain preconceptions: one trait would dominate the contrasting character of a cross; there would be a definite, predictable ratio of types.

Mendel was most successful with *monohybrid crosses,* in which the inheritance of a single trait is followed through several generations. In his most famous experiment, he crossed a variety averaging 6 ft in height with one about 1 ft high. When seeds from this cross were planted, they produced 6-ft plants, not 3.5-ft intermediates. This and similar results with crosses involving the other six traits led Mendel to propose a *law of dominance:* in a cross between parents with contrasting characters, one will predominate in the offspring; the other trait will recede, be masked, or unshown. Mendel called the strong trait *dominant* and the weaker trait *recessive.* The recessive type reappeared in the second generation of offspring (Fig. 5-2). From this Mendel concluded that the hereditary factors existed in pairs in an individual organism, and that only one member of the pair was passed on to the offspring by the parent. The hereditary factors are segregated during the process of sexual reproduction. Mendel called this principle the *law of segregation.* In modern terms, only one member of a gene pair gets into a sex cell. Mendel had observed the results of meiosis, even though he knew nothing—nor did anyone else then, for that matter—of chromosomes. Before continuing our discussion of Mendel's discoveries, let us depart for a moment to briefly introduce the shorthand of genetics.

A capital letter of the alphabet is chosen to represent the dominant gene of a pair of contrasting traits; the lowercase of the same letter stands for the recessive gene. Thus, T is the symbol for tall, and t represents the gene for shortness. The cells of an individual contain pairs of genes. Therefore, a pure-breeding tall plant has the genetic formula TT,

[2]M. J. Sirks and Conway Zirkle, *The Evolution of Biology* (New York, N. Y.: The Ronald Press Company, 1964), p. 300.

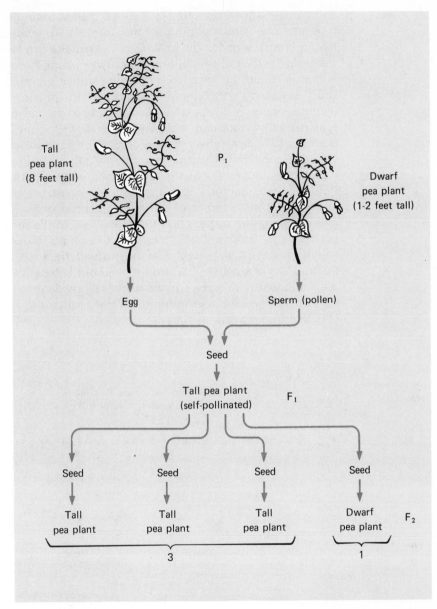

Fig. 5-2   This diagram shows the results of Mendel's most famous experiment, in which he crossed two distinct strains of peas: one tall, one dwarf. The F_1 generation were all tall, every bit as tall as their tall ancestor, but when these first-generation tall plants self-pollinated, their seeds produced both tall and dwarf plants in a ratio of 3:1 in the F_2. Compare this cross with that in Fig. 5-1. How do the results of the two differ?

and can produce only one type of sex cell, containing the gene T. A true-breeding short plant has the formula tt, and produces only t-containing sex cells. The hybrids, Tt, can make two types of sex cells: T-holding cells and t-including cells. The genetic formula for an individual is called the *genotype*. The appearance of an organism (tall or short) is called the *phenotype*. Sometimes, the phenotype represents a definite genotype. A recessive phenotype betrays a pure-breeding genotype (tt), but a dominant phenotype can be the result of either a true-breeding (TT) or a hybrid (Tt) genotype. The parental generation is designated P_1; the first offspring generation, F_1; the second, F_2.

A simple device for analyzing crosses and predicting results is the Punnet square method (Fig. 5-3). A square is constructed and subdivided to represent the number of different potential sex cells from the two parents along the upper and left-hand sides. In the monohybrid cross between true-breeding tall and dwarf pea plants, only one type of sex cell with regard to height can be formed by each parent (T- and t-containing sex cells). The square need not be subdivided. The genes are combined within the square. This is equivalent to fertilization and the conception of a new individual. The resultant F_1 hybrid genotype (Tt) determines a tall phenotype.

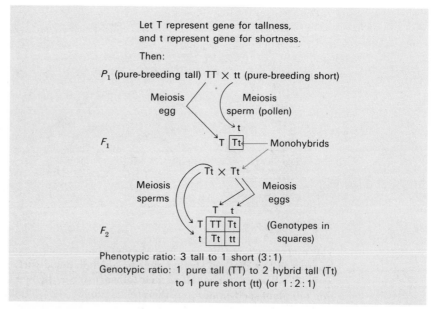

Let T represent gene for tallness,
and t represent gene for shortness.

Then:

P_1 (pure-breeding tall) TT × tt (pure-breeding short)

Phenotypic ratio: 3 tall to 1 short (3 : 1)
Genotypic ratio: 1 pure tall (TT) to 2 hybrid tall (Tt)
to 1 pure short (tt) (or 1 : 2 : 1)

Fig. 5-3 **The monohybrid cross in genetic shorthand and the square method of analysis.**

When F$_1$ hybrids self-pollinate or are crossed with each other, two types of sex cells can be formed by each parent: T and t. The square must be divided in fourths. The potential F$_2$ genotypes are three: TT, Tt, and tt. The *phenotypic ratio* is $\frac{3}{4}$ tall to $\frac{1}{4}$ short (or 3:1), the typical F$_2$ phenotypic ratio of a monohybrid cross involving sharply contrasting characters. Of the tall plants, only 1 in 3 will breed true if allowed to self-pollinate. Two of 3 tall plants in the F$_2$ are not pure-breeders, but rather hybrids, which produce both tall and short offspring. The *genotypic ratio* in the F$_2$ of a monohybrid cross is 1:2:1 ($\frac{1}{4}$ true-breeding tall, $\frac{2}{4}$ hybrid tall, $\frac{1}{4}$ pure-breeding short).

To determine if a dominant organism (A?) is true breeding (AA) or hybrid (Aa), it can be subjected to a *test cross* with a recessive individual. If the test cross yields both phenotypes in a ratio of 1:1 (half dominant and half recessive), it can be concluded that the genotype of the tested individual is hybrid. If all offspring show the dominant phenotype, it is clear that the test individual is pure breeding (see Fig. 5-4).

By simply allowing the F$_2$ pea plants to self-pollinate, Mendel determined which were true breeding and which were not. He saw that the laws of probability readily explained the results. The crossing of hybrids is like a game of chance in which two coins with two possibilities (heads and tails) are tossed simultaneously. For each coin, the chance of heads in 1 in 2; the probability of heads is $\frac{1}{2}$. The probability of tails is likewise $\frac{1}{2}$ for each coin tossed. Although the coins are thrown simultaneously, they are independent, and the probabilities for one are unaffected by the other. The probability that two independent events will coincide is the product of their individual probabilities. This rule is called the *product law* of probability. Thus, when 2 coins are tossed at the same time, the chance that both will turn up heads is $\frac{1}{2} \times \frac{1}{2}$, or $\frac{1}{4}$, which means that with a sufficient number of double tosses, 2 heads or 2 tails will land uppermost about 1 out of every 4 trials. The probability that coin A will land heads-up and coin B tails-up is also $\frac{1}{2} \times \frac{1}{2}$, or $\frac{1}{4}$, and the probability that coin A will be tails and coin B heads is likewise $\frac{1}{2} \times \frac{1}{2}$, or $\frac{1}{4}$. The total probability of heads-tails combinations is $\frac{1}{4} + \frac{1}{4}$, or $\frac{1}{2}$.

The coins of a hybrid cross (Tt × Tt) are the diploid monohybrids themselves, and meiosis is equivalent to the tossing of the coins. The genes for tallness and shortness are the opposite sides of the coin. The chance that a sperm nucleus will be allotted the gene for tallness (T) is 1 in 2, or $\frac{1}{2}$. The probability that a spermatazoan nucleus will receive the gene for shortness (t) is also $\frac{1}{2}$. In the other sex, the chance that an egg nucleus will be given the dominant gene (T) is again $\frac{1}{2}$; the probability that an egg nucleus will receive the recessive gene (t) is likewise $\frac{1}{2}$. Note that each sex cell nucleus contains only a single gene from the pair as

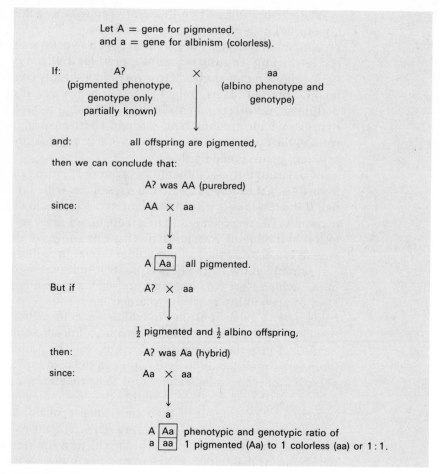

Let A = gene for pigmented,
and a = gene for albinism (colorless).

If:　　　　　　A?　　　　　×　　　　　　aa
　　　　(pigmented phenotype,　　　　　　(albino phenotype and
　　　　　genotype only　　　　　　　　　　　genotype)
　　　　partially known)

and:　　　　　　all offspring are pigmented,

then we can conclude that:

　　　　　　　　　A? was AA (purebred)

since:　　　　　AA × aa

　　　　　　　　　　　↓
　　　　　　　　　　　a
　　　　　　　　A │Aa│　all pigmented.

But if　　　　　A? × aa

　　　　　　　　　　　↓

　　　　$\frac{1}{2}$ pigmented and $\frac{1}{2}$ albino offspring,

then:　　　　　A? was Aa (hybrid)

since:　　　　　Aa × aa

　　　　　　　　　　　↓
　　　　　　　　　　　a
　　　　　　　A │Aa│　phenotypic and genotypic ratio of
　　　　　　　a │aa│　1 pigmented (Aa) to 1 colorless (aa) or 1 : 1.

Fig. 5-4　　　**Monohybrid test cross, a method of determining whether an individual is pure-bred or hybrid for a trait.**

a rsult of the reduction division of meiosis. When the egg and sperm nuclei unite in fertilization, the double number of genes and chromosomes is restored. Fertilization is like the double toss of coins, and the probability of the resulting genetic combinations can be calculated in the same way. The chance that two dominant genes (T) will come together in fertilization is $\frac{1}{2} \times \frac{1}{2}$, or $\frac{1}{4}$. Of the F_2, 25% will be true-breeding tall (TT); likewise, a quarter of the offspring will be pure-breeding short (tt). The chance of a t-containing sperm nucleus uniting with a T-containing egg nucleus is $\frac{1}{2} \times \frac{1}{2}$, or $\frac{1}{4}$; the probability of a T-containing

sperm nucleus fusing with a t-containing egg nucleus is $\frac{1}{2} \times \frac{1}{2}$ or $\frac{1}{4}$. Therefore, the total chance of getting hybrid offspring (Tt) is $\frac{1}{4} + \frac{1}{4}$, or $\frac{1}{2}$. The
ratio of probable offspring is $\frac{1}{4}:\frac{1}{2}:\frac{1}{4}$, or $1:2:1$. The Punnett square method
is merely a means of visualizing probabilities or possible genetic combi-
nations that can be predicted by the algebraic method (Fig. 5-5). In following the inheritance of a single trait, the square method and the probability technique are equally useful, but, when two or three traits are
followed at once (dihybrid and trihybrid crosses), the mathematical
method is superior, for the square becomes cumbersome with 16 and 32
spaces.

Mendel reported the results of several *dihybrid crosses,* in which the
inheritance of two traits was studied simultaneously. His most famous
dihybrid experiment involved both plant height and seed texture.
Smooth pea (S) is dominant to wrinkled (s). Mendel crossed pure-
breeding tall plants with smooth seeds (TTSS) and short plants with
wrinkled seeds (ttss). (Why is it unnecessary to describe the latter as true
breeding?) All the first generation (F₁) dihybrids were tall, smooth (TtSs).
When these dihybrids were crossed (or allowed to self-pollinate), $\frac{9}{16}$ of
the F₂ offspring was tall, smooth, and $\frac{1}{16}$ of them was short, wrinkled.
The parental types (tall, smooth and short, wrinkled) had reappeared in
the F₂. In addition, $\frac{3}{16}$ of the second generation was tall, wrinkled, and $\frac{3}{16}$
of them was short, smooth. These last two phenotypes show a new com-
bination of traits not seen in the parental generation. The phenotypic

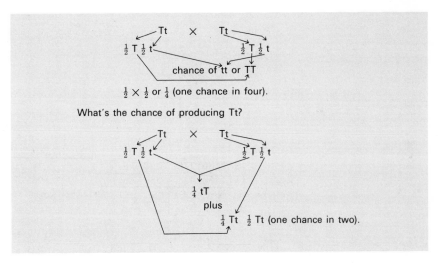

Fig. 5-5   **The product law of probability and the monohybrid cross. What is the probability or chance of pure tall (TT) or pure short (tt) offspring?**

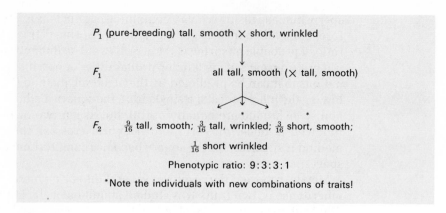

Fig. 5-6   **Phenotypes in a dihybrid cross.**

ratio of a dihybrid cross is $9:3:3:1$ (Fig. 5-6); the genotypes are shown in Fig. 5-7. Note that $\frac{1}{16}$ of the offspring should be pure breeding, tall, smooth, and $\frac{1}{16}$ should be short, wrinkled. The genotypic ratio of a dihybrid cross is unwieldy ($1:2:2:4:1:2:1:2:1$). What is the most frequent genotype, accounting for $\frac{4}{16}$ or $\frac{1}{4}$ of the offspring?

The new combinations of traits (tall, wrinkled and short, smooth) led

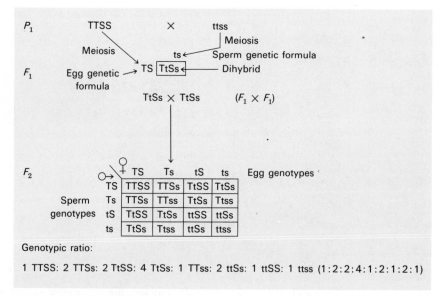

Fig. 5-7   **Genotypes in a dihybrid cross.**

Mendel to propose the *law of independent assortment*. In this law he
stated that the way in which one pair of genes is segregated in the formation of sex cells is independent of the assortment of the other pair. It is
like the game of chance, the double coin toss. The direction in which the
genes for height (tallness and shortness) went was independent of the
path the genes for seed texture (smoothness and wrinkledness) followed.
New combinations of them were formed. If the tall and smooth genes
were physically bound together (*TS*), then they couldn't separate and
combine with members of the other pair of genes as they obviously did
in the new combinations. The same rule applies to short and wrinkled
genes. If they were tied together (*ts*), then they would remain together
rather than going their separate ways. Completely tied, or bound, genes
for the two traits would produce only two phenotypes in the F$_2$ in a ratio
of 3 tall, smooth to 1 short, wrinkled. Even though this is a dihybrid
cross, the phenotypic ratio would be that of a monohybrid cross (3:1).
In addition, the genotypic ratio would be 1:2:1 (a monohybrid genotypic ratio) rather than the dihybrid genotypic ratio of 1:2:2:4:1:2:1:2:1
(study Fig. 5-8). All tall plants would have smooth peas and all short

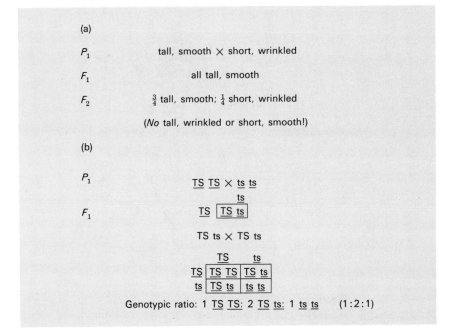

Fig. 5-8   **Dihybrid cross with absolutely linked genes. (a) Phenotypes. (b) Genotypes
(linked genes are underlined).**

plants would have wrinkled seeds. All the dihybrid crosses that Mendel reported showed independent assortment of the two pairs of genes involved.

After mitosis and meiosis had been studied in some detail and Mendel's work rediscovered in 1900, Thomas Morgan (Fig. 5-9), the great American geneticist, and his student Charles Sutton proposed the *chromosome-gene hypothesis,* in which they pointed out the parallel behavior of the chromosomes in meiosis and Mendel's hereditary factors (genes) in segregation. Only one chromosome of a pair went into a sex cell; only one member of a gene pair was passed from the parent to offspring. Mendel's law of segregation was a reflection of meiosis. Morgan and Sutton concluded that genes were carried on chromosomes, and, since there were literally thousands of genes and relatively few chromosomes, more than one gene must be carried on a chromosome.

Fig. 5-9   **Thomas Hunt Morgan (1866–1945), American genius of genetics, who, with his students, dominated post-Mendelian genetics throughout the entire first half of the twentieth century.**
From E. J. Gardner, *Principles of Genetics,* 3rd ed., John Wiley & Sons, Inc., New York, 1968. Drawing by Everett Thorpe.

It was not long before their hypothesis was amply confirmed by numerous experimental results. Many, many pairs of genes for different traits tended to travel together from generation to generation, as though they were physically *linked*. Genes carried on the same chromosome are said to belong to the same *linkage group*. Some genes tend to be linked more strongly than others, but even weakly linked genes do not produce the results predicted by the "law" of independent assortment. Unlinked genes, carried on different chromosomes, obey the law of independent assortment. Linked genes do not sort independently. The meiotic phenomenon of crossing over tends to break linkage groups and restore the law of independent assortment to a degree of validity. Biologists have used the frequency of crossing over as a means of mapping genes, since the further two linked genes are apart, the greater their percentage of cross overs. Closely linked genes (in close proximity) hardly ever cross over.

Most species have characteristic chromosome numbers. Man has 46 chromosomes, 23 pairs of homologues. The common garden pea has only 7 pairs of chromosomes. The seven traits that Mendel studied were located on different chromosomes; his chances of missing linkage were almost 1 in 300. In other words, the odds were 300 to 1 that he would encounter linkage. If Mendel had studied dihybrid crosses involving another, eighth character along with each of his seven, he would have found that the eighth character's genes would have tended to travel along with one of the other seven characters' genes, thus being an exception to independent assortment.

Many authorities find it hard to believe that Mendel had the very good luck of choosing precisely seven, completely independent characters. It has been suggested that he must have studied the inheritance of other traits, but ignored them when he wrote his paper. Alternatively, he might have sought only those results which obeyed the laws of probability.

Most likely, Mendel's first dihybrid cross results obeyed the rule of independent assortment and the laws of probability, thus prejudicing his viewpoint. Then subsequently, results contrary to his expectations might have seemed questionable to him. Perhaps, he discarded contrary results. One can imagine the thrill of finding that the results of a daring biologic experiment conformed with one's hypothesis. Most scientific investigation is cautiously undertaken with only a hair's breadth between reality and the unknown. The truly great experiments very often break with tradition, and their results are less predictable. After going out on the hypothetical limb, one's enthusiasm is redoubled if the results are confirmative.

Mendel had successfully gathered considerable proof of his hypothesis that heredity—in pea plants, at least—followed the laws of chance, was quite like a dice game. No one had done it before. Almost half a century would pass before anyone would do it again, and then it would be professional Germanic scientists, who found an "ignorant" monk a half-century "ahead" of them. Gregor Johann Mendel is justly ranked as one of the two greatest geniuses of biology; alongside Darwin, he stands hardly surpassed, since or before. His was the key that unlocked heredity. It seems strange that a simple monk, studying peas, could have glimpsed so much.

Mendel not only sent his work to a czar of science, the famed German botanist Carl Nageli, but also attempted to clarify the principles his researches revealed in several letters. Many historians of science feel that Nageli simply didn't understand. Perhaps, he was too familiar with breeding experiments in which crosses between extremes produced intermediates in the first generation and continuous variation from one extreme to the other in the F_2. If so, he might have felt that Mendel's results and conclusions were not of general or universal application. This explanation is a possibility, even though Nageli certainly was aware of distinct ratios in some plant hybridizations. Mendel himself was not sure that his scheme applied to much more than peas, and perhaps bees, but he did find it quite interesting that the laws of probability could be applied to a case of inheritance. Without a doubt, Mendel knew of the continuously variable type of inheritance which lacked distinct ratios of either "black or white." It is often stated that Mendel's genius lay in choosing "sharply contrasting characters." Others had crossed extremes. At any rate, Mendel did not approach the problem of variable inheritance.

After his experiments with garden peas, Mendel crossed varieties of hawkweeds at Nageli's suggestion, and was surprised to find that the offspring gave no evidence of the segregation of genes from both parents, but rather were all like their female parents. He must have come to feel that his earlier results were merely peculiarities of the garden pea. It has been suggested that Mendel's discovery was neglected because the biologists of the day were blind to everything but Darwinism. However, the biology of Mendel's time was quite without mathematics and qualitative or descriptive rather than quantitative or measuring. The time might not have been ripe for Mendel. Less than 40 years later, three men made essentially the same discoveries almost simultaneously. All cited the priority of the prescient monk.

BEYOND MENDEL

One of Mendel's rediscoverers, Carl Correns, discovered a fourth "law" of inheritance. He found that when he crossed four-o'clock plants having red flowers with plants having white flowers, all the monohybrid generation developed pink flowers. When the pink F_1 hybrids were crossed, $\frac{1}{4}$ of the F_2 offspring were red, $\frac{1}{2}$ pink, and $\frac{1}{4}$ white flowered (Fig. 5-10). The phenotypic ratio of this cross is $1:2:1$, the genotypic ratio of simple Mendelian inheritance in a monohybrid cross. Unfortunately, Correns called this incomplete dominance, or blending. Today such breeding results are termed *intermediate inheritance*, since neither gene dominates or recedes. Note that the phenotypic and genotypic ratios are the same. Every phenotype betrays a genotype. The hybrids are recognizable (pink) without resorting to a test cross. The pure-breeders are red or white.

Snapdragons also inherit flower color in this fashion. Another pair of snapdragon genes produces three leaf shapes similarly, broad, intermediate, and narrow. The hybrids are intermediate. A red-flowered, narrow-leaved plant crossed with a white-flowered, broad-leaved plant will produce all pink-flowered, intermediate-leaved F_1 plants (dihybrids). When these dihybrids are crossed, the phenotypic ratio of F_2 offspring is $1:2:2:4:1:2:1:2:1$. Do you recognize this ratio as the genotypic ratio of a simple Mendelian dihybrid F_2? Work out the square for yourself. Once again we see that the genotypic and phenotypic ratio are the same in intermediate inheritance. A good phrase to describe simple Mendelian genetics is *dominant inheritance*. Intermediate inheritance is just a simple modification of it.

MENDELIANS VERSUS DARWINIANS

Many historians have speculated as to what would have happened if Mendel had communicated his findings to Darwin. Luckily, he didn't. Darwin believed in gradual, continuous variation, and Mendelian discontinuous inheritance might have shaken his confidence. Darwin's nephew, Sir Francis Galton, was one of the founders of biostatistics or biometrics, the science of biological measurement. Galton applied statistics to the study of the inheritance of human intelligence. The genetics of human intellect seemed to follow the bell-shaped curve of continuous distribution. Galton proposed two principles, following his studies of

Fig. 5-10   **The blending of traits in four-o'clocks. In this case, neither gene is dominant over the other. Parental types are P; F_1 is the first offspring generation and F_2, the second offspring generation, obtained the cross $F_1 \times F_1$.**
From N. D. Buffaloe and J. B. Throneberry, *Principles of Biology*, Prentice-Hall, Inc., Englewood Cliffs, N. J., 1967.

family trees. Each parent contributes about $\frac{1}{4}$ of the inherited traits, each grandparent $\frac{1}{16}$, and each great-grandparent $\frac{1}{64}$. The offspring of unusual parents tended to be more nearly average than their parents. Galton was born in the same year as Mendel (1822) and published his findings a few years after Mendel's unheralded paper appeared. The results of Galton's studies confirmed the traditional belief of blending inheritance, which contrasted so sharply with Mendelian dominance.

One of the rediscoverers of Mendel, Hugo de Vries, published a book entitled *The Mutation Theory* in 1901. Darwin's theory of the natural selection of gradually changing continuously variable traits did not appeal to him. He proposed an alternative theory of sudden change. De Vries held that suddenly appearing "mutants," sharply departed from the norm, were new species, and that evolution took place by such jumps, or large changes (macromutations) rather than by the accumulation of minute, gradual changes (micromutations). De Vries had studied inheritance in the evening primrose and found that distinctly new types were rather frequently produced. One of these "macromutants" was a giant, another a dwarf. In their zeal of discovery, the school of Mendelian geneticists questioned Darwin's entire theory on the basis of his complete ignorance of genetics, but also because of the discontinuous nature of Mendelian variation. The biostatisticians, proponents of continuously variable inheritance, stood behind Darwin and gradual evolution. During the first decade of this century, the argument raged. It seems ridiculous in retrospect that it took so long for the parties to the debate to come to the realization that they were studying two different patterns of inheritance in different organisms. The biometricians felt that the discontinuous variation and simple arithmetic ratios of Mendelian genetics applied to minor exceptions rather than the rule of continuous variation, which could be analyzed only by more complex mathematical procedures. Biometrics was older than the newborn Mendelian genetics, and many biometricians felt that the Mendelians were radical upstarts. Biometrics was closely allied with natural history, and much of its data came from observation rather than experiment. The Mendelians were experimentalists and scoffed at the "armchair biologists." The Englishman William Bateson became the chief spokesman for the Mendelian school. A title to one of his books, *Mendel's Principles of Heredity—A Defense* (1902), reflects the academic wars that attended the birth of genetics. The position of the biostatistical geneticists seemed to support the most archaic ideas about heredity, such as blending. The striking discoveries of the Mendelian geneticists outshone the biometricians. In reality, neither side won or lost. The Mendelians rose to power, but they still had to explain continuous variation.

Simple Mendelian inheritance involves a single pair of genes with two

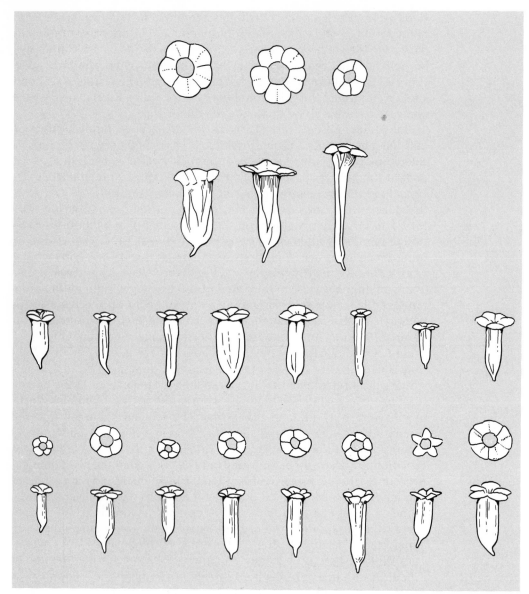

Fig. 5-11   E. M. East's repetition of the tobacco variety cross-breeding experiment of Joseph Koelreuter, nineteenth-century botanist. (a) Top left: short, broad flower of one variety. Top right: narrow, long flower of another variety. Between them: a flower from the F_1 hybrid generation. (b) Some flowers from some F_2 plants (from $F_1 \times F_1$). Note that there is continuous variation from one extreme to the other, with most falling, however, in between. Multiple genes are involved.
From E. M. East, *Genetics*, **1**:164, 1916.

alternative forms, the dominant and recessive. In 1907, Yule suggested that continuous variation might be explained if it were assumed that large numbers of Mendelian genes acted in an additive or cumulative fashion. In 1910, E. M. East repeated botanist Joseph Kolreuter's 1760 study of tobacco flower variation and confirmed his results. He found again that when two extreme parents, which differ in a quantitative character, such as size, are crossed, the F_1 are all more or less intermediate between the parental extremes, and the F_2 show many gradations from one extreme to the other (Fig. 5-11).

In 1908, the Swedish plant breeder H. Nilsson-Ehle crossed a wheat variety with very dark red kernels and another variety with pure white grains. The F_1 grains were of a uniform intermediate shade; the color of the grains in the F_2 ranged from very dark red through various intermediate shades to white. Out of every 16 grains, 1 was as dark as the very red parent, and only 1 out of 16 was pure white.

What does this suggest to you? It reminded Nilsson-Ehle of the results from a simple Mendelian (dominant) dihybrid cross. He proposed that two pairs of genes were involved here and that each pair contained one gene that produced color and one that did not. If two pairs of genes are assumed—A and a, and B and b, with capital letters representing color-producing genes—the results are explained in the diagram in Fig. 5-12.

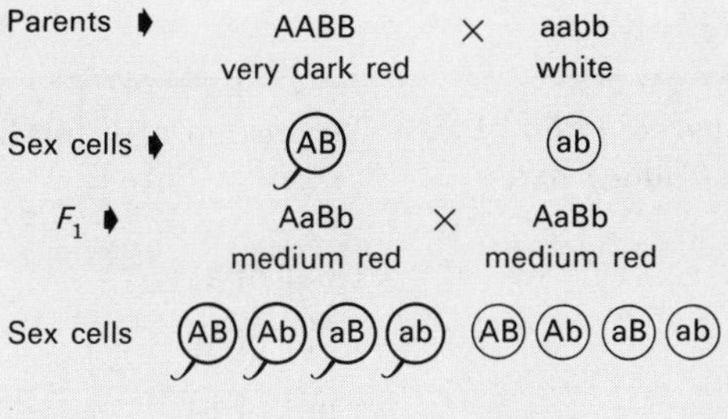

Parents ◗	AABB very dark red	×	aabb white	
Sex cells ◗	(AB)		(ab)	
F_1 ◗	AaBb medium red	×	AaBb medium red	
Sex cells	(AB)(Ab)(aB)(ab)	(AB)(Ab)(aB)(ab)		

F_2 ◗		AB	Ab	aB	ab
	AB	AABB very dark red	AABb dark red	AaBB dark red	AaBb medium red
	Ab	AABb dark red	AAbb medium red	AaBb medium red	Aabb light red
	aB	AaBB dark red	AaBb medium red	aaBB medium red	aaBb light red
	ab	AaBb medium red	Aabb light red	aaBb light red	aabb white

Fig. 5-12   **The multiple-gene explanation of the inheritance of kernel color in wheat.**

On the basis of his studies of corn and tobacco, East proposed a similar hypothesis. The hypotheses of Nilsson-Ehle and East became known as the *multiple-gene hypothesis*. More than a single pair of genes is involved, and they are assumed to be cumulative in their action. This type of inheritance produces many more than the two or three phenotypes of simple Mendelian (dominant) and intermediate inheritance. Multiple gene inheritance is also called polygenic or cumulative gene action. If only 1 out of 64 offspring is as extreme as either parent, then three pairs of genes are supposed to be involved (see Table 5-1).

The heredity of skin color is probably the most famous case of polygenic inheritance in man. In 1913, Dr. Charles B. Davenport studied the offspring of Negro-white crosses in Jamaica and Bermuda, where racial intermarriage is common. The quantity of the brown pigment melanin determines skin color. All people except albinos possess at least some melanin in their skins. Negroes have the most, large amounts of concentrated melanin imparting black color to the skin. The offspring of black-white crosses are light brown, or mulatto. Mulatto marriages can produce children whose skin pigmentation varies considerably, from extreme black to extreme white, but most offspring will be intermediates, like their parents. Some will be slightly lighter, others just slightly darker. The offspring would be distributed along a bell-shaped curve. Among 32 children of mulatto parents, Davenport found two that fell

Table 5-1

Proportion of F_2 as extreme as each parent in multiple or cumulative gene inheritance

Number of Pairs of Genes Concerned	*Proportion of F_2 as Extreme as Either Parent*
1	1 out of 4
2	1 out of 16
3	1 out of 64
4	1 out of 256
5	1 out of 1,024
6	1 out of 4,096
7	1 out of 16,384
8	1 out of 65,536
9	1 out of 262,144
10	1 out of 1,048,576
20	1 out of 1,099,511,627,776

within the white extreme of his scale and two in the black extremity.
This was quite like Nilsson-Ehle's wheat F_2 discussed previously. Of the
offspring, $\frac{1}{16}$ were white and $\frac{1}{16}$ were black. Davenport considered that
two pairs of genes were most likely involved in the inheritance of skin
pigmentation and that they behaved like multiple or cumulative genes.

If we let A and a, B and b represent genes for pigment control, the capital letters standing for pigment-producing genes, Davenport's hypothesis would work out as follows:

AABB	produces black
AABb AaBB	produce dark brown
AaBb AAbb aaBB	produce medium brown
Aabb aaBb	produce light brown
aabb	produces white

What types of offspring (and in what proportions) would you expect
from a light-colored (Aabb) and white (aabb) cross? All the white's sex
cells will carry the genes ab. Half the light colored's sex cells will be Ab
and half ab. Half the offspring should be light colored: $\frac{1}{2}$ Ab $\times$ 1 ab
equals $\frac{1}{2}$ Aabb; half should be white: $\frac{1}{2}$ ab $\times$ 1 ab equals $\frac{1}{2}$ aabb. The ratio
of offspring should be 1:1, or 50:50. Davenport studied 24 families of this
kind with a total of 99 children. Of the offspring, 42 were white, 56 were
light brown, and one was medium brown. Since the expectation (50:50)
was 49.5 (99:2) in each category and none in the third, it is obvious that
there was good agreement between the expected (hypothetical) and
actual findings. Recently, it has been concluded that four, five, or six
gene pairs best explain the range of color variation in the American
Negro population.

Medium color is probably determined by such genotypes as
AaBbCcDdEe or AABBCcddee, black by AABBCCDDEE, light color by
AaBbCcddee, and dark brown by a genotype such as AABBCCddEe.
Olive "whites," or brunettes, are probably the result of such combinations as AaBbccddee, or even AaBbCcddee, rather than aabbccddee,
which may be the Nordic or Scandinavian blonde. "Light colored" and
"dark complected" are really the same. Some Caucasians in India are
nearly black. We shall discuss the adaptive significance of skin pigmentation in the chapter on evolution in man (Chapter 9). In the past, a whole
series of terms, mulatto, quadroon, octaroon, etc., were used to indicate
the percentage Negro in a person's ancestry. One whose grandparent

was Negro, or one born of a mulatto-white marriage, falls easily within the color range of "white." It was considered important to keep track of the family trees of such people, since popular belief held that one who could "pass for white" might have a black child. From our study of multiple-gene inheritance, you should be able to see that this notion is myth. Several other racial qualities seem to have a cumulative gene basis.

It is also commonly believed that brown eyes are dominant to blue. Human eye color was regarded as a classic example of Mendelian inheritance, based upon a single pair of genes. Brown-eyed people were supposed to have the genotypes BB or Bb, and blue-eyed people, bb. The brown pigment melanin causes eye color as well as skin pigmentation. In a blue eye, the superficial layers of the iris (the colored portion of the eye) have no observable amount of brown pigment, and the blue color is produced by the same effect that causes the sky to appear blue. The scattering of light by dust particles causes blue light to be reflected back to the observer. Likewise, colorless cells in the superficial layer of the iris (against the dark background of deeper layers) scatter light so that blue is reflected. There are many shades other than brown and blue: grey, green, hazel, light brown, dark brown, and black. This range suggests multiple gene inheritance.

No subject has been more hotly debated than that of human intelligence. It is the old question of "nature versus nurture," heredity against environment. Does genetics or experience determine learning potential? We live in times of great social revolution, and the sociologists and psychologists have justifiably emphasized the role of environment as a determiner of intellect. No social scientist would, however, subscribe to the view that environment is more important than heredity in determining intelligence. Yet it is a most popular belief that nearly everyone is born with the same intellectual potential and that the differences in intellectual development that do arise are the sole result of environmental differences. Although most people believe that genetics plays some role in determining intelligence and recognize the existence of genius, many feel that environment is more important than heredity in determining mental prowess. The most beautiful constitution in the world says that all men are created equal. Thomas Edison said: "Genius is one percent inspiration and ninety-nine percent perspiration." Can it be that nature has been less than democratic in dispensing this vital trait?

A graph of the intelligence of a large number of children of average parents describes the bell-shaped curve of normal, continuous distribution. The greatest number are average. A fairly large number are low average and high average. There are fewer, but almost equal, numbers of near-geniuses and high-grade mental defectives. Geniuses and low-grade mental defectives are almost equally rare, and these extremes contain the fewest individuals in the sample. There are transitional individuals between the borders of all groups. These are the results that Kolreuter saw in the second generation of tobacco plants and that Davenport described for mulattos' children.

An early study of feebleminded inheritance was that of the Jukes family, descended from supposedly feebleminded sisters. Of 1258 surviving and traceable descendants, 110 were described as mentally defective. A soldier of the Revolution, Martin Kallikak, is said to have had an illegitimate son by a feebleminded girl. A total of 480 descendants were traced from the son. Among them, 143 were allegedly mentally defective and 292 of "uncertain intelligence." Martin Kallikak later married a girl of normal intelligence. A total of 496 descendants were traced from this union. Only one was feebleminded. The families of Darwin and Huxley have often been cited as the opposite extreme, supporting the argument for the hereditary basis of intelligence. It has been noted above that Sir Francis Galton, the first great student of human intelligence and a founder of the science of biostatistics, was a nephew of Charles Darwin. On the basis of Galton's childhood development and later accomplishments, the famous modern psychologist and student of intelligence, Lewis M. Terman, estimated his I.Q. at 200. Thomas Huxley, "Darwin's bulldog," left more than the legacy of evolution. His grandson Julian Huxley is a famous biologist and writer. Aldous Huxley, another grandson, was one of the most famous novelists of the twentieth century. Another descendant, Andrew Fielding Huxley, won the Nobel prize for his studies of the nerve impulse. Just as Galton grew up in a cultured home with the best possible opportunities for education, so the "bad" line of Kallikaks were raised in deprived environments. It is impossible to conduct a controlled experiment in the inheritance of human intelligence. The environment would have to be precisely the same for the various experimental groups, which is the great weakness in the study of family trees for clues as to the heritability of intellect.

The correlation in intelligence is very high between identical twins—even those reared in different environments—and becomes progressively lower, proceeding from brothers and sisters, parents and children, cousins, and finally unrelated children, where the correlation becomes zero. Parents of average intelligence usually have children of average mental abilities, but they have the potential to produce geniuses as well as idiots, infrequently, of course. If AaBbCcDdEeFf is a genotype for average intelligence and AABBCCDDEEFF is genius, then it is easily seen that mediocre parents can produce geniuses as well as mental defectives, perhaps of the genotype aabbccddeeff. If six pairs of genes are involved in the determination of human intelligence, then only 1 child out of 4000 should be born mentally defective. The odds against normal parents having idiots should be at least 4000 to 1. *There are many causes of mental retardation other than genetic alone.* At present, it can only be said that what is called intelligence seems to have a multiple-gene basis.

No nation has ever carried out extensive breeding of mentally select. The Russian practice of selective coeducation makes interbreeding of the intelligentsia more likely, and, perhaps, after several generations of intermarriage followed by academic selection of offspring, a significant increase in the frequency of genius in this class will emerge. However, no way has yet been found to determine inherent intelligence independent of cultural background and educational experience. Many other quantitative, continuously varying human traits, such as stature and weight, seem to be determined at least in part by multiple genes.

MULTIPLE ALLELES

Those two genes that are located at the same place on a pair of homologous chromosomes are called *alleles*. In simple Mendelian inheritance with dominance or in intermediate inheritance, there are no more than two alternative forms of the gene. In dominant inheritance, the maximal number of phenotypes in the F_2 descended from pure-breeding dominant and recessive parents is two. The number of F_2 phenotypes produced by intermediate inheritance is three. More than a single pair of genes is involved in the additive determination of a multiple-gene trait. In this type of heredity, the phenotypic variety in an F_2 descended from a cross of parental extremes is largest, from at least five phenotypes (two pairs of genes) to thousands (five or more pairs of genes). Although two or more pairs of genes are usually involved in multiple-gene inheritance, each pair seems to have no more than two different forms of the gene, two alleles.

There is a type of heredity in which more than two variations on the gene of a given location occurs. This type of inheritance is known as *multiple allelism*. No individual organism's cells ever contain more than two genes each, a pair. The choice of combinations is greater than that offered by the two forms of genes in dominant and intermediate inheritance. There are at least three alleles (rather than two). The various combinations of two from three possibilities produces more phenotypes than are obtained in intermediate inheritance, but usually much fewer than result from the cumulative action of several pairs of genes in multiple-gene heredity. The simplest example of mutiple allelism is the inheritance of human ABO blood types.

There are four phenotypes in this group: A, B, AB, and O. Red blood cells have present on their surface antigens (proteins) A, B, both, or none. The plasma contains antibodies, anti-A, anti-B, both, or none. In the presence of the corresponding antigens, these antibodies will cause the red cells to clump or adhere to one another. This agglutination reaction of the red cells results from the bridging of the corresponding antigens (agglutinogens) on the red cell membranes by means of linking antibodies (the agglutinins). A person's blood can contain antibodies against antigens other than his own. The four blood types discussed are listed in Table 5-2, along with their reactions with test antisera. Plus (+) indicates agglutination and minus (−) means no reaction.

Type AB contains both antigens but no antibodies, of course, for it would clump itself. Type AB reacts with both antisera, but a person of blood type AB can receive limited quantities of other blood types in transfusion. Therefore, an AB individual is called a "universal recipient."

Type O blood contains neither antigen but both antibodies. It can be given in limited transfusion, however, and a type O person is termed a "universal donor." Nevertheless, the "universal donor's" blood will agglutinate the recipient's red cells if it is given in extensive transfusion. In the same way, "universal recipient's" blood cells will be clumped by the donor's serum antibodies in large doses.

Study Table 5-3, on the inheritance of the blood groups, based upon the study of thousands of families. What kind of a genetic mechanism would explain the inheritance of the blood types? At first, it was proposed that two pairs of genes, A and a, and B and b, were involved. This hypothesis is outlined in Table 5-4.

In Table 5-3, note that type O parents seem to be true breeding, and that the gene for type O is apparently recessive. On this basis, type O was assigned the genotype aabb. One would think that this double recessive

Table 5-2

Blood type reactions with antisera

Type	Anti-A	Anti-B
A	+	−
B	−	+
AB	+	+
O	−	−

Table 5-3

Inheritance of ABO blood groups

Blood Groups of Parents	Blood Groups Which May Occur in Children	Blood Groups Which Do Not Occur in Children
O × O	O	A, B, AB
O × A	O, A	B, AB
A × A	O, A	B, AB
O × B	O, B	A, AB
B × B	O, B	A, AB
A × B	O, A, B, AB	—
O × AB	A, B	O, AB
A × AB	A, B, AB	O
B × AB	A, B, AB	O
AB × AB	A, B, AB	O

would be rare in the population, yet, it occurs to the extent of about 47%. Mathematically, 14.2% of the population should also belong to group AB, but only a few percent did. An alternative hypothesis was proposed: three gene forms were postulated. The genotypes are summarized in Table 5-5. Notice that, although there are three types of genes, no individual gets more than a pair. This last hypothesis of multiple alleles fits the data from the population very well. There are medico-legal applications of the blood groups. Suppose a man with blood type B, one of whose parents had blood type O, married a woman with blood type AB, and their first child was type O. Could the man sue his wife for divorce on the grounds of infidelity? A man accused of fathering an illegitimate can be proven innocent by this method but not guilty. Why not?

Another multiple allele trait is the Rh factor of human blood, which

Table 5-4

The first explanatory hypothesis for ABO blood group inheritance

Blood Groups	Genotypes
A	AAbb, Aabb
B	aaBB, aaBb
AB	AABB, AaBB, AABb, AaBb
O	aabb

Table 5-5

**The accepted
genotypes for
ABO blood
types**[a]

A-B Group	Genotype	Reactions with Antiserum	
		Anti-A	Anti-B
O	ii	−	−
A	I^AI^A, I^Ai	+	−
B	I^BI^B, I^Bi	−	+
AB	I^AI^B	+	+

[a]The three genes involved in ABO blood group inheritance are designated I^A, I^B, and i.

was discovered by testing human red cells with antiserum produced against Rhesus monkey red cells. Some human bloods reacted with these antisera and others did not. It seemed that a protein antigen was present on some human blood cells but not on others. Genetic studies indicate many different Rh alleles. A relationship has been established between the Rh factor and the blood cell destroying anemia of newborn.

Matings of an Rh+ father and an Rh− mother can produce an Rh+ embryo. Although maternal and fetal circulations remain completely separate most of the time, microscopic, hemorrhagic accidents do occur in the placenta, and a few fetal Rh+ red cells can pass into the maternal blood stream. The mother's blood contains neither Rh antigen nor antibody, but, once her immunological machinery is exposed to the fetal Rh protein, Rh antibody will develop in her blood. Maternal antibodies can cross the placenta and agglutinate and destroy fetal red cells. The first-born is usually unharmed, but the reaction is more probable in the second and third pregnancies. Knowledge of the Rh compatibility of future parents is extremely important, since alerted physicians can completely exchange the newborn's blood and thus save his life.

There are eight Rh phenotypes that can be demonstrated with three antisera. Four are Rh positive and four Rh negative. No person has more than two genes for the Rh factor. The theory of explanation considers eight Rh alleles: r, r′, r″, r^y, R^0, R^1, R^2, R^3.

SEX-LINKED GENES

Genes carried on the sex chromosomes are said to be sex linked. The chromosomes other than the sex chromosomes are called *autosomes*. Most genes are autosomally linked, carried on chromosomes other than the sex chromosomes. Recall that almost all the chromosomes (with the notable exception of the sex chromosomes) are present in body cells in similar pairs, homologous pairs of chromosomes. For every gene present on the chromosome at one place, there is an allele at the corresponding location on the matching chromosome of the pair. The two genes are not necessarily the same. If one is recessive and the other dominant, the recessive allele will remain hidden. The recessive gene is only expressed in purebreds, where the two alleles are the same. Therefore, the frequency of phenotypic expression of a recessive gene remains low in a breeding population. The dominant gene is most frequently expressed. Since the sex chromosomes of men and flies are unmatched in males (vice versa in birds, amphibians, and butterflies), a recessive gene on one sex chromosome is often without an allele on the other and the recessive trait is expressed, even though there is not a double dose of it. In those sexes where the two sex chromosomes are similar or homologous (human females, roosters, and male toads), recessive genes are most frequently masked by dominant alleles. In man, X-linked recessive genes show up with a greater frequency in males (XY), because much of the Y chromosome is not homologous to the X and therefore does not carry comparable genes. Female birds (ZW) exhibit recessive traits more frequently than males (ZZ), for the same reason. The few Y-linked genes find their expression in males, of course.

Most human sex-linked genes are carried on the larger X chromosome, since the Y chromosome is so small and large areas of it seem genetically "silent," or at least remain unknown. Red-green color blindness, hemophilia ("bleeder's disease"), myopia (nearsightedness), night blindness, ichthyosis ("fish skin"), defective tooth enamel, and a type of muscular dystrophy are all human traits dependent upon X sex-linked genes. Queen Victoria was a carrier of the gene for hemophilia that became widespread throughout the royal families of Europe, appearing in 10 princes but in no descendant princesses. Why? There was no history of bleeder's disease in the royal family before Victoria (Fig. 5-13). A mutation is thought to have occurred in a germ cell of her father. The Russian monk Rasputin wielded undue influence in the late czarist court apparently because of his uncanny ability to stop the bleeding of the hemophiliac prince of Russia, Czarevitch Alexei.

The author of this book is nearsighted (in more ways than one, perhaps). His brother is also nearsighted, but his sister has normal vision. Neither parent is nearsighted. From whence did the defective gene come? Diagram the cross. What was the probability of nearsighted boys? Of nearsighted girls? (Multiple factors are involved in the determination of much nearsightedness.)

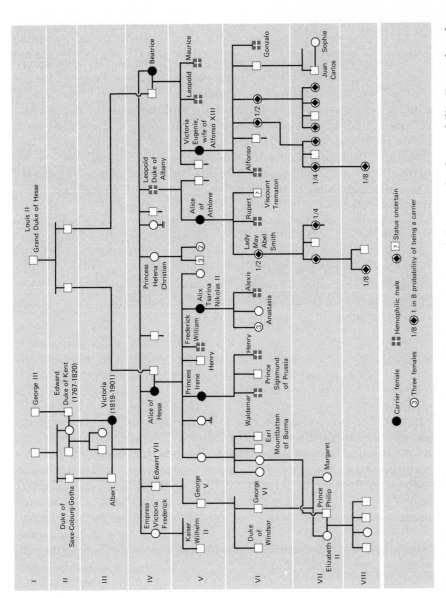

Fig. 5-13 The family tree of European royalty and the transmission of the gene for hemophilia. Every female carrier and every hemophilic male can be traced to Queen Victoria of England, yet there was no history of the disease in the royal family before her. The mutation either originated in her or in one of her parents. The gene for hemophilia is sex linked, carried on the X chromosome. Why do so many males show the trait whereas hardly any females exhibit the disease? Recall the sex chromosomal situation in man.

SEX-INFLUENCED INHERITANCE

Sex can affect the expression of genes for certain traits. The most famous example of sex-influenced heredity is pattern baldness. The gene for this trait behaves as a dominant in the male and a recessive in the female. If the gene for baldness is designated B and the gene for normal hair retention b, BB and Bb genotypes should produce baldness. They do in males, but in females only the pure-breeding (BB) individuals grow bald. The female hybrids (Bb) never grow bald (unless unusually high levels of male sex hormones from the adrenals are present). Thus, you can see it is really impossible to represent the inheritance of the trait by following the strict rules of genetic shorthand. Male castrates, or eunuchs, never lose their hair by pattern balding, even if they are purebred for the gene (BB). Virile males with well-developed secondary sexual characteristics bald quicker than less masculine individuals. Neither will lose their hair unless at least one gene for baldness is present. Apparently, the level of the male sex hormone testosterone regulates the expression of the gene for baldness. Men with abundant body hair often grow bald in their twenties.

MOLECULAR GENETICS

The discoveries considered so far belong to the realm of classical genetics. In many ways, the concept of the gene held by geneticists in 1930 hardly differed from that of Mendel. The gene had been defined three ways: (1) as a unit of segregation and recombination, (2) as a unit of mutation, and (3) as a unit of function. The chromosome was conceived to be a string of beads, the genetic particles, and nuclear proteins were thought to be the biochemicals of the genes. What the genes were and how they exerted their effects were perhaps the most perplexing questions in biology at the time. Flashes of foresight preceded the actual birth of genetics at the molecular level.

In 1901, a year after the rediscovery of Mendel's labor, Sir Archibold Garrod, professor of medicine at Oxford, proposed that the genes ultimately controlled metabolism, the biochemical reactions. Garrod hypothesized that the genes produced enzymes which, in turn, regulated the chemical reactions of life. This was the first statement of what later came to be known as the *gene-enzyme hypothesis.*

Dr. Garrod studied the "black urine disease," a harmless disease in which the urine turns black after exposure to air. The abnormal constituent of the urine responsible for this change was shown to be homogentisic acid, or alkapton, and the disease was called alkaptonuria in medical jargon. By studying families of people with the disease, Garrod showed that the proportion of victims was quite the same as recessive phenotypes in a simple Mendelian monohybrid cross. He further suggested that homogentisic acid appeared in the urine of alkaptonurians because they lacked an enzyme, homogentisic acid oxidase, required for the breakdown of the compound into simpler products that could be converted into the respiratory end products of CO_2 and water. In normal individuals, the amino acid phenylalanine from the diet is converted into a number of other compounds, including homogentisic acid, which is successfully degraded into CO_2 and water.

In 1908, Professor Garrod delivered a series of lectures entitled "Inborn Errors of Metabolism" before the Royal Society of Physicians, in which he considered three other inherited metabolic defects besides alkaptonuria. Garrod's work created no stir. Once again, the world was not ready. Fortunately, in the words of George Wald, a prophet of modern biology, "No great idea is ever lost. Like Antaeus, it is overthrown only to rise again with renewed vigor. It is dismissed only to return, yet never quite the same. Its rejection is only a step in its further development."[3] The idea that genes generated enzymes did not pass away with Garrod. Sometimes, it seems that science is an iron maiden. An idea is only an idea, speculation is "mere speculation," until it is supported by experimental fact.

In the 1930's, Boris Ephrussi and George Beadle set out to investigate the biochemistry of the variety of fruit fly eye pigments, which were known to be controlled by specific genes. They had little success. B. O. Dodge, of the New York Botanical Gardens, recommended the pink bread mold, *Neurospora crassa,* to Thomas Morgan as a genetic experimental organism, potentially superior to the fruitful fruit fly. Early in this century, the capitol of genetics shifted from Europe to the United States, where Morgan and his students proposed the chromosome-gene theory and mapped the genes of the chromosomes of the fruit fly, an amazing feat. For once, students came from Europe to study with an American master. Morgan's academic family was one of the most creative ever produced. They made the fruit fly *Drosophila* the animal of genetics (Fig. 5-14). When Morgan left Columbia University for the California Institute of Technology, he took with him several strains of the pink bread mold. It was from this source that George Beadle and Edward Tatum obtained the organism when they commenced their historic studies in biochemical genetics in the 1940's.

[3]George Wald, "Innovation in Biology," *Scientific American,* Vol. 199, No. 3 (Sept., 1958), p. 100.

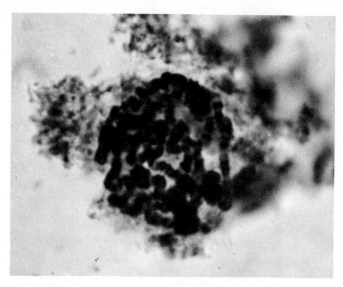

Fig. 5-14   The giant chromosomes of the fruit fly and its relatives are
unusual, not only in terms of their size, but also in that they
are found in an interphase nucleus. Genetic maps of these
chromosomes have been constructed using the frequency of
crossing over as a measure of distance between linked genes.
The giant chromosomes are found in the salivary gland, hindgut, and kidneys of flies and their relatives. Each consists of
500 to 1000 duplicates with homologues synapsed. Chromosomal mutations have been visualized along their lengths, and
sites of genetic activity have been correlated with chromosome puffs along these banded structures. It is as though the
chromosome explodes or is disbanded at the sites of genetic
activity. The blueprint molecule of the DNA genetic code,
mRNA, is made here and makes its way subsequently out into
the cytoplasm.

From M. J. Zbar and M. D. Nicklanovich, *Cells and Tissues*, Saturn
Scientific, Inc., Fort Lauderdale, Fla., 1971.

Neurospora normally grows on a minimal medium, containing nitrate, phosphate, sulphate salts, the sugar glucose, and a single vitamin,
biotin. With these materials the mold can synthesize all its own amino
acids, proteins, carbohydrates, lipids, nucleic acids, and all other vitamins. Like most other microorganisms, the mold is a wondrous biochemist, not as great as the green plant (which starts with less) but certainly much more independent than man with his host of dietary needs.

During most of its life cycle, *Neurospora* is monoploid. The mold grows in the form of interweaving filaments. Some vertical filaments fragment to form asexual spores, which are disseminated like dust. These spores are formed by mitosis, and the nuclei of the vegetative filaments and spores contain a single set of chromosomes, and therefore a single set of genes. Following a brief diploid stage, a fruiting body is formed and monoploid "sexual" spores are produced by meiosis in saclike structures called asci (ascus, singular). These meiotically formed spores germinate to form a typical threadlike mass of monoploid mold. The life cycle of *Neurospora* is outlined in Fig. 5-15. For a recessive trait

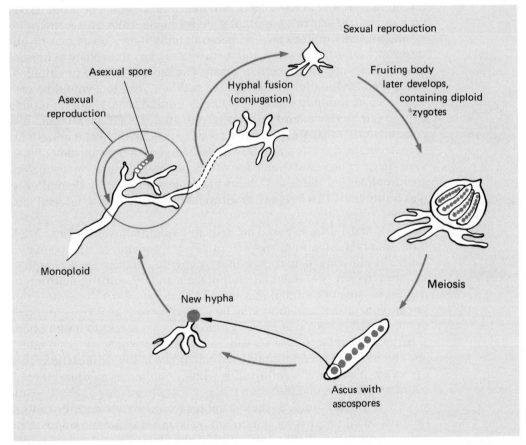

Fig. 5-15   **Life cycle of the pink bread mold. A diploid zygote is formed by thread (hyphal) cell fusion (conjugation); each zygote then undergoes meiosis in a sac (ascus), forming monoploid spores (ascospores), which germinate to produce the thread masses (mycelia) of the mold.**

to appear in the diploid organisms, such as the fruit fly, there must be a double dose (two recessive genes), as one dominant member of a gene pair can completely hide the presence of a recessive gene in hybrids, whose carrier genotype can only be exposed by test cross. In the dominant monoploid phase of the mold's life cycle, even recessive genes are expressed, since only a single representative of each gene is present, without its mate. This is one of the major advantages in the use of *Neurospora* in genetic experiments. Of course, the mold multiplies faster than does the fruit fly.

One of the ways to find out how a gene works is to see what happens when it does not. A mutation is a change in a gene. Every gene has a mutation rate, low in most cases. Studies indicate that 1 in about 33,000 human sex cells carries a mutant gene for hemophilia. Genes are quite stable, as one would expect; in general, mutations are random or undirected, harmful, recessive, and seem to occur spontaneously. A normal gene is functional in a positive sense. The mutant gene is functionless most often, or functions in a negative fashion. This, too, would be predicted for a randomly changed, formerly normal gene. Natural mutants appear rarely. Hermann Muller, a former student of T. H. Morgan, discovered in 1927 that X rays could be used to produce artificial mutations, increasing the mutation rate tremendously. Now it became almost possible to produce mutations for study at will. Before, investigators greedily hoarded the few mutants that appeared. Shortly thereafter, it was found that "black light," or ultraviolet radiation, also induced mutations.

Beadle and Tatum exposed normal, monoploid spores to X rays. They then obtained large numbers of mitotic descendants of the irradiated spores by allowing them to germinate, grow, and multiply on a complete medium, which contained all amino acids and vitamins in addition to the ingredients of minimal medium. Thus supplied with countless duplicate genetic descendants of the irradiated spores, the investigators set out to identify specific *deficiency mutants* by a process of elimination.

If the descendants of the exposed spores subsequently germinated and the mold grew on minimal medium, presumably no mutations had occurred. If no growth followed seeding on minimal medium, then it could be assumed that mutation had occurred.

Next, it was determined whether spores would germinate and grow on minimal medium plus all amino acids, or minimal medium enriched with all vitamins. If the mold did not develop on a vitamin supplemented medium but did grow on complete amino acid medium, it was concluded that the mutant mold had lost the ability to synthesize some amino acid. Growth on total vitamin medium and failure to grow on amino acid medium was interpreted as indicating a mutation in the ability to synthesize some vitamin.

Then it was determined which amino acid or vitamin synthesizing ability had been lost. Spores were seeded in tubes of minimal medium plus one amino acid or one vitamin. If the spore germinated and the mold grew in a culture tube of minimal medium plus lysine (an amino acid), it was concluded that the mutation had blocked lysine synthesis (study Fig. 5-16).

Since *Neurospora* has a sexual and diploid phase in its life cycle, Beadle and Tatum were able to prove that the deficiency mutations had a genetic and Mendelian basis. Certain monoploid filaments of two mating strains form sex cells by mitosis. These unite, forming a diploid nucleus like that of the fertilized egg. The diploid cell undergoes meiosis, yielding four monoploid cells that then divide mitotically producing a total of eight monoploid spores. All these spores are contained within a sac. (Review Fig. 5-15.) These monoploid sexual spores are genetically equivalent to the sex cells of higher organisms.

It is possible to remove each spore individually from the sac and attempt to germinate it upon experimental media to determine its genetic identity. When two different, single-deficiency mutants were so crossed, new combinations of traits were sometimes observed in the offspring. Neither parent strain would grow on unsupplemented minimal medium, but some of the spores descended from their union would grow on minimal medium alone, whereas other spores would germinate only on a medium supplemented with *both* parental deficiency substances.

If the new combinations of traits account for 50% or more of the descendant spore types, the genes are unlinked. If the recombinations are considerably less than 50%, linkage is assumed and the recombinants are considered to result from crossing over during meiosis. When a normal strain is crossed with a deficiency mutant, four of the spores in the sac will grow on minimal medium and four will not. Thus, the deficiency mutations segregate like Mendelian hereditary factors or genes.

Beadle and Tatum found several closely related mutants. One mutant (*A*) would grow on minimal medium supplemented with any of three amino acids: ornithine, citrulline, or arginine. Another mutant (*B*) would grow on minimal medium enriched with either citrulline or arginine, but it could not grow on minimal medium plus ornithine. The third related mutant (*C*) would grow on minimal medium plus arginine *alone*: it would not grow on either ornithine or citrulline medium. An explanatory hypothesis linked all three molecules in a multistep biochemical synthetic pathway:

$$\text{gene } A \qquad \text{gene } B \qquad \text{gene } C$$
$$\downarrow \qquad\qquad \downarrow \qquad\qquad \downarrow$$
$$\text{precursor} \longrightarrow \text{ornithine} \longrightarrow \text{citrulline} \longrightarrow \text{arginine}$$

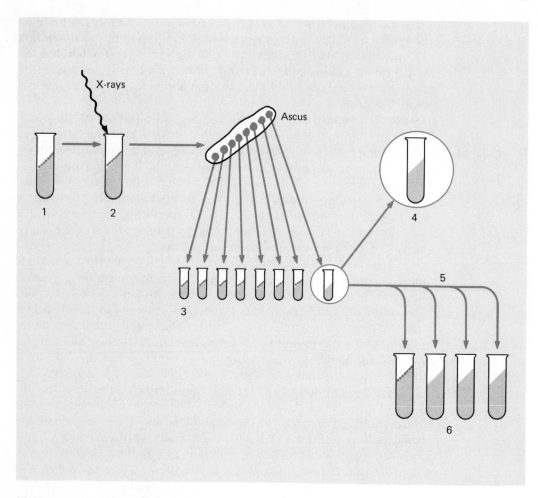

Fig. 5-16   Technique for inducing and identifying mutations in *Neurospora crassa* (the pink
bread mold). (1) Wild type or normal, nonmutant mold grows on an agar slant
of "minimal medium" (carbohydrate, the vitamin biotin, and inorganic nutrients, including NH_4^+). (2) Wild-type mold is irradiated with X rays. (3) Spores are
dissected out of asci and individually cultured on enriched medium [containing
all amino acids and all vitamins in addition to the normal components of minimal medium listed in (1)]. (4) For mutant identification, portions of each culture
are now transferred to minimal medium; if any fails to grow, a mutation is most
likely responsible. (5) Spores of the suspected mutant strain are then transferred to a series of tubes, each containing minimal medium plus one possible
additional requirement (e.g., one amino acid or one vitamin). (6) This mutant
survives on minimal medium plus whatever is in the first tube besides minimal
medium. If it were the amino acid arginine, we could conclude that the mutation
causes a loss of the capacity to synthesize arginine.

Apparently, mutation of the three genes blocked the synthesis at different stages. A mutation of gene *B* would explain why either citrulline or arginine would satisfy the deficiency of mutant *B*. How would a mutation of genes *A* and *C* explain the similarities and differences in the nutritional needs of mutant strains *A* and *C*?

Similar, enzymatically controlled pathways involving the preceding compounds were known to occur in the cytoplasm of certain animal cells. Beadle and Tatum proposed that the genes produced the enzymes catalyzing the biochemical steps of arginine synthesis:

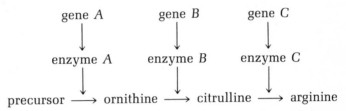

Mutations of the genes deactivated them or resulted in the production of defective enzymes, incapable of effectively catalyzing reactions in the sequence.

Beadle and Tatum had proved that mutation could result in nutritional deficiency and that the explanation of the defect lay in the loss of the organism's power to execute certain biochemical reactions. They had shown that the loss was a heritable or genetic trait. Finally, they proposed that the function of one gene was the direction of the synthesis of a specific enzyme. The idea was called the *one gene-one enzyme hypothesis*. In 1959, for their work, George W. Beadle and Edward L. Tatum were awarded a Nobel prize.

THE HEREDITARY CHEMICAL

In 1938, Richard Goldschmidt criticized the conception of the chromosome as a string of individual genes and proposed an alternative view of the chromosome as a long chain, nucleoprotein molecule. Goldschmidt hypothesized that spatial arrangements along the normal molecule were the crucial properties of the genes and that a mutant gene was a point where rearrangement of a portion of the molecule had occurred. "Prescience" is another word for foresight of this kind, and it means "before knowing." Goldschmidt lived to see the vindication of his idea that the gene was a chemical rather than a physical or structural unit.

Was the genetic chemical protein, nucleic acid, or a combination of the two, nucleoprotein? What was the genetic chemical? These were the questions biologists asked then and answered subsequently.

In 1928, F. Griffith found experimental evidence for a bacterial genetic transforming substance. He was working with two strains of pneumonia bacteria. The phenotype of one included smooth colonial growth on agar. This strain was designated S (for smooth). Another strain (R) formed rough colonies. Microscopic examination revealed that the individual bacteria of the smooth variety were encapsulated with a mucous sheath of carbohydrate nature and that the rough were not; otherwise, they were nearly identical. The smooth strain was virulent or pathogenic (disease causing) when injected into mice. The rough strain was harmless (avirulent). When strain S was killed by heating and then injected into mice, no disease followed, but when live strain R bacteria were injected into mice along with dead strain S organisms, a significant percentage of the mice so treated developed the disease and died. The blood of these doubly injected mice always contained *live encapsulated* bacteria. Somehow, the dead, pathogenic bacteria had transformed the live, harmless, rough organisms into the disease-producing smooth variety. In 1944, three scientists at the Rockefeller Institute identified the transforming principle of the bacterium as a "nucleic acid of the deoxyribose type" (see Fig. 5-17).

The chromosomes were known to be composed of nucleoproteins, the nucleic acid DNA and certain basic nuclear proteins, the histones in particular. About 1870, the German Friedrich Miescher isolated what was later to be called "nucleic acid" for the first time from pus cells. At infection sites, vast numbers of white blood cells converge, phagocytose bacteria and cellular debris, then die. White cells have a small volume of cytoplasm, so that the "pus" is loaded with nuclear chemicals. In 1924, Robert Feulgen and coworkers devised a color test for "thymonucleic acid" and showed that it only occurred in the chromosomes of the cells. This test became known as the "Feulgen reaction"; it was historically important because it specifically located one of the nucleic acids (later shown to be DNA) in the nucleus alone.

Between 1926 and 1930, James Sumner of Cornell University and John Northrop of the Rockefeller Institute proved beyond doubt that enzymes were proteins. In 1941, the Belgian cell biologist and embryologist Jean Brachet and the Danish cytologist Torbjorn Caspersson showed that secretory cells, specialized for protein synthesis, possessed cytoplasms rich in the second nucleic acid, RNA, and postulated that RNA was involved in protein synthesis. The coincidence was great: all protein secretory cells were rich in RNA. Shortly thereafter, biochemists found that radioactive amino acids were rapidly incorporated into protein

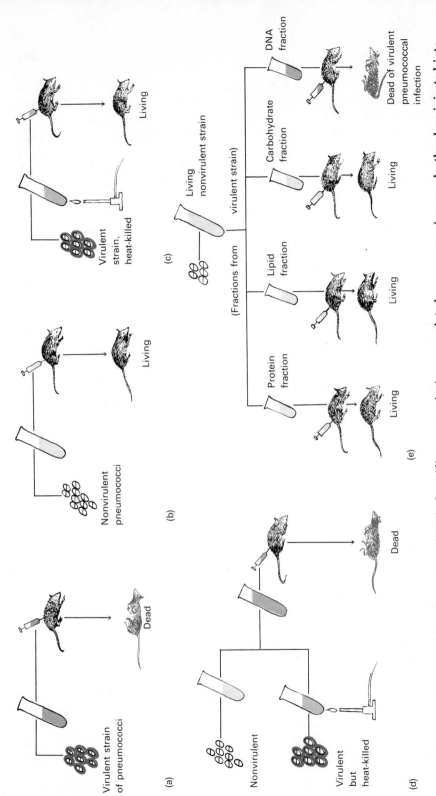

Fig. 5-17 Bacterial transformation. (a) Virulent (disease-causing), encapsulated pneumococci cause death when injected into mice. (b) Nonvirulent, unencapsulated pneumococci do not cause death when injected into mice. (c) Heat-killed virulent pneumococci do not cause death. (d) A mixture of living nonvirulent and heat-killed virulent pneumococci are fatal. Living, virulent, encapsulated bacteria are recovered from the victim, which implies that heredity of the live, nonvirulent strain was altered by some factor from the dead bacteria that allowed the otherwise capsuleless variety to build capsules and become virulent again. (e) Living nonvirulent bacteria are injected into mice along with each of several fractions of killed virulent bacteria. Only the DNA fraction proves capable of transforming the nonvirulent strain.

From N. M. Jessop, *Biosphere*, Prentice-Hall, Inc., Englewood Cliffs, N. J., 1970.

structure in the RNA-rich microsomal fraction of the cell. Beadle and Tatum had shown that the genes directed enzyme synthesis. Enzymes were proteins. How did the genes direct protein synthesis? What was the connection between the nucleic acids (DNA and RNA) and protein structure?

By 1950, there was little doubt that DNA was the genetic chemical, although a few still held to the proteinaceous gene hypothesis. A great deal of information about DNA composition had accumulated. It remained for someone with a synthetic mind to organize the countless observations into a hypothesis that could be tested experimentally.

In 1950, the great chemist Linus Pauling, of the California Institute of Technology, had proposed a spiraled or coiled arrangement of the polypeptide chains in certain proteins. The helical arrangement was offered as an explanation for patterns produced by passing X rays through molecules.

When a ray of light is passed through a narrow slit, the edges of the beam are bent at the edges of the opening, and it is as though the beam is broken into several rays, none of which is traveling in the same direction. This fragmentation of a ray is called diffraction. Some parts of the beam interfere with others; they cancel each other. The result is a dark band upon a screen held up beyond the slit. Other parts of the beam add to or reinforce others, creating bright line bands. The pattern of light and dark bands produced by passing light through a narrow slit is called a diffraction pattern. Light rays have relatively long wavelengths, wider than most molecules, and the narrowest slits that man can make are very large compared with the distances between the atoms (ions) in a crystal.

X rays have short wavelengths, and, in their passage between the ions of inorganic crystals and the molecules of biologic compounds, such as proteins, they are diffracted like light waves and produce patterns of light and dark on X-ray plates. The spaces between adjacent molecules act like the slits of light diffraction.

After a study of the X-ray diffraction patterns produced by several proteins, Pauling and R. B. Corey suggested that the polypeptide chain of amino acids in many proteins was coiled like a spring or the turns of a screw. Hydrogen bonds held the turns of the coil together. Pauling astounded the world of biologists with his successful three-dimensional analysis of the protein molecule. Until this time, the study of biological molecules had been largely two dimensional. To attempt to visualize a third dimension had seemed like an impossible task. Pauling speculated that the structure of DNA might also be helical. He thought that three chains were wound up in the DNA molecule.

In the 1920's, Oscar Levene had shown that nucleic acids' chains could be broken into simpler units called nucleotides. A nucleic acid is a poly-

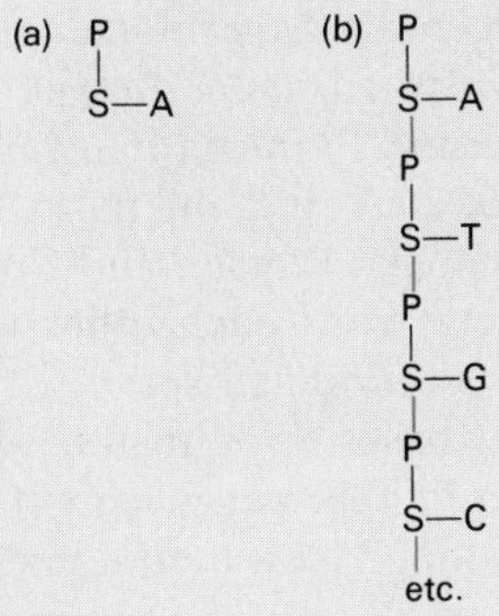

Fig. 5-18

A nucleotide (a) and a polynucleotide chain (b). Phosphate is represented by P; S means sugar; and A, T, G, and C stand for nitrogen bases adenine, thymine, guanine, and cytosine, respectively.

nucleotide, just as a protein is a polypeptide chain of amino acids. A DNA nucleotide is composed of the 5-carbon sugar deoxyribose bound to both a phosphate and a nitrogenous base (Fig. 5-18a). In the chain, phosphates alternate with sugars, and the nitrogenous bases stick out to the side (Fig. 5-18b). Four types of nitrogenous bases were found in most DNA's: double-ringed adenine and guanine, and single-ringed thymine and cytosine. How did DNA duplicate itself as the chromosomes and genes did during cell division? What was the spatial arrangement of the DNA molecule? How did DNA direct protein synthesis? These were the next questions.

An English group of X-ray diffractioners led by Maurice Wilkins had made a number of molecular measurements of DNA fibers. Pauling's Cal Tech group and the London scientists both favored the idea that there were three chains in the molecule. Pauling thought that the nitrogen bases stuck out of the molecule, but Rosalind Franklin of Cambridge felt that the bases projected into the center of the molecule.

In 1951, James Watson, an American postdoctoral fellow, met Francis Crick, a doctoral candidate in biophysics at Cambridge. Crick and others had proposed that helical (spiral) molecules produced a crossways diffraction pattern (Fig. 5-19). Rosalind Franklin's X-ray diffraction analyses of DNA revealed this pattern. The biologist Watson was most aware that the preeminent number of biology was two, as evidenced by chromosome and gene pairs, formed by the union of two sex cells from two parents. Watson wished to build a DNA model with two chains twisting around each other, in contrast to Pauling and others, who favored three as the number of strands. He preferred Franklin's view that the nitrogen bases projected into the molecular center rather than pointed outward, as Pauling had suggested.

Wilkins and Franklin had found that the DNA molecule was a long, thin rod of 20 A in diameter (1 A = 1/250,000,000 in) with a regular repeating pattern. Watson and Crick deduced that two double-ring bases projecting toward each other from opposite strands would make the molecule too wide to agree with actual measurements. Likewise, two single-ring bases bridging the two chains would result in a molecule too narrow to fit the measurements of Wilkin's crystallographs. The distance produced by a model matching one large (double-ring) molecule with one small (single-ring) molecule satisfied the observed measurements. Jerry Donahue of Cambridge showed Watson how to write the most probable formulas for the nitrogen bases that would allow hydrogen bonding. In 1950, Erwin Chargaff had shown that the amount of adenine (double ring) always equalled the amount of thymine (single ring) in any particular DNA, and that the quantity of guanine (double ring) invariably was the same as that of cytosine (single ring). No one

Fig. 5-19   **An X-ray diffraction pattern of a crystalline fiber of a DNA salt. It shows the famed crossways pattern (from 11 o'clock to 5 o'clock and 1 o'clock to 7 o'clock), which provided James Watson, Francis Crick, and others with evidence for a helical (spiral) arrangement of DNA.**
Courtesy of M. H. F. Wilkins and W. Fuller.

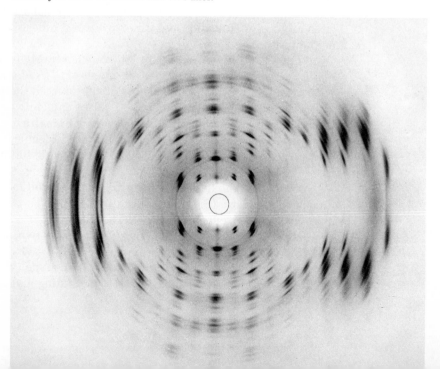

will ever know whether Watson and Crick were aware of Chargaff's rule, but, nevertheless, this information was in keeping with their subsequent proposal.

In 1953, James Watson and Francis Crick proposed "A Structure for Deoxyribose Nucleic Acid" in the British scientific journal *Nature* (Fig. 5-20).[4] They hypothesized that the DNA molecule was a double helix with two polynucleotide chains coiled about each other like a gradually spiraling staircase or a twisted ladder. The sides of the ladder were composed of alternating sugar and phosphate molecules, and the rungs were made up of two nitrogenous bases, one large and one small per rung, tied together by hydrogen bonds. Adenine and thymine were complementary or matching bases, as were guanine and cytosine (see Fig. 5-21).

Watson and Crick wrote "that the specific pairing we have postulated immediately suggests a possible copying mechanism for the genetic material."[5] The two strands of the double helix could separate by breaking

[4] James D. Watson and Francis H. C. Crick, "A Structure for Deoxyribose Nucleic Acid," *Nature,* London, 171, p. 737.

[5] Ibid., p. 737.

Fig. 5-20   **James Watson (left) and Francis Crick (right), proposers of the double helical model of DNA, probably the most celebrated biologists of the twentieth century.**

Bettman Archive.

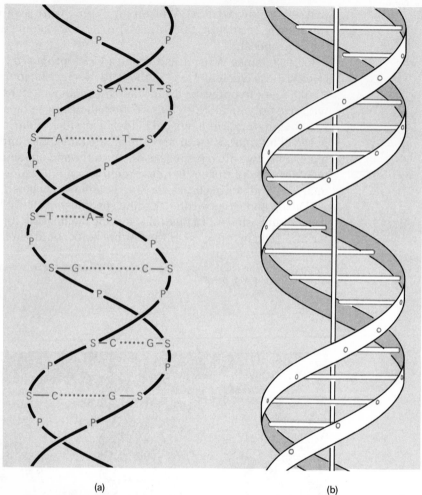

(a)   (b)

Fig. 5-21   **Simple representations of DNA. (a) Phosphate is P, S is sugar (deoxyribose), T is thymine, A is adenine, C is cytosine, and G is guanine. Note that A always bonds with T, and C always goes with G. (b) The "spiral staircase" of DNA.**

the weak hydrogen bonds between bases, and the exposed sequences of bases along the single strands could serve as a pattern for the alignment of matching bases in the construction of replicate strands. Adenine (A) would align a matching thymine (T) nucleotide, and guanine (G) would bring a cytosine (C) nucleotide into place, and so on down the line. Where there were two strands, there would be four (Fig. 5-22). One strand of each daughter double helix would be old and one strand would

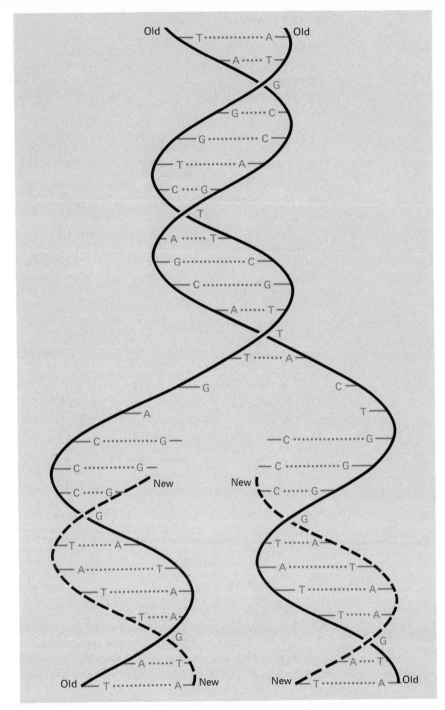

Fig. 5-22   **The replication of DNA according to the Watson-Crick hypothesis.**
From W. R. Breneman, *Animal Form and Function,* Blaisdell Publishing Co., Waltham,
Mass., 1966.

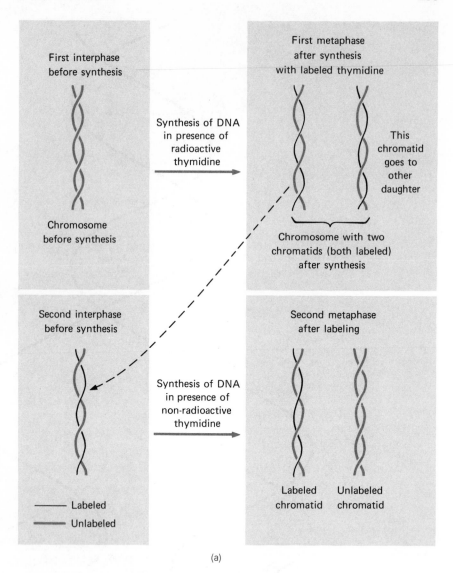

(a)

be newly assembled under the direction of the old. This proposal for DNA replication seemed to explain how genetic information was preserved, duplicated, and transmitted through generations.

Wilkins and Franklin checked the predicted X-ray scattering properties of the Watson-Crick model for DNA with their actual X-ray photographs of DNA and found nothing inconsistent. Proof of the DNA replication hypothesis was not long in coming. Herbert Taylor labeled the chromosomes of dividing onion root cells with radioactive thymine

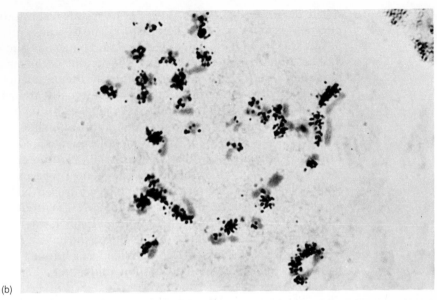

(b)

Fig. 5-23   **(a) OPPOSITE PAGE: The experiment of H. J. Taylor demonstrated that the method of replication diagrammed in Fig. 5-22 is correct for chromosomes as well as for double helices of DNA. Taylor used onion root tip cells, but the results at the second metaphase can be seen in the radioautograph of hamster chromosomes (b). One chromatid of each two part chromosome is hot, owing to the uptake of radioactive thymidine in the replication of DNA. Why was a radioactive compound of thymine used instead of one of the other bases?**
Courtesy of Dr. T. C. Hsu, M. D. Anderson Hospital and Tumor Institute, Houston, Texas.

nucleotide. Thymine nucleotide (thymidine) is found only in DNA. (Its place is taken in RNA by uridine, uracil nucleotide.) Thus, thymidine uptake is a precise reflection of DNA synthesis. After Taylor grew parental, dividing onion root tip cells in radiothymidine medium, all the hybrid chromosomes of the first division generation were radioactive. These cells with labeled chromosomes (normal plus radioactive DNA) were then grown upon a nonradioactive medium. Half the chromosomes of the second generation were labeled, and half were not. These results are in keeping with the Watson-Crick hypothesis. Unlabeled DNA strands must have constructed radioactive complementary strands, making hybrid double helices, composed of the old, unlabeled strands and new radioactive strands. When the hybrid helices unwound and replicated using non-radioactive thymidine, only half of the new double helices remained labeled. The original unlabeled chains organized new unlabeled strands and joined with them to form wholly nonradioactive double helices (see Fig. 5-23a).

The technique by means of which the results were studied in this experiment is known as radioautography. The dividing cells are placed in a "hot" medium, removed, fixed, mounted on a slide and covered with a photographic emulsion or film. Silver grains in the emulsion will be darkened where radioactive particles strike them, and, in this way a cell will take its own picture, an autograph. Actually, it was the chromosomes that took their own picture (see Fig. 5-23b). Taylor's experiment was the first step in verifying the DNA duplication hypothesis.

M. Meselson and F. Stahl grew bacteria upon a medium with heavy nitrogen isotope (N^{15}) until all the bacterial DNA contained nothing but heavy nitrogen in its nitrogen bases (N^{15}/N^{15}). The investigators then plated these bacteria upon a lighter, normal nitrogen (N^{14}) medium, long enough to divide once, and found that the DNA so produced was hybrid (N^{14}/N^{15}). This hybrid DNA (heavy N plus light N) was of a weight intermediate to that of the lightest (N^{14}/N^{14}) and the heaviest, purely isotopic (N^{15}/N^{15}) DNA. Extracted DNA molecules settle to a certain level or band in a density gradient solution under very high-speed centrifugation. Three distinct bands were recognized in the centrifuge tubes. The Watson-Crick DNA duplication hypothesis predicted and explained such results. Intermediate weight DNA could only be the result of a light (normal) strand remaining intact and synthesizing a heavy, isotopic complement (Fig. 5-24). This experiment and Taylor's confirmed the hypothesis.

TOWARD THE CODE

The major reason that proteins had remained so long the favored choice as hereditary biochemicals was that their variety was seemingly infinite, with more than 20 amino acids that could be arranged in any number of orders. This degree of variety seemed a necessary prerequisite for the genetic chemical, and protein was found in the chromosomes. In addition, the enzymes that controlled almost all the known biochemical reactions were protein in nature, and the properties of genes were compared to those of these biologic catalysts. The ability of the genes to reproduce themselves led to their description as "autocatalysts," agents that controlled their own duplication reaction. Although the genes directed the production of substances and structures, they themselves remained unchanged in the process most of the time. In this way, they were like inorganic catalysts or biocatalysts (enzymes), which remain unchanged at the end of a reaction they have promoted. Genes, like en-

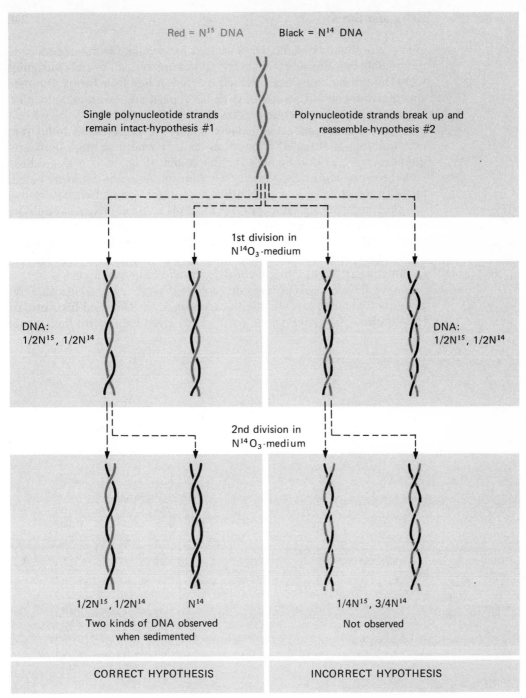

Fig. 5-24   The Meselsohn-Stahl proof of the Watson-Crick hypothesis of DNA replication made use of the fact that DNA synthesized using a heavy isotope of nitrogen could be separated from DNA containing the lighter isotope of nitrogen in a centrifuge. Three weights of DNA were identified. Can you find them in the following diagram?

zymes, are highly specific. DNA's major drawback for the genetic role at first sight was the lack of variety of its components. The only variables were the nitrogenous bases, which existed in just four forms. Biologic specificity depended upon the synthesis of particular protein molecules. How could four bases code for 20 amino acids?

In 1954, American astrophysicist George Gamow speculated that combinations of three DNA bases, or triplets, would be more than adequate to serve as code words for the 20 amino acids. If a single base coded for one amino acid, then only four amino acids could be coded ($4^1 = 4$), which is not enough. With a combination of two bases, or a couplet, representing each amino acid, 16 ($4^2 = 16$) code words could be formed, still not enough. With three bases, 64 ($4^3 = 64$) code words were possible (Fig. 5-25). This figure is more than enough to code for the 20 odd amino acids, and, even before it was experimentally proven, Gamow predicted that many code words would be synonyms.

A new decade came before cell biologists were ready to test this hypothesis. Then in 1961, Marshall Nirenberg of the National Institutes of Health broke the genetic code with a key experiment. Until then, prog-

Singlet code (4 words)	Doublet code (16 words)				Triplet code (64 words)			
					AAA	AAG	AAC	AAU
					AGA	AGG	AGC	AGU
					ACA	ACG	ACC	ACU
					AUA	AUG	AUC	AUU
					GAA	GAG	GAC	GAU
					GGA	GGG	GGC	GGU
A	AA	AG	AC	AU	GCA	GCG	GCC	GCU
G	GA	GG	GC	GU	GUA	GUG	GUC	GUU
C	CA	CG	CC	CU	CAA	CAG	CAC	CAU
U	UA	UG	UC	UU	CGA	CGG	CGC	CGU
					CCA	CCG	CCC	CCU
					CUA	CUG	CUC	CUU
					UAA	UAG	UAC	UAU
					UGA	UGG	UGC	UGU
					UCA	UCG	UCC	UCU
					UUA	UUG	UUC	UUU

Fig. 5-25   **The possible number of code words based on singlet, doublet, and triplet codes are shown here. It is now believed that the triplet code is correct.**

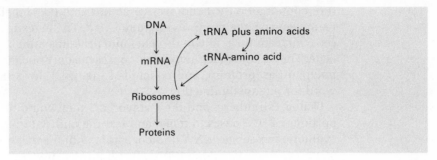

Fig. 5-26   **Relationship of DNA, RNA, and protein. DNA synthesizes messenger RNA (mRNA), which carries blueprints to ribosomal sites, where amino acids are also brought by transfer RNA (tRNA) molecules. The ribosome puts the pieces (amino acids) together following the blueprint (mRNA). Protein is the result.**

ress had been primarily made by observation (the DNA molecular measurements by the Wilkins' group), synthesis of observed facts into theory (the Watson-Crick model), and speculation (the Gamow hypothesis).

Meanwhile, Jacques Monod and Francois Jacob of the Pasteur Institute in Paris had coined the phrase "messenger RNA" (mRNA) to describe a type of RNA that was complementarily synthesized by DNA and carried genetic messages from DNA to the ribosomes, those ribonucleoprotein particles that had been identified as the cytoplasmic sites of protein assembly. Small molecules of another type of RNA, "transfer RNA" (tRNA), had been discovered. The number of forms of transfer RNA equalled the variety of amino acids. Transfer RNA apparently picked up amino acids activated with ATP and transported them to the protein assembly sites (Fig. 5-26). Several investigators had demonstrated that a soup of disrupted bacterial cells minus their walls still carried on protein synthesis. The biochemical machinery of protein synthesis, ribosomes, transfer RNA's, amino acid activating enzymes, and ATP, could function outside the cell in these "cell-free extracts." The incorporation of amino acids into protein can be followed in such systems by using amino acids containing the radioactive carbon isotope C^{14}. In 1955, Severo Ochoa and Marianne Grunberg-Manago of the New York University School of Medicine isolated an enzyme that would join RNA nucleotides together in the absence of a DNA template.

Nirenberg and Heinrich Matthei used this enzyme to combine uracil nucleotides (uridine molecules) into a synthetic messenger RNA that they called "poly-U" (polyuridylic acid). Poly-U was added to a bacterial cell-free extract along with a mixture of the 20 amino acids. Each mixture contained only 1 radioactive amino acid (C^{14} label); the other 19 amino acids in the mixture were unlabeled. The investigators wanted to see which amino acid was directed into protein synthesis by the poly-U code. They found that phenylalanine was the only one incorporated into polypeptide (protein), and concluded that UUU was the triplet code word for phenylalanine (Fig. 5-27).

Ochoa, Nirenberg, and their coworkers then combined various proportions of two bases to make and test the ability of other varieties of synthetic messenger RNA to incorporate amino acids into protein. For example, Nirenberg mixed 70% U (uridine) and 30% C (cytidine) with the enzyme for making RNA chains. The probability of getting the triplet sequence UUU is $0.7 \times 0.7 \times 0.7$, or 0.34; 34% of the triplets are expected to be UUU. The probability of getting CCC is $0.3 \times 0.3 \times 0.3$, or 0.27; that is, 27% of the triplets are expected to be CCC. Thus, the probability of getting UUC triplets is $0.7 \times 0.7 \times 0.3$, or 0.147 (14.7%). The relative amount of an amino acid directed into protein compared with another depends upon the ratio of bases in the synthetic RNA, since this determines the probability of their triplet combination.

H. G. Khorana of the University of Wisconsin devised a method of synthesizing chains of DNA and RNA in which two nucleotides were repeated over and over again in known sequence. It was thus possible to make a DNA strand of TCTCTC, which would form a double helix with the strand sequence AGAGAG. Then, by using RNA polymerase (the enzyme that makes mRNA on DNA templates), synthetic messenger RNA strands of UCUCUC could be made. When DNA serves as a template for mRNA synthesis, uracil (U) takes the place of thymine (T), which is found only in DNA. Therefore, AGAGAG (DNA sequence) would align UCUCUC (mRNA complementary sequence). A DNA se-

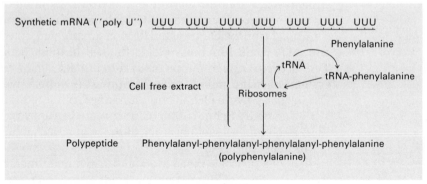

Synthetic mRNA ("poly U") UUU UUU UUU UUU UUU UUU UUU

Phenylalanine

Cell free extract

tRNA

tRNA-phenylalanine

Ribosomes

Polypeptide   Phenylalanyl-phenylalanyl-phenylalanyl-phenylalanine
(polyphenylalanine)

Fig. 5-27   **Nirenberg's genetic code cracking experiment.**

quence of TCTCTC would synthesize an mRNA sequence of AGAGAG. Each of these synthetic mRNA's, when placed in a cell-free system, produced protein chains of two alternating amino acids. Poly-UC produced polypeptides of the amino acids serine and leucine; poly-AG incorporated arginine and glutamic acid. A polynucleotide that began CUC directed the incorporation of leucine into the polypeptide (protein) chain first. An mRNA strand that started UCU placed serine first in the amino acid chain. This was the first proof that the code was read sequentially and that the codon mRNA consists of an uneven number of nucleotides (triplets). Khorana succeeded in synthesizing nucleotide chains consisting of known sequences of three different nucleotides, such as GAUGAUGAU, which incorporated the same amino acid (aspartic acid) in polypeptides, just as poly-U (UUUUUU) directed polyphenylalanine synthesis.

Only 61 of the 64 triplets code for amino acids. There are many synonyms, as Gamow suggested: UCU, UCC, UCA, and UCG all code for serine. Three combinations (UAA, UAG, and UGA) code for no amino acids. They might be the punctuation marks signaling the break between polypeptide messages, perhaps the beginning and end of a "gene." The genetic code is shown in Table 5-6.

Nirenberg and Khorana received the 1967 Nobel prize in medicine and physiology for breaking the genetic code. Maurice Wilkins, James Watson, and Francis Crick were awarded the same for their monumental studies of the DNA molecule in 1961. Their momentous discoveries along with the great work of the electron microscopists gave birth to the "new biology," molecular biology.

A SUMMARY OF DISCOVERY

DNA has been established beyond any reasonable doubt as the biochemical of heredity. The answers to the questions as to how the genetic chemicals duplicate themselves and direct protein synthesis are remarkably similar. In genetic duplication, one strand of the double helix of DNA serves as a template for the synthesis of a complementary chain of DNA. In the *transcription* of DNA's genetic code, a strand of DNA serves as a template for the synthesis of a complementary chain of messenger RNA (mRNA), which leaves the nucleus for the ribosomes, the cytoplasmic sites of protein synthesis.

The triplets of DNA are the genetic *code* words. The triplets of mRNA are the transcribed *codons*. Transfer RNA molecules bear a matching

Code Triplets	Amino Acid	Code Triplets	Amino Acid
AAA	lysine	CAA	glutamine
AAG	lysine	CAG	glutamine
AAC	asparagine	CAC	histidine
AAU	asparagine	CAU	histidine
AGA	arginine	CGA	arginine
AGG	arginine	CGG	arginine?
AGC	serine	CGC	arginine
AGU	serine	CGU	arginine?
ACA	threonine	CCA	proline
ACG	threonine	CCG	proline?
ACC	threonine	CCC	proline
ACU	threonine	CCU	proline
AUA	isoleucine?	CUA	leucine
AUG	methionine	CUG	leucine
AUC	isoleucine	CUC	leucine
AUU	isoleucine	CUU	leucine
GAA	glutamic acid	UAA	gap (comma)
GAG	glutamic acid	UAG	gap (comma)
GAC	aspartic acid	UAC	tyrosine
GAU	aspartic acid	UAU	tyrosine
GGA	glycine?	UGA	tryptophan
GGG	glycine?	UGG	tryptophan
GGC	glycine?	UGC	cysteine
GGU	glycine	UGU	cysteine
GCA	alanine?	UCA	serine
GCG	alanine?	UCG	serine
GCC	alanine?	UCC	serine
GCU	alanine	UCU	serine
GUA	valine?	UUA	leucine
GUG	valine	UUG	leucine
GUC	valine?	UUC	phenylalanine
GUU	valine	UUU	phenylalanine

Table 5-6

The amino acid code

sequence of bases that complement the triplets of mRNA. These triplets of transfer RNA (tRNA) are called *anticodons*. The ribosomes are the sites of *translation,* or the synthesis of protein according to the genetic code (DNA) in the form of mRNA (the transcribed code). The ribosomes move along the mRNA strands. They carry along the growing polypeptide chains at the same time that they hold together the codons of mRNA and the anticodons of tRNA, while additional amino acids are added to the growing protein chain. Transfer RNA molecules are released after their carried amino acids are attached to the lengthening polypeptide. At the end of the message, the polypeptides are released, and the ribosome is free to function as a tool (jig) of translation again (Fig. 5-28). Ribosomes seem to function in groups, the polysomes. Usually several polypeptides associate and fold to form more complicated configurations of the actual proteins. Often, a vitamin derivative (coenzyme or prosthetic group) is added to the protein before its synthesis is finished.

THE MODERN GENE

The old definitions of the gene as a unit of recombination and crossing over emphasized the particulate conception of the hereditary factor. The discoveries of molecular biology clarified the definition of the gene as a unit of function and mutation. The work of Beadle and Tatum, the "one gene-one enzyme theory," stressed the functional role of the gene. The gene as a unit of duplication was very well explained by the Watson-Crick model of DNA, which stressed the chemical rather than the physical definition of the gene.

The gene has been the biologic unit of greatest stability, yet it is not completely stable; therefore, it is difficult to define absolutely. The journalistic technique of education depends heavily upon the supposition that the more closely a problem is examined, the simpler it becomes. Like the atom, the gene can be subdivided, and some of the old definitions of the gene apply to subdivisions, which are not entire genes. Although Occam's rule that the simplest explanation is most likely correct is almost a commandment of science, at the same time it seems that the rule (rather the exception) is that the closer one examines a problem the more complicated it becomes. Man tries to answer unanswered questions with models from his past experience. New words and models must be added to his vocabulary. So it was with the attempts to precisely define the gene.

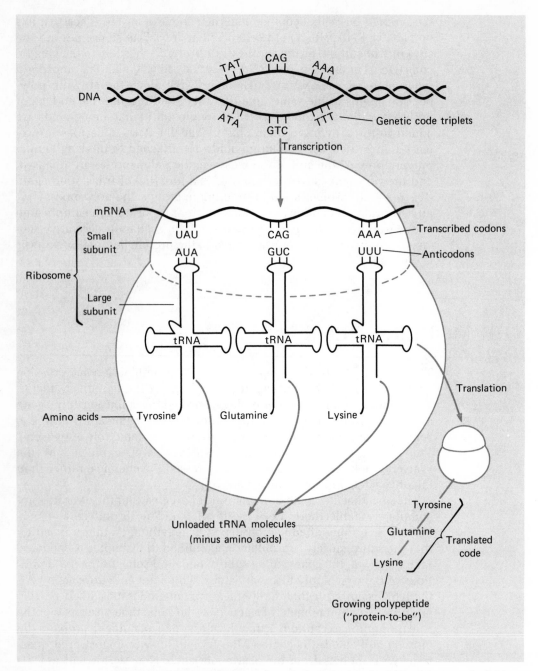

Fig. 5-28   **The transcription of the genetic code (mRNA synthesis) and its translation into the amino acid sequence of protein. See text for additional description.**

Since many proteins are made up of more than one polypeptide, the one gene-one protein hypothesis had to be altered slightly. The one gene-one enzyme hypothesis became the "one gene-one polypeptide" hypothesis. In addition, not all genes directed the production of enzymes. The respiratory protein hemoglobin is composed of four polypeptide chains, two alpha and two beta strands. Together, the four chains contain a total of 574 amino acid units. One might speculate that there would be 1722 (574 × 3) nucleotides in the hemoglobin gene sequence, but, since the protein consists in two pairs of polypeptides and the two chains of each pair are identical, the number is perhaps half as great—861. There must be a nucleotide sequence for the alpha polypeptide and another for the beta polypeptide. If we strictly followed the one gene-one polypeptide concept, then the "gene" for hemoglobin would really be two genes. A mutation is known that replaces normal hemoglobin with an altered form of the molecule, which is found in the red blood cells of victims of sickle cell anemia. Sickle cell hemoglobin differs from normal hemoglobin by a single amino acid substitution in each of the two beta chains. Valine replaces glutamic acid. The DNA code words for glutamic acid are CTT, or CTC, and those for valine are CAT or CAC. The difference between them is a single base, or nucleotide. The unit of change is a single nucleotide, yet one nucleotide codes for nothing. Obviously, the old definition of the gene as both a unit of mutation and a unit of function is no longer quite adequate.

Seymour Benzer and other molecular geneticists encountered this discrepancy early in their mapping studies of the interior of bacterial virus genes. The term *muton* was suggested for a unit of mutation, the smallest element within a gene able to change, which can be as small as one nucleotide. A DNA segment that specifies a particular protein is the *cistron,* or functional gene of the one gene-one enzyme hypothesis. Cistrons vary in size, but the smallest are composed of hundreds of nucleotides, and the largest include a thousand nucleotides or more. The unit of recombination cannot be as small as a single nucleotide. The probability of two nucleotides crossing over from different chains is almost zero, or practically impossible. Yet, the *recon* (unit of recombination) can be smaller than the functional gene (cistron).

MUTATION: A FINER LOOK

Since mutation is a change in a gene and the gene is chemical (DNA), mutations must be the result of chemical changes in DNA. The word mutation simply means change, and biologists also consider chromosomal abnormalities to be mutations. A portion of one chromosome can be lost (deletion), turned around (inversion), become attached to the homologous chromosome resulting in a duplication of genes, or be translocated to a nonhomologous chromosome. Mutation at the genetic level is called *point mutation,* and the single-base changes (mutons) that we have studied are the smallest possible examples of such changes.

X rays cause chromosomal breakage, but also produce single gene mutations. X rays are said to be ionizing radiation. They eject electrons from the atoms or molecules through which they pass, thus ionizing the atoms or molecules, leaving them positively charged. Organic ions are extremely reactive, or short lived, entering into chemical reactions with surrounding or neighboring molecules after a few milliseconds' existence. These reactions change the chemical identity and spatial arrangements of the reacting molecules and therefore the identity and informational value of the gene, which so depends upon the spatial arrangement and nature of its chemical components. X rays also change cell water molecules into reactive ions and molecular species, such as H_2O_2. The other forms of ionizing radiation, alpha, beta, and gamma rays, produce

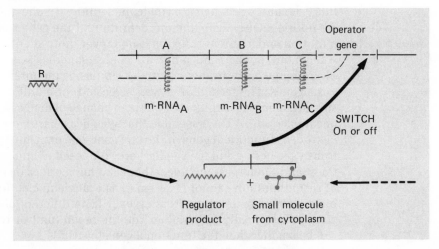

Fig. 5-29   **Genetic regulation. The regulator gene is denoted R; A, B, C are linked genes determining the enzymes of a synthetic pathway; mRNA$_A$ is the messenger RNA made by gene A, etc. Production of mRNA is either permitted or inhibited by the operator gene, which, in turn, is turned on or off by the reaction between the regulator product and a small molecule from the cell cytoplasm or environment.**

From D. M. Bonner and S. E. Mills, *Heredity,* Prentice-Hall, Inc., Englewood Cliffs, N. J., 1964.

similar results. Although ultraviolet light is a nonionizing radiation, it is absorbed strongly by the nucleic acids, excites their electrons to higher energy levels, and increases the likelihood of reaction considerably.

The chemical mutagens (mutation-producers) include the war gases, the nitrogen mustards, whose derivatives have been used in the treatment of cancer, formaldehyde, nitrous acid, and several steroids that are carcinogenic (cancer-producing). Nitrogen mustards react with the phosphoric acid groups of nucleic acid (DNA), producing spatial distortions of the sides of the ladder. Nitrous acid can change one code base into another, and therefore one code triplet into another. Some mutagenic chemicals are similar to the nitrogen bases of DNA and can substitute for them, yet destroy the informational value of the gene. Two such compounds, 5-fluorouracil and 2-aminopurine, are synthesized in the laboratory, but some naturally occurring molecules, such as caffeine and theophylline (found in coffee and tea), may also be mutagenic for the same reason. LSD causes chromosomal breakage. Nicotine and the hallucinogenic alkaloids are also nitrogen-carbon ring compounds often similar to the code bases.

REGULATORY GENES

Jacob and Monod found that certain bacterial genes regulate the activities of other genes rather than direct cytoplasmic protein (enzyme) synthesis. Regulator genes produce repressor substances that can repress ("shut off") operator genes, which, in turn, "switch on" the structural genes that direct protein synthesis (Fig. 5-29).

It appears at present that different genes are active in cells specialized for different functions. Even though each body cell nucleus contains the same genes, it is illogical to assume that all genes are active in each cell. Interestingly, environmental substances have been found that activate and repress genes.

RELEVANCE

One of Darwin's greatest clues as to the hereditary plasticity of the species was the amazing variety of domestic plants and animals produced by plant and animal breeding experiments. By inbreeding livestock and plants, it has been possible to develop largely pure-breeding strains.

Fig. 5-30   **A Santa Gertrudis cow in south Florida. This breed was developed by crossing the heat and insect-resistant Indian humped cattle, the Brahman breed, with the superior meat-producing English breeds, the Shorthorn and Hereford. Brahman cattle are poor meat producers and the Herefords do poorly in hot, humid, insect-ridden climates. Crossbreeds such as this one combine the desirable features of both and eliminate the undesirable traits to some extent.**
From M. D. Nicklanovich and M. J. Zbar, *Introductory Genetics*, Saturn Scientific, Inc., Fort Lauderdale, Fla., 1972.

Undesirable characteristics due to recessive genes are revealed, and, by preventing the reproduction of individuals possessing such traits, the frequency of the harmful genes in the population can be reduced to nearly zero.

Crossbreeding of inbred lines has often led to new strains, combining the desirable traits of both parental lines. Several European and American cattle breeds, Herefords and Shorthorns, have superior meat-producing traits but poor heat, drought, and insect resistance. Indian Brahman and Zebu breeds have poor meat, but great heat, drought, and insect resistance. By crossing European beef breeds with heat-resistant Brahmans, the Santa Gertrudis breed, which combined the valued features of both parental strains, was developed (Fig. 5-30). This breed has been very successful in the southern and southwestern United States. Geneticists of the United States Department of Agriculture are now crossing the high milk-producing Jersey with the drought-resistant Zebu in an attempt to produce a new breed with the desirable qualities of both strains.

In addition to yielding new, highly desirable combinations of traits, crossbreeding of inbred lines can produce superior offspring with "hybrid vigor." The crossbreeding of inbred lines of corn is practiced regularly to obtain hybrid seed, from which grow plants superior in ear production and size (Fig. 5-31). Perhaps the hybrid condition is itself beneficial. It may be that different "vigor genes" from both inbred strains act in a complementary fashion when combined. Some think that unidentified, harmful, recessive genes from the purebred parent strains are masked in the hybrid offspring of crossbreeding.

Rarely, desirable mutants appear spontaneously in domestic organisms. The hornless mutant is one such in cattle. The gene responsible is recessive to the normal, horned condition. The procedure for developing a hornless strain involved breeding a hornless bull with his horned daughters. Why?

ON BEING WELL BORN

Man is no longer subject to natural selection, and it is doubtful that he will improve biologically in future generations. By consciously selecting among his own genes, man can undoubtedly improve his future genera-

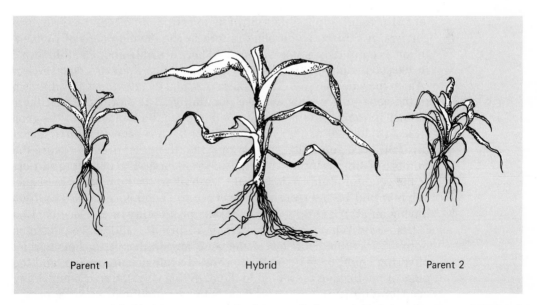

Parent 1   Hybrid   Parent 2

Fig. 5-31   **Hybrid vigor in corn. Each of the small plants on the left and right are the inbred parents of the hybrid corn shown in the middle.**

tions. Sir Francis Galton coined the term *eugenics,* which he defined as "the science which deals with all influences that improve the inborn qualities of a race; also, with those that develop them to the utmost advantage." The "race" meant here is the human race. Eugenics has been called "the science of being well born."

Eugenic societies were organized in Europe and the United States early in the present century. Eugenists desired to improve the genetic composition of the human species. They advocated encouraging reproduction by the genetically superior and discouraging or actually preventing the reproduction of undesirable genotypes. Unfortunately, the eugenics movements frequently attracted prejudiced persons, ignorant of the science of genetics. The racist propaganda and genocide practiced by the Nazis probably retarded the progress of eugenics for a quarter-century or more. Even today, most laymen warily eye anyone mentioning human genetic selection. The social revolutions of today emphasize environmental rather than hereditary roles in the determination of human phenotypes. The disciplines that deal with the improvement of mankind through the control of environmental factors have been grouped under the term euthenics, and it should be distinguished from eugenics, which is now reserved specifically for those processes that would result in the improvement of genetic material.

The elimination of harmful genes from the population of man is desirable, but immensely difficult. Most deleterious genes are recessive and at present impossible (with few exceptions) to identify in hybrid carriers. Albinism, a complete failure in the development of pigmentation, is caused by a recessive gene. There are only about 5 albinos (aa) in 100,000 people, but it has been calculated that there are 1420 carriers (Aa) in the same sample. Thus, there is a total of 1430 genes for albinism in the gene pool of this average population. The sterilization of the 5 albinos in the population would only reduce the incidence of the gene from 1430 to 1420, an insignificant amount. It has been computed that at this rate, 5000 years (100 generations) would be required to reduce the number of albinos by half. The only current method of detecting carriers of the gene for albinism is with the birth of an albino child to a couple. What would be their probable genotype? The chance of two carriers having an albino child is 1 in 4, but the probability of carrier offspring is $\frac{1}{2}$. How so? What percentage of their offspring would you expect to be normal? Identified carriers of the gene for albinism and their normal offspring could be genetically counseled against reproducing, and the odds of misfortune presented to them. In this way, the frequency of the gene for albinism could be reduced in the population in a shorter period of time, but it could never be brought to zero, since the gene permitting the development of normal pigmentation (A) mutates to the defective gene for albinism (a) at a certain low rate.

Not all harmful genes are recessive. In some cases, dominant deleterious genes can be eliminated in a single generation. Huntington's chorea is a degenerative nervous disease inherited as a simple dominant. Unfortunately, the symptoms do not appear in the carrier until he is in his thirties or forties, by which time children usually have been produced. The disease begins with twitchings of the limbs and face, progresses to a severe spastic condition, accompanied worst of all by the complete loss of mind, due to brain cell degeneration. The offspring of known victims should probably be forbidden the right to reproduce.

The gene for chondrodystrophic dwarfism is also a dominant, but it would be more difficult to reduce in the population than the gene for Huntington's chorea. In this type of dwarfism, the head and torso are normal size, but the limbs are extremely short. Ten out of 94,000 babies born in Copenhagen, Denmark, have developed into such dwarfs. Two of these babies had dwarf parents; eight did not. Obviously, the latter represent new mutations. The gene involved must mutate to produce chondrodystrophy once in about 12,000 births. Thus, it would be very difficult to significantly reduce the frequency of the gene in this population. Nevertheless, negative eugenic measures (sterilization) could have prevented two cases of disfigurement.

In some cases, it is possible to detect the hybrid carrier of a harmful gene. Cooley's anemia, or thalassemia major, is fatal to those carrying two of the harmful causative genes. Blood tests can reveal carriers of one abnormal gene, since these individuals suffer from a mild form of the disease, thalassemia minor, or microcythemia. In some sections of the Po River Valley of Italy and Sicily, as high as 27% of the total population has the disease. When two people with the mild form of the disease marry, about one-fourth of their children (the purebreds for the trait) die, and about half their offspring are microcythemic (carriers). What percentage would you expect to be normal? Each year Cooley's anemia kills 1000 children. Geneticists and physicians are well on their way to controlling the disease. Those who have the gene are advised not to marry and by all means not to wed relatives or carriers like themselves, with microcythemia.

It has been estimated that perhaps 5% of the general population has the genotype for diabetes. Since the disease is incurable and even patients on replacement therapy develop other conditions, it may be desirable to reduce this seeming genetic trait in the population.

The hottest controversy about potential eugenic practices centers about intelligence. The sterilization of high-grade and low-grade mental defectives is practiced in some states. The mentally defective are not as prolific as is commonly believed. If population curbs are instituted in the future, it is conceivable that intelligence could become a criterion for reproductive "privilege." The intelligent may be allowed to reproduce more than the average. It seems unlikely that such practices would develop in a democracy.

It is possible to store viable sperm by freezing for long periods of time. Perhaps the semen of geniuses will be so preserved. Another possibility is the transplantation of the nuclei of body cells into denucleated eggs and either implanting these into a prenatal nurse's womb or controlling development in an artificial medium. The person so formed would be an identical twin, a genetic duplicate of the source. Of course, he or she would be a different person, growing up in a different environment, but the potential should be the same as the donor.

Under totalitarian dictatorships, it would be possible to produce various "breeds" of men, as has been done with domestic plants and animals. A genetic caste system could be developed with phenotypic classes of geniuses, gigolos and prostitutes, dull slave laborers, etc. The same methods that have produced the various breeds of dogs and cattle, inbreeding, artificial selection, crossbreeding, etc., could produce the same results in the human species.

DISCUSSION QUESTIONS

1. Which of Mendel's "laws" are laws in reality?

2. Why did the Mendelians doubt Darwinism?

3. How are skin color and intelligence inherited?

4. What experimental evidence supported the hypothesis that DNA was the genetic substance?

5. How was the genetic code broken?

6. Why were the proteins thought to be the genetic chemicals?

7. What proof is there of the double helical model of DNA?

8. Is heredity or environment most crucial in inheritance?

9. In rabbits black hair is dominant to white. A black rabbit was crossed with a white, and half the offspring were black and half were white. What was the genotype of the black rabbit?

10. Two extremes were crossed, and their offspring were average. A cross of averages produced a generation, half of which were average, and one quarter of which fit into either extreme. What type of inheritance was this? What were the genotypic and phenotypic ratios?

11. In another case, when two extremes were crossed and the average offspring mated, the second generation showed continuous variation from one extreme to the other with the most frequent phenotype being average and the least frequent types being the extremes. $\frac{1}{16}$ of the offspring fitted into each extreme category. What type of inheritance was this?

12. The normal type of a certain bacterium had no amino acid requirements. Following exposure to ultraviolet light, several mutant strains appeared. Two closely related deficiency mutants were isolated. One could survive on minimal medium plus either of two amino acids: phenylalanine or tyrosine. The other could survive only on minimal medium plus tyrosine. Phenylalanine can normally be converted into tyrosine. How do you explain these results?

13. A normal sighted couple had three nearsighted children. Who was responsible? Write the genotypes of all.

REFERENCES

BONNER, DAVID M., and STANLEY E. MILLS, *Heredity*, Prentice-Hall, Inc., Englewood Cliffs, N. J., 1964.

CRICK, F. H. C., "The Structure of the Hereditary Material," *Sci. Am.*, Oct., 1954.

CRICK, F. H. C., "The Genetic Code: III," *Sci. Am.*, March, 1966.

GORINI, LUIGI, "Antibodies and the Genetic Code," *Sci. Am.*, Apr., 1966.

LEVINE, R. P., *Genetics*, Holt, Rinehart & Winston, Inc., New York, 1968.

MOODY, PAUL A., *Genetics of Man*, W. W. Norton & Company, Inc., New York, 1967.

NIRENBERG, MARSHALL N., "The Genetic Code: II," *Sci. Am.*, Mar., 1963.

SNYDER, L. H., and P. R. DAVID, *The Principles of Heredity*, D. C. Heath & Company, Boston, 1957.

WATSON, JAMES D., *The Double Helix*, Atheneum Publishers, New York, 1968.

OBJECTIVES

1. Contrast evolution and special creation.

2. List three characteristics of the fossil record that can be explained by evolutionary theory.

3. Distinguish between homologous and analogous organs.

4. Contrast the evolutionary theories of Darwin and Lamarck.

5. List four great principles which led Darwin to evolution.

6. Summarize the five points of Darwin's evolutionary mechanism.

7. Define the "modern synthesis."

8. Define the species three ways and criticize each definition.

9. Contrast the Hardy-Weinberg equilibrium and evolution.

10. Explain the potential roles of isolation and genetic drift in evolution.

11. List three sources of variability; cite three types of adaptations.

12. Describe three types of selection.

13. Compare the composition of the primitive and the modern atmosphere; outline Miller's experiment and explain its significance.

14. List five major steps involved in the evolution of land plants.

15. List six major steps in the evolution of land vertebrates.

16. Contrast adaptive radiation and convergent evolution.

17. Summarize the arguments for the fresh water and salt water theories of vertebrate origin.

18. List the seven classes of vertebrates; give three characteristics of each class and two common representatives; construct a vertebrate family tree.

19. List the four major divisions of the plant kingdom and briefly discuss their evolutionary relationships.

20. Describe the plants, animals, and climates of the three last great ages in Earth's history.

21. Diagram and label the basic pentadactyl limb and describe its modification in several vertebrate lines.

22. Briefly describe the evolution of the modern horse.

23. Analyze the famous cases of industrial melanism, DDT resistance, and sickle cell anemia in evolutionary terms.

Evolution:

Perpetual

Revolution

ERWIN SCHRÖDINGER, NOBEL PHYSICIST, proposes that "this age, which delights in calling itself the age of technology, will in some later time be described in terms of its brightest lights. . . . as the age of the evolutionary idea."[1] The idea of evolution has been more potent than any other idea in all areas of modern knowledge and life. Appropriately, the noted historian of evolutionary thought, Loren Eiseley, called his major work *Darwin's Century.* The nineteenth-century intellectual world belonged to Darwin and Darwinism has not yet surrendered the twentieth-century world of thought.

One alternative to evolution has been *special creation,* the doctrine that the species are unchanging, being today as they were at the time of Genesis, in the beginning, fresh from the hand of God; in other words, there are no new species. Evolution is the generally accepted theory that all present species of plants and animals have developed from earlier forms by the genetic transmission of slight variations in successive generation. Above all, the types of life are changeable. They are descended with modification from countless common ancestries, all eventually converging upon the one or the few, rudimentary, original forms, which, in turn, had their beginnings in the nonliving world. The species is anything but holy or inviolate. Its twisted genealogy, the family tree, is scarred with numerous extinctions and the transformations of old species into the new.

---

[1] Erwin Schrödinger, *My View of the World,* trans. C. Hastings (Cambridge, London: Cambridge University Press, 1964), pp. 5–6.

It is nearly inconceivable that a man with a choice would prefer common ancestry with apes to his divine creation. Evolution seems quite inescapable. The history of conflict between evolution and religion is long and by no means ended. There are many who maintain a reconciliation, making of God the director of biologic change, but it must be confessed that the grandly piloted evolution so frequently envisioned is a far cry from the seeming facts of the history of life on earth. Then, there are those who perhaps rightly hold that science and religion are entirely separate realms, since the former is concerned with strictly natural phenomena and the latter deals with the supernatural. We shall sample the views of both schools in the concluding chapter.

Julian Huxley has stated that evolution is almost scientific law to the biologist. Still, science is basically conservative in that there are few laws. Perhaps, it is more accurate to say that the theory of evolution is the only plausible explanation of the present and past varieties and commonalities of life. German physicist Ernst Mach has pointed out that the process of science itself is evolutionary, a continual, gradual adjustment of thoughts and facts, a selection of the most useful explanatory hypotheses (the fittest) and a rejection of those least in keeping with the facts (unfit).

The term evolution is derived from the Latin word *evolutio,* meaning "an unrolling." To evolve is to change in modern parlance. The geologic evolution of the earth was but a fraction of the cosmic evolution of the universe. *Organic evolution* refers to the unrolling of the present panorama of life, the history of biologic change. Evolution explains the seeming paradox of why living things are so alike and yet so different.

A HISTORY OF EVOLUTIONARY THOUGHT

The intellectual environment in which evolution was born was anything but sterile. In the Platonic vein, the species as an absolute type was considered to be a misleading appearance rather than the ideal, final reality. Sensory impressions were false; truth lay behind the scene. The Enlightment had left heresy in its wake. Similarities of organisms were considered proofs of the ideal design. In 1699, Edward Tyson compared the anatomies of monkeys, apes, and men. In 1809, French biologist Jean Baptiste Pierre Antoine de Monet, the Chevalier de Lamarck (Fig. 6-1), proposed a theory of evolution in his *Philosophie Zoologique* (A Zoologic Philosophy). Lamarck (as he is commonly called) expressed his doubts as to the immutability of the species:

Fig. 6-1   **Jean de Monet, Chevalier de Lamarck, first great evolutionist, father of invertebrate paleontology, author of** *Philosophie Zoologique* **(1809).**

The name species has been given to every collection of similar individuals which have been produced by other individuals like themselves. This definition is correct. . . . But to this definition has been added the supposition that the individuals which make up the species never vary in their specific characters and that consequently the species has an absolute constancy in nature. It is exactly this supposition that I propose to combat.[2]

Lamarck perhaps implied an ape origin for man. He did not believe in the extinction of species, but rather that the "extinct" forms had been altered to form present types. Lamarck powerfully hinted at an evolutionary tree. Although he had been a plant biologist for most of his life, he became an animal biologist late in his life and pursued the study of

---

[2]Jean Baptiste Pierre Antoine (de Monet) De Lamarck, *Zoological Philosophy*, trans. Hugh Elliot (New York, N. Y.: Hafner Publishing Company, 1963), p. 35.

animals without backbones (the invertebrates) with such a youthful passion that he became the father of invertebrate paleontology.

Rightly so, Lamarck emphasized the change in the external environment, climate, geography, etc., as the prompters of biologic change. He proposed that alterations of the physical environment produced changes in physiologic "needs," leading to modifications of behavior and "habit," and finally, in consequence, transformations of bodily structure itself. Lamarck supposed that the "use or disuse" of parts led to their hypertrophy (enlargement or development) or atrophy (withering and disappearance). The reason for the waxing and waning of structures was the unconscious recognition of need by the organism in its changing environment (see Fig. 6-2). Lamarck had expressed a fanatic belief in the superplasticity of life, an almost unlimited ability to change by "involuntary will." Nevertheless, the evolution he envisioned was not instantaneous.

Today, it is believed that the capacity of life to change is not unlimited, but restricted over long periods of time to the variety already present, the "hand dealt by chance and life's tolerability," and that the final source of truly new variety (other than new combinations of genes) is random genetic mutation. Thus, over short periods of time, life is relatively rigid, static. Lamarck was excusably ignorant of genetics; this science was not born until 1900. He believed (like most others of his time) in the inheritance of acquired characteristics, individual changes induced by environmental deviations, a view no longer defensible.

The Napoleonic dictator of French education, the Baron Georges Cuvier, publicly defeated the evolutionist Lamarck and his colleague St. Hilaire in a debate before the French Academy of Sciences. Oddly enough, Cuvier opposed evolution, even though he founded comparative anatomy and vertebrate paleontology, the two sciences providing the most fundamental evidence for evolution. Cuvier divided the animal kingdom into four great groups, the Mollusca, the Radiata (animals with wheel symmetry, such as starfish and jellyfish), the Articulata (animals with jointed segments, such as the annelids and arthropods), and the Vertebrata. He could see no similarities in their anatomies and therefore could not imagine a common ancestry for them. In this, he was right: Hardly any definite anatomical similarities exist between mollusks and men. Yet, it was just such an absurd comparison that St. Hilaire strove to defend in debate with the greatest anatomist of his day, Cuvier. Unfortunately, the meritorious points of Lamarck's theory fell along with the ridiculous. The public birth of evolution was delayed for a half-century, and France would be the last western civilization to embrace it. To explain the extinctions of the fossil record, Cuvier proposed that the earth had suffered a series of awesome catastrophies, of which the Flood was the most recent.

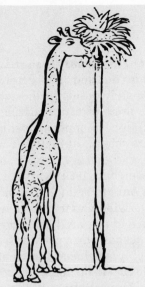

— Dis donc, papa, *pourquoi* que les palmiers sont si grands?

— C'est pour que les girafes puissent les manger, mon enfant, car. . .

. . .si les palmiers étaient tout petits, les girafes seraient treś embarrassées.

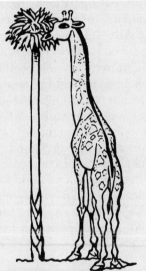

— Mais alors, papa, *pourquoi* que les girafes ont le cou si long?

— Eh bien! c'est pour pouvoir manger les palmiers, mon enfant, car. . .

. . .si les girafes avaient le cou court, elles seraient encore bien plus embarrassées.

Fig. 6-2   A French caricature of Lamarck's theory of evolution by Caran d'Ache. A child asks his father why the palm is so tall. The father explains that the height of the palm permits the giraffe to eat it, for, if the palm were small, the giraffe would be embarrassed. The child then asks why the giraffe's neck is so long. The father replies: ". . . if the giraffe had a short neck, it would be even more frustrating."

A form of Lamarckianism survived until recently in the Soviet Union. Lysenko, the czar of Russian biology, insisted that simply exposing wheat to cold would almost immediately result in the formation of cold-resistant varieties, even though none of his "experiments" were repeatable. The "idea" had a "revolutionary appeal," and fitted party dogma well. The advance of the entire science of biology was retarded behind the iron curtain.

Cuvier's brain was preserved and weighed (1830 g versus 1350 g average); Lamarck is disinterred in texts only to ridicule the more foolish aspects of his evolutionary theory. Darwin, who followed him, did not clearly distinguish between inherited and acquired characteristics, but his mechanism of change was more mechanistic, "needing no need."

Nineteenth-century England probably produced the greatest "amateur" scientists in all of history. Charles Lyell, the father of modern geology, was by profession a lawyer, and Charles Darwin, graduate in theology and medical school dropout, never preached. The wealthy families of the day sent their boys off to study "a learned profession" (law, medicine, or the ministry), much as they sent their girls off to finishing school. Many never practiced their professions afterward. It would be unfair to hobbyists to call them amateurs, for they were anything but that, pursuing their "avocations" nearly full time. The countryside seems to have been alive with pigeon fanciers, horticulturists, and "rock hounds." At Cambridge, Darwin took up with several beetle collectors who introduced him to Professor Henslow, a man of wide scientific curiosity and a plant breeder. The professor led an informal gathering of students and faculty passionately interested in nature. Many English ministers were famous naturalists who found evidence of God's handiwork in the designs of nature. The works of one such famous theologist-naturalist writer, William Paley, influenced Darwin considerably. After graduation, Henslow secured for young Darwin a position as naturalist (without pay) on board the H. M. S. Beagle, bound on a 5-year mapping expedition around the world. The year was 1831 and "the sun never set on the British Empire." Darwin sailed around the world and "back in time." Of the voyage Darwin later wrote,

The voyage of the Beagle has been by far the most important event in my life and has determined my whole career; . . . I have always felt that I owe to the voyage the first real training or education of my mind. I was led to attend closely to several branches of natural history, and thus my powers of observation were improved, though they were already fairly developed.[3]

---

[3]Charles Darwin, *The Autobiography of Charles Darwin,* (New York, N. Y.: W. W. Norton & Company, Inc., 1969), pp. 76–77.

Darwin was quite familiar with the plants, animals, and rocks of the British Isles. His journal of the voyage reveals that he was a competent geologist, meteorologist, and biologist. Last, but not least, Darwin was an accomplished writer, capable of powerful yet calm, elegantly simple description. He noted and recorded the geographic variation of plant and animal life as he proceeded from the tropical rain forests of northern South America to the Argentine pampas, the deserts of Chile, and the Andean highlands.

Upon his return to England, Darwin published his *Journal of Researches;* his appearance at that time is recorded in Fig. 6-3. On July 1, 1837, he opened his first notebook on the "species problem" and continued to work on it for the next 20 years. In 1842, Darwin's small book *Coral Reefs* was published, and his *Geological Observations on South America* appeared in 1846. For 8 years, he worked on barnacles, finally publishing two thick volumes on these aberrant crustaceans. Darwin's curiosity and patience seem to have been limitless. His interest in the unfamiliar and "insignificant" was one of his greatest assets. On his barnacle work, Darwin remarked: "I do not doubt that Sir E. Lytton Bulwer had me in his mind when he introduces in one of his novels a Professor Long, who had written two huge volumes on Limpets." Even after he had gained great fame for his work on evolution, Darwin continued his studies of this nature. In 1862, his book *Fertilisation of Orchids* was published; *The Formation of Vegetable Mould through the Action of Worms* was printed in 1881. Of this study of earthworms, Darwin wrote, apologetically, "This is a subject of but small importance; and I know not whether it will interest any readers, but it has interested me."

These titles represent no more than half of his studies and works. Darwin's powers of absorption and concentration were extreme. He himself confessed that his mental constitution made it difficult for him to turn from one subject or labor to another. Darwin drew the keys to his success thusly:

Therefore, my success as a man of science, whatever this may have amounted to, has been determined, as far as I can judge, by complex and diversified mental qualities and conditions. Of these the most important have been—the love of science—unbounded patience in long reflecting over any subject—industry in observing and collecting facts—and a fair share of invention as well as of common sense. With such moderate abilities as I possess, it is truly surprising that thus I should have influenced to a considerable extent the beliefs of scientific men on some important points.[4]

---

[4] Ibid., pp. 144–145.

Fig. 6-3   **Charles Darwin (1809–1882).**
Bettmann Archive.

To this list we might add that he loved nature, as the following prose
written at the age of 65 reflects:

*The glories of the vegetation of the Tropics rise before my mind at the
present time more vividly than anything else. Though the sense of sub-
limity, which the great deserts of Patagonia and the forest-clad mountains of Tierra del Fuego excited in me, has left an indelible impression
on my mind. The sight of a naked savage in his native land is an event*

which can never be forgotten. Many of my excursions on horseback through the wild countries, or in the boats some of which lasted several weeks, were deeply interesting; their discomfort and some degree of danger were at that time hardly a drawback and none at all afterwards.[5]

Darwin said, "If we choose to let conjecture run wild, then animals, our fellow brethren in pain, disease, suffering and famine—our slaves in the most laborious works, our companions in our amusements—they may partake of our origin in one common ancestor—we may be all melted together." This statement inspired Loren Eiseley to write:

If he had never conceived of natural selection, if he had never written the Origin, *it would still stand as a statement of almost clairvoyant perception. There are few youths today who will pause, coming from a biology class, to finger a yellow flower or poke in friendly fashion at a sunning turtle on the edge of the campus pond, and who are capable of saying to themselves, "We are all one—all melted together." It is for this, as much as for the difficult, concise reasoning of the* Origin, *that Darwin's shadow will run a long way forward into the future. It is his heritage from the parson-naturalists of England.*[6]

Alfred Russell Wallace, who independently proposed the same theory of evolution, spent years collecting biologic specimens in the Amazon and Southeast Asia. In contrast to Darwin, he was not a member of the leisure class, a gentleman explorer like Alexander von Humboldt or Charles Darwin. Wallace actually made a living selling butterfly collections, stuffed birds, etc., to gentlemen back in England. Before that he had been a surveyor. In 1858, Darwin received an essay from Wallace, *On the Tendency of Varieties to Depart Indefinitely from the Original Type,* which contained exactly the same theory as that of Darwin. Darwin proposed to step aside, but several of his scientific friends persuaded him to prepare a brief extract of his materials. Darwin's extract and Wallace's essay were published together in the journal of the Linnaean Society. Darwin then condensed his projected several volume work into one manuscript, which was published in 1859 under the title *The Origin of Species.* The means or mechanism of speciation (species formation) that Darwin and Wallace proposed was *natural selection,* or "the preservation of favoured races in the struggle for life." There were several ideas and bodies of fact which led inevitably to the synthesis of organic evolution.

---

[5]Ibid., p. 80.
[6]Loren Eiseley, *Darwin's Century* (Garden City, N. Y.: Doubleday, 1958), p. 378.

Geology was reborn in Darwin's youth. We have seen that he was an avid geologist. On the 1831 voyage, one of the few books that Darwin took with him was the first volume of Lyell's *Principles of Geology,* and he received the second volume a year later along the coast of South America. The theory of the perpetual evolution of the Earth's surface, first proposed by the Scotsman James Hutton in 1785 and reintroduced by Charles Lyell in 1831, opposed the theory of catastrophism and the naïve concept of an Earth literally as old as the few biblical generations, no more than a few thousand years. The theory of Hutton and Lyell was called uniformitarianism, since it was assumed that the events in the geologic history of the Earth were the product of the same natural forces active today. The same agents that wear away mountains today were said to be operative in the past. Wind and sand, ice and stream, made sediments of mountains, and rivers carried the sediments to the sea floor, where they could be pressed into stone and changed by heat into metamorphic rocks that could be thrust up again with the volcanic eruptions of the ocean floor. The sea erodes the shore; one continent falls, another rises. The changes are so gradual as to be hardly recognized by a man in his brief lifetime. Because of the nearly imperceptible nature of its change, the theory of uniformitarianism has also been called gradualism. Hutton saw "no vestige of a beginning,—no prospect of an end." He felt that the Earth was as "eternal as the heavens." Where the Earth's crust had buckled and upthrust or been cut by stream or shovel, innumerable layers or strata were exposed. Nearly fathomless time was revealed by examining the strata and estimating the eons required for their formation. The Earth was millions of years old at least. It wasn't until the twentieth century that the age of the Earth was established at several billion years, and then even the timeless conception of the universe was abolished. With modifications, uniformitarianism has become the geologic viewpoint of the twentieth century. What does all of this have to do with Darwin and biologic evolution?

Darwin was as much a "professional" geologist as he was a "professional" biologist. As a student of Lyell, Darwin was well aware of this theory of geologic evolution, which not only paralleled bioevolution, but provided a time scale adequate for gradual organic evolution. If the Earth were only a few thousand years old, then evolution would be improbable and the special creation of all the variations of life likely. Just as the Earth did not change before a man's eyes, neither did the species of life. Wallace, too, had read Lyell's *Principles of Geology.*

The second great principle was the concept of *gradation* in living forms. The idea of a scale of natural perfection, the *scala naturae,* dates back to the ancients. The Greeks and their subsequent worshipers of the Middle Ages recognized that animals could be arranged along a ladder proceeding from the lowest (worms) to the highest (man). The ladder was static. There was no climbing from one rung to the next with one notable and crucial exception: it was believed that the lowest forms were spontaneously generated, that is, arose from nonliving matter. It was not until late in the nineteenth century that Pasteur disproved the myth that contemporary microscopic plants and animals were spontaneously generated. Nevertheless, students of the plant and animal kingdoms saw that organisms could be arranged according to their grades of organization. There was a gradation from simple to complex.

William "Strata" Smith (1769–1839), an English canal engineer, was primarily responsible for recognizing that each of the different layers of rock or strata had its own characteristic types of fossils and that the lower the layer the less its fossils resembled living forms. The shallower the stratum, the more closely its fossils looked like living forms. Vertebrate fossils were found only in the highest strata. Fossils of the invertebrates ("lower animals") dominated the lower strata. The examination of the fossil record reveals the extinction of species, whose fossils are absent in the higher layers, and the appearance of new species, whose fossils are missing in lower strata. Thus, there is a geologic succession of the forms of life.

The creationists held that the progression evident in the fossil record could be explained by assuming that a succession of more advanced creations had followed each catastrophic geologic disturbance. This doctrine did not imply descent from one form to another. Man was believed to be the goal of the process.

Darwin saw "descent with modification," evolution, in the geologic succession and the living scale of organization:

We can understand how it is that all forms of life ancient and recent, make together a few grand classes. We can understand, from the continued tendency to divergence of character, why the more ancient a form is, the more it generally differs from those now living; why ancient and extinct forms often tend to fill up gaps between existing forms, sometimes blending two groups, previously classed as distinct, into one; but more commonly bringing them only a little closer together. The more ancient a form is, the more often it stands in some degree intermediate between groups now distinct; for the more ancient a form is, the more nearly it will be related to, and consequently resemble, the common progenitor of groups, since become widely divergent. Extinct forms are seldom directly intermediate between existing forms; but are intermediate only by a long and circuitous course through other extinct and different forms. We can clearly see why the organic remains of closely consecutive formations are closely allied; for they are closely linked

together by generation. We can clearly see why the remains of an inter-
mediate formation are intermediate in character.

The inhabitants of the world at each successive period in its history
have beaten their predecessors in the race for life, and are, in so far,
higher in the scale, and their structure has become generally more spe-
cialised; and this may account for the common belief held by so many
palaentologists, that organisation on the whole has progressed.[7]

Darwin thus recognized that a true classification of the forms of life
would be an evolutionary history including extinct as well as living organisms. He saw in the fossil record a continuous story of gradual
change, yet abruptly punctuated with extinctions in many places.
Throughout the *Origin,* Darwin speaks of the "great principle of gradation." This idea undoubtedly formed one of the bases of his theory.

A third thought that entered into the making of Darwinism was
Malthus' *principle of population.* Thomas Malthus was a peculiar blend
of clergyman and economic theorist, who became amazingly influential
in political economics and unintentionally in biology. David Ricardo,
the "champion of rising capitalism," completely agreed with what Malthus had to say about population and so did Charles Darwin and Alfred
Wallace. Malthus' thesis, the "Malthusian Doctrine," also came to be
called the "dismal theorem." In his "Essay on the Principle of Population," he stated that there is a "constant tendency in all animated life to
increase beyond the nourishment prepared for it." The population always outruns the food supply. The checks on population "which repress
the superior power of population, and keep its effects on a level with the
means of subsistence, are all resolvable into moral restraint, vice, and
misery." Malthus' "positive checks," which had operated throughout
history were war, disease, infanticide, poverty, and famine, the four
horsemen of the Apocalypse and more. The minister believed that the
inevitable lot of mankind was misery. He neither advocated vice nor
condoned misery, but held them inescapable, since man would not
practice "moral restraint" (sexual abstinence). The essay was almost a
sermon. It has been reprinted many times, especially today, when
human population looms as the primary problem of civilization.

In his autobiography, Darwin relates the following:

In October 1838, that is, fifteen months after I had begun my system-
atic enquiry, I happened to read for amusement Malthus on Population,
and being well prepared to appreciate the struggle for existence which
everywhere goes on from long-continued observation of the habits of

---

[7] Charles Darwin, *The Origin of Species* (New York, N. Y.: Crowell-Collier Publishing Company, 1963), pp. 362–363.

animals and plants, it at once struck me that under these circumstances favourable variations would tend to be preserved, and unfavourable ones to be destroyed. The result of this would be the formation of new species. Here, then, I had at last got a theory by which to work. . . .[8]

In the *Origin,* Darwin wrote of the struggle for existence: "It is the doctrine of Malthus applied with manifold force to the whole animal and vegetable kingdoms; for in this case there can be no artificial increase of food, and no prudential restraint from marriage." Wallace said that the reading of Malthus catalyzed his essay on natural selection. Thus, both Darwin and Wallace attributed their insight into the struggle for existence to Malthus. Darwin spoke of Malthus as a "great philosopher."

The ability of man to modify wild animals and plants by selective breeding to produce the amazing variety of domestic plant and animal races, strains, or breeds, deeply impressed Darwin. From this *artificial selection* Darwin saw the possibility of a parallel natural selection. Of course, he realized that the variations or traits valued by man were not the same that nature selected, and the rate at which nature produced new races was much slower than the human process. Nevertheless, Darwin saw that the "hand of nature" was like the hand of man. The title of the first chapter in the *Origin of Species* is "Variation under Domestication." One need only contemplate the varieties of domestic dogs, all derived from a few wild types, to appreciate the potential of selection, artificial in this case, since many of the bizarre domestic breeds are not fit to survive in the "state of nature."

Let us summarize the principles that stimulated and contributed to the evolutionary theory of Darwin and Wallace, in the order in which they were presented:

1. *geologic gradualism*
2. the *gradation* of living and fossil forms
3. the *principle of population*
4. *artificial selection*

The evolutionary mechanism in Darwin's theory is natural selection. A brief outline of it follows.

1. More individuals are born in each generation than can possibly survive, thus "precipitating the struggle for existence."
2. There is variation within the species.
3. Individuals with favorable variations have a greater chance of surviving to reproduce.

---

[8]Charles Darwin, *The Autobiography of Charles Darwin* (New York, N. Y.: W. W. Norton & Company, Inc., 1969), p. 120.

4. If the variations are heritable, they will be passed on to the off-spring.

5. In this way, the characteristics of the descendants will gradually depart from the ancestral type, forming first subspecies, and finally new species.

Overpopulation and its consequent *competition* lead to a *differential survival* (differential reproduction) of individuals in the spectrum of *intraspecific variation,* which eventually culminates in *speciation,* the "origin of species." Although Darwin and Wallace were not the first to say that evolution had occurred, they were the first to propose the natural selective mechanism of evolution. With few modifications, it remains the modern theory.

Darwin was entirely ignorant of genetics, and so he was incapable of distinguishing between acquired and heritable variations, the same failing as Lamarck's. It has been noted that Darwin's "genetics" was essentially the same as that of the ancient Greeks. Fortunately, he was unaware of Mendel's work. The sharply contrasting characters of Mendelian inheritance might well have shaken his belief in the gradual, continuous nature of evolution. Darwin referred to Lamarck as a "justly celebrated naturalist" in the historical sketch of evolutionary thought that precedes the introduction to the *Origin.* In numerous passages of the *Origin,* Darwin employs the Lamarckian concepts of "use and disuse" and "habit," as evidenced in the following paragraph of the first chapter:

Changed habits produce an inherited effect, as in the period of the flowering of plants when transported from one climate to another. With animals the increased use or disuse of parts has had a more marked influence; thus I find in the domestic duck that the bones of the wing weigh less and the bones of the leg more, in proportion to the whole skeleton, than do the same bones in the wild-duck; and this change may be safely attributed to the domestic duck flying much less and walking more, than its wild parents. The great and inherited development of the udders in cows and goats in countries where they are habitually milked, in comparison with these organs in other countries, is probably another instance of the effects of use. Not one of our domestic animals can be named which has not in some country drooping ears; and the view which has been suggested that the drooping is due to the disuse of the muscles of the ear, from the animals being seldom much alarmed, seems probable.[9]

Yet, on the very next page, Darwin writes: "Any variation which is not inherited is unimportant for us." Darwin's lack of genetic knowledge is

---

[9] Charles Darwin, *The Origin of Species* (New York, N. Y.: Crowell-Collier Publishing Company, 1962), p. 33.

completely forgivable. As he himself wrote, "The laws governing inheritance are for the most part unknown." They were then.

With the development of genetics, it was not long before the species was redefined genetically and natural selection reinterpreted in terms of changing gene frequencies. The fusion of Darwin's theory (natural selection) and genetics became known as the *modern synthesis,* and it is the current theory of evolution. There was no reason to discard the mechanism of natural selection just because Darwin was ignorant of genetics.

THE SPECIES UNDEFINED

Dictionaries still define a species as "a single, distinct kind of plant or animal, having certain distinguishing characteristics: a category of biological classification." The word species is derived from Latin and can be translated as "a seeing," "an appearance." The species were first classified according to appearance. Members of the same species look alike. As a group, they have an astronomical number of common characteristics. To the naked eye, these shared attributes take the form of *homologous structures,* which are similar in anatomic detail and development. Comparative anatomy is the study of homologous structures, and it forms one of the foundations of evolution. Genetically speaking, homologous structures are phenotypic expressions of a common genotype. Possession of common genes is the ultimate criterion of kinship.

The father of modern biologic classification, Carolus Linnaeus, was the first influential figure who systematically gave each organism two names, the *genus* and *species* names. This system of naming is called *binomial nomenclature,* the assignment of two names. Linnaeus seems to have believed in the unchangeability of the species. Nine of the ten editions of his *System of Nature* (1735) contained the statement: "There are no new species." To paraphrase in terms of special creation: all species are old species, present since the Genesis. Notably, this statement was omitted from the last edition of the *System.* Perhaps, by then, Linnaeus had come to doubt the immutability of the species. Elsewhere in his writings, Linnaeus speaks of "freaks" or "sports of nature" that departed considerably from the ideal type. Darwin and Wallace were well aware of intraspecific variation. Wallace wrote, "When a sufficient number of individuals are examined, variations of any required kind can always be met with." Herein lay the "species problem." The species was not an absolute type. There was wide variation within it. A typical member of a species represented the average.

The English clergyman and biologist John Ray (1627–1705) was the first to give a breeding definition to the species. He emphasized the interfertility or inbreeding of members of the same species and the intersterility or failure of breeding between members of different species. This definition is also far from absolute. The horse and the mule belong to the same genus but different species. By the preceding definition they should be intersterile, yet a horse mare and a donkey stud will successfully mate. Their interspecific hybrid offspring, the mule, is sterile, since the horse and donkey chromosome sets are not compatible in sex cell formation. Mule sex cells are not viable. The mule is a domestic creation. In the natural state, horses and donkeys virtually never interbreed. Neither do the horse and the zebra (a type of horse) normally interbreed. Wild burros frequently run in mixed herds with mustangs, yet they do not interbreed. A horse mare in heat will choose a horse stallion every time she has a choice. The interbreeding of close species is often unsuccessful: most of the time hybrid offspring do not survive; if they do, they are frequently sterile. Interspecific hybrids are rare in nature; intergeneric hybrids (crosses between genera) are even rarer. The members of a species have a high degree of species recognition. The various patterns of coloration and shape as well as behavior all serve as a means of species identification. The confused often fail to reproduce: they cast their seed upon the waters. It is the reproductively successful that inherit the earth, persist as a species in time.

Yet all the species of ducks are potentially interfertile, although in a state of nature they remain actually separate. Interspecific hybrids and even intergeneric hybrids have been recovered from natural communities. Thus, the interbreeding definition of the species is no more absolute than that of type.

It has been noted that homologous structures denote common genes. Phenotype is often but not always an accurate reflection of genotype. The possession of common genes is the final criterion of relationships: organisms with genes in common are related by "blood," descended from a common ancestor. The members of the same species have a myriad of genes in common, the basis of their numerous common characteristics. The species has been defined as a natural biologic unit, whose members share a common gene pool and interbreed. Yet, the genetic composition of widely separated populations of a species may differ considerably. Only a few species are composed of single populations. The *endemic* species, which are found in only one place on the earth's surface, are the only species that satisfy all three definitions of the species, morphologic, interbreeding, and genetic. Most species consist in many populations distributed over as wide a geographic range as several continents in several cases and differing in genetic composition and appearance at least slightly. The greatest uniformity is found within a single breeding population. Therefore, a single population species is most specific genetically. Only a single population species can be defined as a collection of genes, a *gene pool*. So, the genetic definition of the species is no more satisfactory than the morphologic or breeding. All agree that the genetic population is the unit of evolution, but a population does not a species make.

Perhaps "species" is an obsolete term. Only the exception, the single population species, satisfies all three of the attempted definitions. We should not be disappointed: the species is of but a single time and place. It has no absolute definition. The species is mutable, or changeable. If we must define the species, then we can temporarily say that the members of the same species are anatomically and genetically alike and actually interbreeding. In centuries, they may not be any of these.

A species can be divided into subspecies, varieties, or races. The species, like life itself, is plastic, capable of adaptive response. Therefore, it finally eludes our absolute defining attempts. In terms of geologic time, the species is ephemeral, transient, short lived, appearing, and disappearing. Less than a handful of species have remained unchanged over millions of years. The species has been—like many concepts of man—imperfect. The species originates in a place where the parent population remains, daughter populations migrate, fragment into new populations, which when separated subspeciate at first, and finally transform into new species. The species is but a tiny segment of the many curves of life. Variation, genetic and morphologic, is maintained within the species, even within its populations. Variation is the evolutionary wealth, necessity of the species. As many seemingly cynical philosophers have said, "The only constant is change itself." The branches of the tree of life bend in the winds of time and they never straighten. Some break. Many species have become extinct. Others have been transformed into new species. An individual organism is capable of limited adaptive response to small changes in the environment. Over the generations, life responds considerably to larger environmental changes. Evolution is the adaptive response of generations. Not infrequently, the identity of the species is altered in the process. Some long-lived species have persisted apparently unchanged in changed environments. A species that does not change in a changing environment is usually fated to extinction. Perhaps you can see what Darwin and others meant by the "species problem." Nature is continually at work "undefining the species," and we call the process evolution.

THE CIRCLE OF THE RACES

Populations of a species distributed over large geographic regions show a continuous gradation of differences. Adjacent, or neighboring populations are most alike, and widely separated populations differ most.

The common grass or leopard frog (*Rana pipiens*) is the most widely distributed frog in North America, ranging from Canada to Mexico and beyond. Populations from Florida and Georgia successfully interbreed, as do frogs from Maryland and North Carolina. However, frogs from Florida crossed with frogs from Vermont produce deformed, unviable offspring. Interbreeding fails even though frogs from New England and the deep South are morphologically similar enough to be placed in the same species. The gene pools of the populations at the extremities of the range have departed from each other to such a degree that successful interbreeding is no longer possible. Two of the three definitions of the species have been undone. Only the anatomic definition remains semi-valid. The members of the two widely separated populations retain many morphologic and genetic similarities as the result of gene flow north and south through the intervening populations, which thus serve as a "genetic bridge" between the extremes. If the connecting populations were wiped out, then the resultant, geographically isolated populations (which are already reproductively isolated) could drift apart genetically and morphologically until they would be recognizably different species. Several authorities have already suggested that more subspecies of *Rana pipiens* should be recognized.

The populations of a species covering a large area often show minor form differences. Such species are said to be *polytypic* ("many types"), and the different forms are classified as subspecies or races. Even the morphologic unity of the populations of these species is breaking down. The graded series of morphologic or physiologic (usually both) variations exhibited by a species' populations along a line of geographic or environmental transition is called a *cline*.

In some instances, the populations of a polytypic species are arranged in the form of a circle, a *rassenkreis* ("circle of race"). The common harbor gull, *Larus argentatus,* forms a rassenkreis encircling the lands around the north pole. Ornithologists believe that the original home of this species was eastern Siberia, west of Alaska. Let us call this original subspecies *A*. Races *B*, *C*, and *D* are distributed eastward across North America into Greenland. Populations of races *E* and *F* are scattered westward across Siberia into northern Europe and Scandinavia. European subspecies interbreed with western Siberian races, and these last with parent population *A*. Also, *B* interbreeds with *C*, and *C* with *D*; *A* and *B* cross successfully. North American and western European races both occur in the British Isles and along coastal Europe, but they do not interbreed. The circle is broken at this point. The genes flow westward and eastward, clockwise and counterclockwise, back and forth around the circle's circumference except where the American and European races meet. If the species were only defined in interbreeding populations, then the European and American races would seem to be separate species, but the Siberian race would belong to both. If the Siberian populations were wiped out, the genetic connections between the European and American races would be destroyed, and they could drift further apart. A rassenkreis is like a cline whose extremities are drawn together in the process of opposite migrations from the parent populations. In the case of the harbor gull, the European and American races are new neighbors and recent races, both equally descended from the theoretical parent population of Siberia. Supposedly, the circle was drawn by two lines proceeding in opposite directions from eastern Siberia.

One further example of a rassenkreis may be cited. The Buckeye butterfly, *Junonia lavinia* (*Precis lavinia*), is widely distributed from Florida along the Gulf Coast into Mexico and Central America, down into South America and up through the West Indies of the Caribbean (see Fig. 6-4). The populations gradually change as one proceeds around the ring, but neighboring populations are similar and interbreeding. This ring of races is broken in Cuba, where butterflies like those from Florida coexist without interbreeding with Buckeye races like those to the south. In Cuba, the two populations behave like distinct species, yet there is no one place around the rest of the ring where it is clear that one species ends and the next begins. This rassenkreis, too, must reflect the routes of dispersal of the populations of the parent species. Extremities of distribution have finally overlapped without interbreeding.

The rassenkreis and the cline demonstrate how an originating species disperses and subspeciates, finally to the point of intersterility, which is the verge of speciation. They represent the geographic or environmental (ecologic) anatomy of a splitting species. These subspecific circular and linear gradations of the species have evolved over thousands of years.

Over short periods of time, geologically speaking (centuries), the characteristics and frequencies of genes in a population remain constant. Over millenia (thousands of years), the genes of a population change. This is the essence of evolution, and the mechanisms of change are the methods of evolution.

There is a law of population genetics that is, in a sense, the diametric opposite of evolution. The *Hardy-Weinberg law* states that if (1) there is an equal survival rate (no natural selection), (2) the rate of mutation

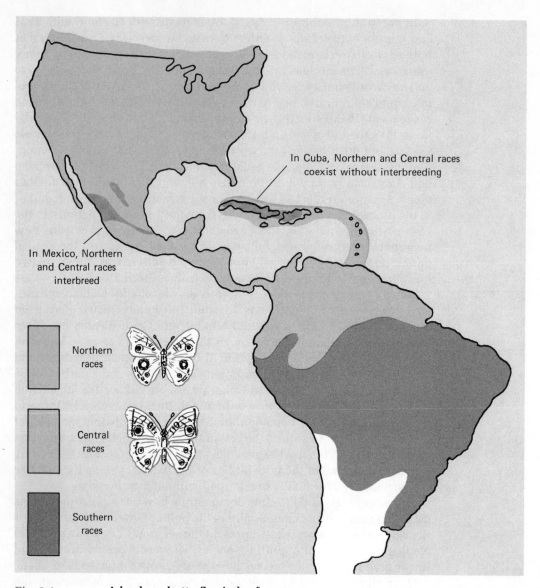

In Cuba, Northern and Central races coexist without interbreeding

In Mexico, Northern and Central races interbreed

Northern races

Central races

Southern races

Fig. 6-4   **A buckeye butterfly circle of races.**
From D. J. Merrell, *Evolution and Genetics,* Holt, Rinehart, & Winston, Inc., New York, 1962.

is zero or nearly so, (3) the population is large (4) with purely random mating and (5) no migration into or out of the population, and (6) sex cells carrying the two forms of a gene are produced in equal numbers, then the relative frequencies of genes in a population tend to remain constant from generation to generation. These conditions describe a genetic equilibrium. All prerequisites of the Hardy-Weinberg law are the opposites of the mechanisms of evolution. The Hardy-Weinberg equilibrium is true over short periods of time, but evolution eventually overthrows it. In calculus, smaller and smaller portions of a gradual curve are assumed to be more and more nearly a straight line for the purposes of triangular solution. So the Hardy-Weinberg equilibrium is an instantaneous value of the lengthy evolutionary curve. It is an approximation, valid for a short distance.

THE MODERN SYNTHESIS

The modern view of evolution attributes evolutionary change to four major causes: mutation and sexual reproduction, natural selection, genetic drift, and isolation. Mutation and the new genetic combinations produced in sexual reproduction are the source of variety. For a variation to have evolutionary possibilities, it must have a genetic basis. Nature can only select from genes in the existing pools. Natural selection alters the frequencies of genes in the genetic pools of populations. The frequencies of genes determining desirable traits are raised, whereas the incidence of genes underlying less favorable characteristics tends to decrease. Certain genetic combinations are promoted over others. Once the average genotype of a population departs far enough from the frequency of genes in the parent pool, interbreeding between ancestral and descendant populations may fail, and then there are two species where there was one.

Speciation seldom occurs in adjacent populations. As long as the members of the various populations of a species live in similar environments, actually interbreed successfully, and exchange genes, they remain essentially alike. Occasionally, the old species is transformed into the new in time. In other cases, the parent populations of the species remain unchanged in the old environments while new races and finally species arise from migrants in new environments. When individuals of closely related but different populations fail to interbreed, they are said to be *reproductively isolated,* and the evolutionary paths of these groups will not cross again. For all practical purposes, they have become separate species. Populations can be separated in time and space. Obviously, contemporary populations of a species cannot interbreed with their ancestors. Historic and current populations are reproductively isolated in time.

Almost all evidence suggests that spatial or *geographic isolation* precedes actual reproductive isolation and speciation. Two recently separated populations are effectively, reproductively isolated, but potentially interfertile or interbreeding. It has been suggested that reproductive isolation results from the genetic divergence that occurs during the origin of subspecies and species in populations spatially isolated from one another. The reproductive isolation of populations leads to further genetic departure. It is extremely difficult to see how speciation could occur without isolation.

Others have proposed that reproductive isolation is the result of nature's tendency to eliminate hybrids themselves as well as select against genes favoring hybridization. We have seen that many hybrids are inviable, sterile, or untrue-breeding. The survival of crossbreeders is thus imperiled. Natural selection would favor those organisms that breed only with nearly identical types. In this way, natural selection would tend to "purify" the genotypes of subspecies to the point of speciation.

Small migrant populations most often represent nonrandom samples of the large parent population. They are sampling errors. Some genes may be missing from the pool of the small population that are present in the large, and the frequencies of other genes in the small population's genetic pool may well differ from the proportions of genes found in the parent population. These random changes in gene frequencies resulting from sampling errors are said to be due to *genetic drift.*

Sometimes, a population is "forced through a small bottleneck," or selected on the basis of a single criterion rather than many. A large population of ducks can be nearly wiped out in a viral epidemic. The few surviving by chance are not necessarily average representatives of the former population. Perhaps, the survivors are atypical in that they possess a native immunity to the viral disease. They can be inferior to the average in many other respects. The "bottleneck" effect is a type of genetic drift.

SEXUAL SELECTION

Darwin recognized a third type of selection, sexual selection, in addition to natural and artificial:

Inasmuch as peculiarities often appear under domestication in one sex and become hereditarily attached to that sex, so no doubt it will be under nature. Thus it is rendered possible for the two sexes to be modified through natural selection in relation to different habits of life, as is sometimes the case; or for one sex to be modified in relation to the other sex, as commonly occurs. This leads me to say a few words on what I have called Sexual Selection. This form of selection depends, not on a struggle for existence in relation to other organic beings or to external conditions, but on a struggle between the individuals of one sex, generally the males, for the possession of the other sex. The result is not death to the competitor, but few or no offspring. Generally, the most vigorous males, those which are best fitted for their places in nature, will leave most progeny. But in many cases, victory depends not so much on general vigor, as on having special weapons, confined to the male sex. A hornless stag or spurless cock would have a poor chance of leaving numerous offspring. Sexual selection, by always allowing the victor to breed, might surely give indomitable courage, length to the spur, and strength to the wing to strike in the spurred leg, in nearly the same manner as does the brutal cockfighter by the careful selection of his best cocks. How low in the scale of nature the law of battle descends, I know not; male alligators have been described as fighting, bellowing, and whirling around, like Indians in a war-dance, for the possession of females; male salmons have been observed fighting all day long; male stag-beetles sometimes bear wounds from the huge mandibles of other males; the males of certain hymenopterous insects have been frequently seen by that inimitable observer M. Fabre, fighting for a particular female who sits by, an apparently unconcerned beholder of the struggle, and then retires with the conqueror. The war is, perhaps, severest between males of polygamous animals, and these seem oftenest provided with special weapons. The males of carnivorous animals are already well armed; though to them and others, special means of defence may be given through means of sexual selection, as the mane of the lion, and the hooked jaw to the male salmon; for the shield may be as important for the victory, as the sword or spear.

Amongst birds, the contest is often of a more peaceful character. All those who have attended to the subject, believe that there is the severest rivalry between the males of many species to attract, by singing, the females. The rock-thrush of Guiana, birds of paradise, and some others, congregate; and successive males display with the most elaborate care, and show off in the best manner, their gorgeous plumage; they likewise perform strange antics before the females, which, standing by as spectators, at last choose the most attractive partner.[10]

More shall be said about these matters in the chapter on animal behavior (Chapter 8).

---

[10] Ibid., pp. 97–99.

Darwin's work on the development of man (*The Descent of Man and Selection in Relation to Sex*) was more devoted to sexual selection than human evolution. His hypotheses in this area were largely ignored or rejected until recently, since it was thought that sexual selection was but one aspect of natural selection. Darwin proposed that sexual selection came about in two ways: male competition and female choice. The idea of female choice has been severely criticized as being too anthropomorphic (man-centered). Nevertheless, the idea of sexual selection has risen in recent times as the only explanation of the behavior and characteristics of such widespread species as jewelfish, American chameleons, prairie chickens, sage grouse, birds of paradise, antelopes, wolves, and baboons. In these cases, the female chooses her mate from a number of males often according to rank established by male combat. Males are selected by females on the basis of size, strength, vigor of display, and vividness of secondary sexual characteristics.

The traits produced by sexual selection are not always those that would be favored by natural selection. In fact, the course of sexual selection seems in some cases to be the exact opposite of that which would have been followed according to the dictates of natural selection. It is easy to see how size and strength might be favored by natural selection in certain situations, but the brilliant plumage of many sexually competitive birds is anything but camouflaging coloration, the type of pigmentation that natural selection would logically favor. Sexual selection has bizarrely modified the males of several species, enlarging some secondary sexual characteristics to cumbersome extremes. There is evidence that large male lions are at least partially dependent on the smaller, more agile females to obtain food; most of the kills of a pride are made by the lionesses. The huge antlers of the extinct Irish elk must have been produced by sexual selection. It would seem that any trait that enhanced reproduction would be adaptive, yet in these cases sexual selection seems to have operated not only independently of natural selection but even in contradistinction to it. It has been proposed that sexual selection has speeded up the course of evolution in several animals. We shall explore this topic further in Chapter 8.

A BRIEF HISTORY OF LIFE ON EARTH

Perhaps the most disappointing thing about the birth of life on Earth was the lack of drama attending it. There might have been flashes of lightning, but all else was at the submicroscopic level of molecular reactions.

Earth is about 4.5 billion years old. This age has been established through radioactive dating. Of the naturally occurring radioactive elements, uranium is one of the longest lived. In its radioactive decay it is converted to lead. The half-life of uranium is 4.5 billion years; in that time, half a sample of uranium is converted to lead. By establishing the ratio of uranium to lead in the oldest rocks, geologists have arrived at a figure for the age of Earth that lies between 4 and 5 billion years. What is the ratio of uranium to lead in the oldest rocks?

Life has been present for probably less than half Earth's lifetime, no more than 2 billion years. The fiery conditions of newborn Earth made life impossible, but, not long after the surface had cooled and the oceans had formed, the first life appeared.

It is assumed that the primitive atmosphere of Earth contained methane (CH_4), ammonia (NH_3), hydrogen (H_2), and water vapor. The spectroscopic analysis of the atmospheres of other planets, such as Jupiter, have revealed such a gaseous composition. Free oxygen is assumed to have been absent. Most of this element was either in water molecules or bound up with the metals, iron, silicon, and aluminum, in the form of oxides in the earth's crust. The oxygen present in the atmosphere today exists in two forms: O_2 and O_3, ozone. At present, ozone filters out most of the high-energy, short-wavelength ultraviolet from the sunlight. Only low-energy, long-wavelength ultraviolet reaches the earth's surface today. In the past, it is assumed that high-energy ultraviolet continually poured down through the primitive atmosphere onto the surface of the earth and its oceans. It is believed that ultraviolet (UV) served as the primary source of energy of activation for the reactions that resulted in the formation of the smaller biochemical molecules. The maximum absorption band of methane lies in the UV range. The absorption of energy wavelengths by molecules raises their electrons to higher energy levels, increasing the probability of chemical reaction.

In 1952, Stanley Miller, a student of Harold C. Urey, Nobel chemist, circulated a primitive atmosphere in a closed system past an electric discharge, simulating lightning. At the end of a few days, several amino acids were detected in the condensate (Fig. 6-5). To be sure, it is a long way from amino acids to proteins, as many skeptics of the importance of this experiment pointed out, but it was the first experiment of its kind that showed definitely that some of the most important "small molecules of the living machine" could be formed in the absence of life, abiogenetically. There are those who favor the hypothesis that the highly reactive hydrogen cyanide (HCN) was also a component of the primitive atmosphere. Using various energy sources, such as UV, radiation, and heat, in addition to electric discharge, investigators have synthesized a number of simple "biochemicals" from primitive atmospheres. The nitrogenous

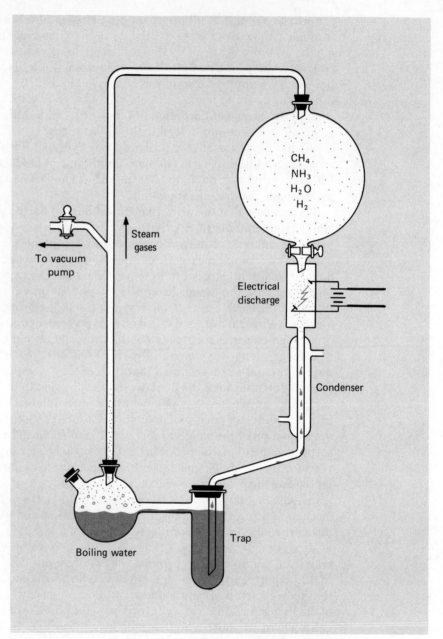

To vacuum pump

Steam gases

CH₄
NH₃
H₂O
H₂

Electrical discharge

Condenser

Boiling water

Trap

Fig. 6-5   **Urey-Miller experiment, which simulated conditions on earth before the appearance of life. Amino acids were found in the condensed vapor in the trap after several days.**

From G. B. Moment, *General Zoology*, Houghton Mifflin Company, Boston, 1967.

bases of the nucleic acids, short-chain fatty acids, and simple sugars have been prepared by these methods, in addition to amino acids. The results of these experiments leave little doubt that the simple molecules of the living system could have been formed under primitive Earth conditions. It thus seems likely that a long period of molecular or "prebiochemical" evolution preceded the birth of life.

The next step would logically have been the formation of large molecules, such as proteins and nucleic acids. It has been experimentally shown that some amino acids combine when heated dry to form proteinlike molecules. Organic substances, such as amino acids, tend to aggregate. This process might have been a means of concentrating these substances in the primitive ocean, which has been described as a "hot, dilute soup." The polymerization of monomers, such as nucleotides and amino acids, to form nucleic acids and proteins was a crucial step in the generation of life. A hotly debated question concerns whether proteins or nucleic acids came first. Nucleic acids are found in all forms of life, but so are proteins. The very same year that Urey and Miller reported their abiotic ("without life") synthesis of amino acids, the building blocks of protein, Watson and Crick proposed the double helical model for DNA. It was then shown how the nucleic acids (DNA and RNA) directed protein synthesis. The eyes of the world, so long focused on the proteins as the ultimate "stuffs of life," shifted to the nucleic acids. Yet enzymatic proteins catalyze DNA and RNA synthesis. The question of the priority of nucleic acids and proteins may never be answered; there are such questions.

The simplest forms of life are the viruses, which contain both protein and nucleic acid. All known viruses are obligatory parasites and display the minimum property of life (reproduction) only within other cells. Viruses have been called "naked genes" as well as "naked chromosomes," since they are analogous and somewhat homologous to these nuclear parts of the cell. It has been proposed that the first forms of life were "free-living viruses," capable of independent reproduction. It has been hypothesized that there may be free-living viruses in existence today unknown to us, since the only way that viruses come to our attention is through their ability to cause diseases.

Perhaps there were free nucleic acids "in the beginning," that could duplicate themselves, or carry out reproduction, which most agree is the primary property of life. According to this view, then, life began when nucleic acid molecules acquired the ability to catalyze the reaction that resulted in the duplication of themselves by utilizing simpler compounds (base nucleotides) from the surrounding sea. At this point, natural molecular selection might have begun: those molecules most efficient in autoduplication would come to dominate the molecular species.

Somewhere along the line, the nucleic acids came to direct protein synthesis, and the association between nucleic acids and proteins began. We could call the free-living, nucleoprotein virus a "protochromosome."

The formation of protective membranes about aggregates of protochromosomes must have been the next step in the evolution of the cell. The evolution of the membrane system and of the energy-transforming organelles has been discussed thoroughly in Chapter 2. It is interesting to speculate that life began to control its immediate environment very early. Compartmentalization of the cellular contents from the haphazard, hazardous external environment must have been essential to the achievement of a steady state.

The next step must have involved the attainment of multicellularity, as the ancient protozoa and protophyta, the unicellular animals and plants, respectively, gave rise to intermediates, and finally, the metazoa and metaphyta, the advanced, multicellular animals and plants. Figure 6-6 summarizes the most important events in the early history of life.

The seas of 500 million years ago (the Cambrian period) already contained the majority of the animal phyla (Fig. 6-7). Most of plant evolution was yet to come. Vertebrate fossils have never been found in Cambrian strata. There was probably no life on the land at this time. The barren landscapes must have been scenes of terrific erosion with scattered boulders, gravel beds, and sand and dust storms, such as occur on Mars today.

The first organisms to permanently invade the land were plants. The early land plants differed from their lower aquatic plant (algal) ancestors in that their exposed portions came to be covered by waxy *cuticles,* retarding evaporation from the deeper tissues. It seems that two different lines of land plants appeared in the transition from aquatic to terrestrial habitat. In one, the mosses and their relatives, no vascular (conducting) or supportive tissue developed. All higher land plants contain considerable vascular tissue. The mosses are apparently a blind alley in plant evolution, not giving rise to any higher forms. The second line of land plants evolved *vascular tissue* and presumably gave rise to the higher plants (see Fig. 6-8).

Although complex marine algae, such as the kelps, have developed bladelike "leaves," tubelike "stems," and nubby-fingered holdfast devices, these are not considered to be *true* leaves, stems, and roots, since botanists reserve those terms for vascularized portions of land plant anatomy. The lower aquatic plants have no need of vascular tissue, since they are bathed by a rich and various salt solution. Neither do they require supportive tissue, since they are buoyed up by the water. Some kelps reach a length of 150 ft or more.

As the land plants rose up from the earth in competition for sunlight,

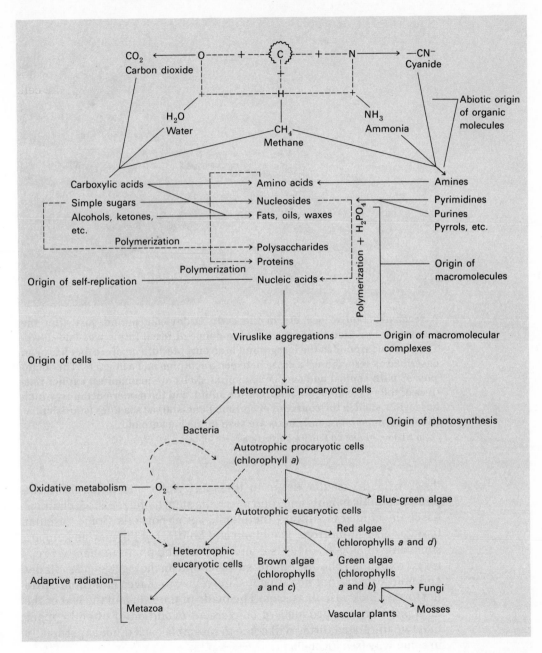

Fig. 6-6   **Summary of important stages in history of life: prebiochemical, biochemical, protocellular, unicellular and multicellular.**
Modified from N. M. Jessop, *Biosphere*, Prentice-Hall, Inc., Englewood Cliffs, N. J., 1970.

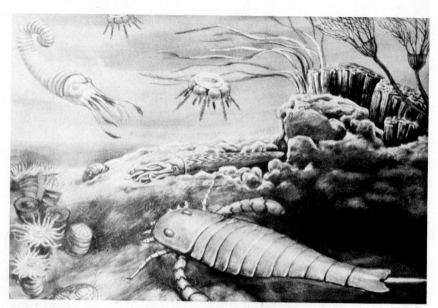

Fig. 6-7   **A reconstruction of sea life in the early Ordovician period, just after the Cambrian, in which the first great abundance of fossil forms was laid down. The giant arthropod in the foreground is an eurypterid, a creature that has been described as resembling "a cross between a scorpion and a lobster." The "octopuses" with conical and semicoiled spiral shells are ammonites, extinct relatives of today's chambered nautilus, the squid, and the modern octopus, which has lost its shell in the course of evolution. Long-stalked sea lilies (starfish relatives), sea anemones, and corals are seen in the background.**
From *History of Life on Earth Charts,* Denoyer-Geppert, Co., Chicago.

they evolved vascular tissue to transport water and minerals to the photosynthetic heights and food down the stem to the roots. Mechanical tissue developed to support the plants' aerial portions. Some vascular tissue is also supportive. The blade and holdfast of the marine kelp are somewhat analogous to the leaf and roots of land plants in that they represent expanded photosynthetic surfaces and anchorage devices. Blades and holdfasts are not homologous, however, to leaves and roots, since the former lack vascular tissue. The blade of the kelp and the leaf of the land plant are the products of *convergent evolution.* It often happens that two unrelated forms will come to resemble each other superficially in some way (see Fig. 6-9).

For eons, the land plants were bound to the water's edge (or at least moist climates), since the transport of their ciliated or flagellated sperms required water. To escape the water or the semiaquatic habitat and to

(a)   (b)

Fig. 6-8   **(a)** *Psilotum,* **the "whiskfern," the most primitive living vascular plant, lacks leaves and roots. (b) In this cross section of its photosynthetic stem, note the vascular tissue (conducting tubes) in the center. The epidermis and cuticle are the outermost layers.**
Photos courtesy of Saturn Scientific, Inc., Fort Lauderdale, Fla.

Fig. 6-9   **This large "seaweed," or kelp, possesses a leaflike blade, stemlike stripe, and rootlike holdfast, all of which differ from the leaves, stems, and roots of land plants in that they lack conducting tissues.**
Courtesy of Ward's Natural Science Establishment, Inc., Rochester, N. Y.

exploit the unoccupied land environments, a new method of sperm transport had to be found.

The first advanced animals to invade the land were probably scorpions and centipedes (Fig. 6-7). The first vertebrates evolved in the waters of 420 million years ago; it would be almost 100 million years before they came haphazardly ashore.

The first animals with backbones were jawless fish, lacking paired fins and covered with a bony scale armor. Due to the latter feature, they have been called *ostracoderms* ("shell skin") (Fig. 6-10a). The lampreys and the hagfishes are the modern-day descendants of these primitive forms (Fig. 6-10b). Of course, the contemporary jawless fishes are considerably departed from the ostracoderms. Neither the lamprey nor the hagfish has scales. The modern jawless fish are parasitic to other fish, attach themselves to the bodies of their hosts by their round, suckerlike mouths, and rasp flesh and blood from their prey with their horned tongues. The ancient ostracoderms appear to have been bottom feeders.

One of the loudest controversies in the silent science of paleontology centers about the freshwater versus the saltwater theories of vertebrate origin. The fossils of other animals and plants, and the chemical composition of the geologic beds in which ostracoderms are found, suggested that these most primitive vertebrates first appeared in the freshwater habitat. Harvard paleontologist A. S. Romer leads the exponents of the freshwater theory. Homer Smith, kidney authority and physiological evolutionist, argued that the structure and function of the majority of fish kidneys strongly support the freshwater theory and proposed

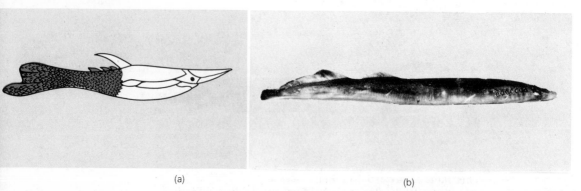

(a)   (b)

Fig. 6-10   **Jawless fishes. (a) Ostracoderms. These ancient, armored, jawless fishes are believed to have been the first of the vertebrate line. (b) Lamprey, a modern jawless fish.**
Courtesy of Carolina Biological Supply Co., Burlington, N. C.

that the armor of ostracoderms was not only a physically protective device, but also a physiologic barrier to the entry of water into the body fluids of these fishes, living in a hypotonic (freshwater) medium. The Chicagoan R. H. Denison and his group support a saltwater theory of vertebrate origin, since some of the features of the geologic beds of ancient jawless fish fossils cannot deny a saltwater origin.

A third viewpoint has emerged—namely, that the ostracoderms arose from primitive, oceanic invertebrate chordates (the phylum Chordata includes invertebrates as well as vertebrates) that took to swimming up estuaries and rivers and lived upon the rich organic debris coming down with the current. Evolution has most often occurred in the *ecotone,* where one ecosystem blends into another, in this case, where the salt water meets the fresh, the estuary. In the ecotone, the change of environment is not abrupt but gradual; the ecotonal environment is not as extreme as it is in the environments to either side. Species that normally dwell in one habitat or the other can very often tolerate the intermediate conditions of the ecotone. Most ecotones abound with a great variety of life, visitors, natives, and invaders. Whether the vertebrates evolved in the fresh or salt water may well remain an undecided point, but it seems logical that they might have evolved in an ectonal situation, in an estuary or a salt marsh. Since their jawless anatomy strongly suggests that the ostracoderms were bottom feeders, it must mean that plants invaded the land long before the vertebrates. The richest muck, laden with the organic matter of plant remains, is deposited in river deltas. Most of the organic matter found there is derived from land plants.

About 360 million years ago, the first jawed fishes, the *placoderms,* appeared. It is thought that they arose from the ostracoderms, as the first gill arch was modified to form the lower jaw. Paired fins also made their first appearance in the placoderms. Many of the placoderms were predators. These primitive jawed fishes gave rise to the class of *cartilaginous fish* (sharks, rays, etc.) as well as the *bony fish* (Osteichthyes).

The bony fish make up the largest class of vertebrates, dominating both the salt and fresh water. Osteichthyans (bony fish) are usually divided into two groups: the *ray-finned fishes* and the *lobe fins,* or the "muscle-finned" fishes. Most common fish (bass, trout, carp, perch, pike, catfish, tuna, cod, mackerel, herring, grouper, etc.) are ray fins. Today, the lobe fins are represented by only six *relict species* (survivors from an earlier period). There are five species of lungfish: the Australian, the South American, and three African. One species of lobe fin, the "coelacanth," still survives in the salt water off Madagascar (see Fig. 6-11a). The lobe fins are thought to have been ancestral to the land vertebrates.

The fresh water is a precarious habitat, subject to drought. Ancient freshwater fishes must have early evolved lungs, blood-rich diverticula

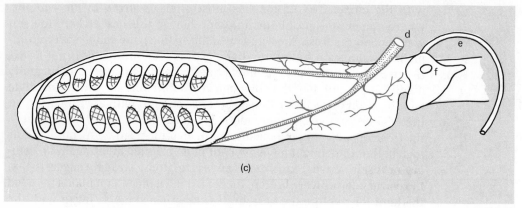

Fig. 6-11   **Lobe-finned, or flesh-finned fishes, also known as Choanichthyes, literally "fish with internal nostrils." The middle portions of their paired hip and shoulder fins contain jointed skeletons with attached muscles. Close relatives of these fish probably gave rise to amphibians in the past. (a) A coelacanth, a lobe fin that was thought to have been extinct, but was recently found living off the coast of Madagascar. (b) Australian lungfish. (c) Lung of lungfish opened in the rear to show its air cells; d, pulmonary vein; e, pulmonary artery; f, glottis, opening from esophagus into lung.**

of the throat, used as accessory respiratory organs compensating for deficient gill respiration in the oxygen-depleted, stagnant waters of drying ponds. Today, the gar and mudfish of the Everglades and the Mississippi drainage basin still possess lungs, although they are not nominal lungfishes; they belong to an ancient group of bony fish. The garfish and the mudfish (as the latter's name implies) survive the stagnant waters of the periodic droughts that visit the Everglades much better than their more recent relatives, bass and bream, whose ancient lungs—like those of most other fish—have been modified to form air or *swim bladders,* mere buoyancy devices. Catfish survive the drought well, because their scaleless skin is highly vascularized and serves as a respiratory accessory to the gills. Some fish that feed on the bottoms, which are frequently oxygen-poor waters, have evolved special blood respiratory pigments (hemoglobins) that are more efficient in stagnant water respiration than the normal. There is more than one solution to the problem of life in stagnant water.

Although lungfish have gills, they cannot survive with gill respiration alone, even in well-oxygenated water. Unless lungfish can get their heads out of water in the barrels in which they are collected, they drown. The lungfish "summer hibernate," or *estivate*, to avoid the drought. As their fresh water dries up, they burrow in the mud, make blowholes to the surface, secrete mucus about themselves, and remain in a state of lowered metabolism until the water comes again, even for as long as several years. This is one solution to the drought, but it's not the only possibility.

The lobed fins apparently evolved as devices for clambering over sandbars and mudflats to get from one drying mudhole to the next, a means of returning to the water rather than leaving it (see Fig. 6-12). The lobed fins differ from the rayed fins of most fish in that muscle masses have moved out onto the bony elements of the fin and certain bony components of the fin have enlarged. Rayed fins have slender bony elements and their muscles of control remain close to the body. The walking catfish (*Clarias batrachus*) from Southeast Asia, which has been recently introduced into Florida and threatens to overrun the Everglades, is neither a catfish nor a lobe fin, but it has arrived at features of both (Fig. 6-13a). Its skin is not scaled, and it has thickened one of the rays in each of its pectoral fins (shoulder fins) for use as a digging and propelling organ for support upon and movement over the land (Fig. 6-13b). In addition, the walking catfish has modified its rear gills as lung-like devices (Fig. 6-13c). Southeast Asia, from whence it comes, is like the Everglades in that there is annual drought and monsoon.

The pattern of thickened bony elements in the fin lobes of fossil lobe-finned fishes is quite similar to the arrangement of bones in the limbs of most ancient amphibians. There is excellent geologic evidence that the Devonian period (300 million years ago), in which the first amphibians appeared, was an age of severe climatic change, of flood and drought. Masses of fossils of early amphibians (Stegocephalians) have been found in several localities, suggesting that the dying animals had been crowded together in some drying mudhole or pond.

The lungs and fin lobes of the lobe fins and lungfish were adaptations permitting survival in the precarious freshwater habitat. Although they would be modified further to form the lungs and limbs of land vertebrates, their evolution in the first place had nothing to do with an "in-

(a)

(b)

Fig. 6-12   (a) Devonian swampscape. These extinct lobe fins had internal nostrils and probably lungs. They resembled the surviving coelacanth and lungfishes. Primitive jawless and jawed fishes swam in the water. Note that some of these first land plants resemble the present-day whiskfern (Fig. 6-8). Other land plants seen here are club mosses and horsetails. Ferns and seed plants were still millions of years in the evolutionary future. (b) *Periophthalmus*, the mud skipper, a fish that chases its prey on land. Its shoulder fins are bent at an angle so that they support the weight of the body, and, as in the ancient lobe fins, muscles have moved on out onto the thickened boney rays of the fins.

(a) from *History of Life on Earth Charts*, Denoyer-Geppert, Co., Chicago. (b) courtesy of CCM: General Biological, Inc., Chicago.

(a)   (b)

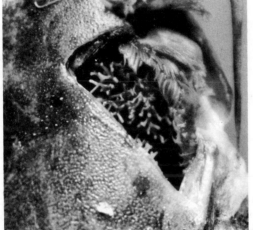

(c)

Fig. 6-13   "Walking" catfish is neither a catfish nor a lobefin. *Clarias batrachus* is a fish
                   from Southeastern Asia shows some superficial similarities to both of the
                   above. (a) Specimen from above. (b) Note thickened ray (spine) in shoulder
                   fin used as balancer. (c) Photo showing rear gill modified to form lunglike
                   device.

                   From M. D. Nicklanovich and M. J. Zbar, *Everglades Ecosystem*, Saturn Scientific, Inc.,
                   1972.

tended" invasion of the land. Anatomical devices acquired by a group of animals in one habitat that fit them for life in a new and different environment are said to be "preadaptations." Mud is really a type of an ecotone between the fresh water and the dry land. The lobe fins and the lungfish (and their descendant amphibians) were fitted out for life "in between."

The *choanae* are the internal openings connecting the nasal passageways and the throat. Most fish lack them, even though they have external nostrils. The lobe fins have such "internal nostrils," and were therefore once called Choanichthyes ("choanae fish"). The imaginative naturalist Loren Eiseley refers to the lobe fin as the "Snout" in the following passage:

In the passage of days the pond became a puddle but the Snout survived. There was dew one dark night and a coolness in the empty stream bed. When the sun rose next morning the pond was an empty place of cracked mud, but the Snout did not lie there. He had gone. Down stream there were other ponds. He breathed air for a few hours and hobbled slowly along on the stumps of heavy fins.

It was an uncanny business if there had been anyone there to see. It was a journey best not observed in daylight, it was something that needed swamps and shadows and the touch of the night dew. It was a monstrous penetration of a forbidden element, and the Snout kept his face from the light. It was just as well, though the face should not be mocked. In three hundred million years it would be our own.

There was something fermenting in the brain of the Snout. He was no longer entirely a fish. The ooze had marked him. It takes a swamp-and-tide flat zoologist to tell you about life; it is in this domain that the living suffer great extremes, it is here that the water-failures, driven to desperation, make starts in a new element. It is here that strange compromises are made and new senses born. The Snout was no exception. Though he breathed and walked primarily in order to stay in the water, he was coming ashore.[11]

Although the land had been colonized by plants and invertebrates, it was devoid of vertebrates. A whole new world of ecologic niches was open to the first land vertebrates. The Devonian period is known as the "Age of Fishes." Amphibians appeared late in the Devonian; their name implies that they were capable of life either on the land or in the water. Actually, the surviving amphibians are not truly land animals. They are still held to the water's edge by the heaviest links in the chain of life—

---

[11]Loren Eiseley, *The Immense Journey* (New York, N. Y.: Random House, Inc., 1957), pp. 50–51.

reproduction. Almost all amphibians must return to the water at least to reproduce. The first amphibians must have looked like "four-legged" fish. The evidence is quite convincing that ancient members of the lobe-finned fish group gave rise to the primitive amphibians. The lobe fins had functional lungs, choanae opening into the throat, and bones in their elongated fins that corresponded with those in the legs of ancestral, stem amphibians. One late Devonian fossil, *Ichthyostega,* is a "missing link" between lobe-fin fish and amphibians, since it shows a combination of characteristics from both groups.

The amphibians reached their peak of diversity in the Carboniferous period (250 million years ago), the "Age of Amphibians." By the end of Carboniferous times, three distinct groups of ancient amphibians had evolved. These varied from a few inches in size to 15 ft in length. Some were aquatic and some terrestrial; a few had lost their limbs and returned to the water permanently. The modern amphibians, the frogs, toads, salamanders, and newts, did not appear until later, in the Age of Reptiles.

The primitive *labyrinthodont* ("complex-toothed") amphibians (Fig. 6-14) gave rise to reptiles, the first truly land vertebrates. The fossil *Seymouria* (Fig. 6-15) is a missing link between amphibians and reptiles, exhibiting a nearly equal mixture of the traits of both classes. Fossil reptilian eggs are known from the end of the Age of Amphibians. Reptiles were present in numbers about 230 million years ago, the Permian period, at the end of which sharp climatic changes occurred, the Appalachian mountains upthrust, and most amphibians became extinct.

In surviving amphibians, it is still possible to see life's plasticity. Apoda ("footless"), one living order of amphibians, has lost all traces of limbs (like snakes among the reptiles) and lives wormlike in the tropical lands of the world (Fig. 6-16). No fossil apodes are known, perhaps indicating that they are of recent origin. Were the legs so long in coming discarded? Varying degrees of limb degeneration are found in other groups of amphibians. The Congo eel (*Amphiuma*) has nearly useless, toothpick legs and an elongate eel-like body (Fig. 6-17). It is an aquatic salamander with gills. Amphibians generally pass through a limbless, gilled, fishlike larval stage (tadpole) of aquatic existence. In metamorphosis, the gills are resorbed and limbs sprouted, as the animal assumes a terrestrial life. Several salamanders remain in the water even as adults, retaining their gills. The salamander body form is in many ways closer to ancestral amphibians than the frogs, which have lost their tails in the course of evolution. The aquatic and even terrestrial salamanders walk with sideways undulations of their bodies brought about by the alternate contractions of segmental muscle masses on opposite sides of their bodies. This method of movement is quite like that of fish. The muscular

Fig. 6-14   A labyrinthodont amphibian, probable reptilian ancestor, stands at the foot of the tree in this Carboniferous swamp. It is the Age of the Amphibians, when they reached their peak in numbers and types. An eel-like amphibian with degenerate limbs and external gills swims in the water, and a giant dragonfly with a wingspan of 5 ft flies overhead. The great forests that flourished during this time formed our present coal deposits. No flowering plants were present. It was still the Age of Ferns. Seed plants had appeared, clearly modified ferns.
From *History of Life on Earth Charts*, Denoyer-Geppert, Co., Chicago.

Fig. 6-15   *Seymouria,* one of the stem reptiles, derived from labyrinthodont amphibians and ancestral to other reptilian groups.

Fig. 6-16   **An apode, a legless, terrestrial amphibian.**
                   Photo courtesy of Professor Lewis D. Ober.

segments of *Siren* (the "mud eel") are visible along the length of its body.
The forelimbs of this animal have disappeared, and the small, hindlegs
have moved forward to a position just behind the external gills. The skin
of the tail is flanged above and below like the tail fin of a fish (Fig. 6-18).

From this short discussion, we see that even the surviving amphibians
show *adaptive radiation,* divergent evolution, fitting them to many
ways of life. Some have converged upon the fishlike body form; others
have assumed a snakelike appearance. The ancestral amphibians were
even more varied. Obviously, there is no unity of plan here, or typical
type, no evolutionary progression toward a fixed goal. "Nature" modified the ancestral amphibian body form (the common ancestor) this way
and that, "for various ends." Some pathways led to the loss of limbs and

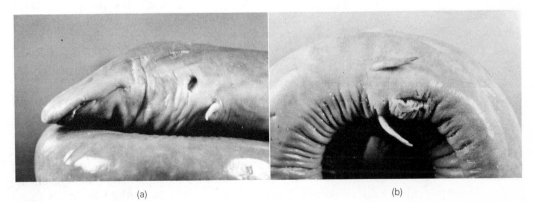

(a)   (b)

Fig. 6-17   **The Congo eel, a permanently aquatic amphibian with degenerate limbs.
                   (a) Head and forelimbs. (b) Hindlimbs.**
                   Courtesy of Professor Lewis D. Ober.

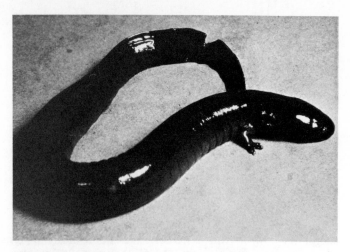

Fig. 6-18   *Siren*, **the mud eel, has one pair of legs and has assumed a considerably fishlike body form in the course of its evolution.**
Courtesy of Professor Lewis D. Ober.

the taking up of ways of life anything but "amphibian." One past line led to the reptiles, which "inherited the earth," only to lose it to their mammalian and avian heirs.

The climate of the continental land masses during the Age of Amphibians was warm and moist, and the earth was covered with lush swamps of primitive plants. Plant biologists call the Age of the Amphibians the "Age of Ferns," since the great swamp forests of Carboniferous times were made up of giant ferns, club mosses, and horsetails (fern allies). The first seed plants, the gymnosperms ("naked seeds"), appeared in this age. Flowering plants were absent. The forests that flourished during these times formed our present coal deposits. The ferns and their allies were vascular plants (tracheophytes), but they were not seed producers. It would be the seed plants that would conquer the land.

The Permian period that ended almost 200 million years ago (see Table 6-1) saw the Appalachians newborn and a greater ice age than the most recent. Marks of this Permian glaciation are still to be found in South America, India, and Australia. There is good evidence to indicate that Africa, South America, and Antarctica were fused in one great southern continental land mass, "Gondwana," until this time. Fossils of reptiles and amphibians have recently been unearthed in Antarctica. The climate of the Permian was cold and dry. The Carboniferous forests of the preceding periods dried up and the amphibians declined. The dominant land vertebrates became the stem reptiles, or cotylosaurs,

Fig. 6-19   **Early Triassic period, the first period of the Age of the Reptiles. Most reptiles have simple peg teeth, all of which are essentially the same. These dog-toothed, mammal-like reptiles possessed incisors, canines, and cheek teeth with several cusps. These lizardlike creatures grew up to 8 ft long. The Age of Reptiles was also the Age of the nonflowering seed plants, the gymnosperms, whose seeds are not encased in fruits (the common state in flowering plants). The pines and the firs that we see in the background are essentially like the modern coniferous (cone-bearing) evergreen gymnosperms. This era was drier than the preceding one, and both the reptiles and the conifers are adapted to the drought, whereas the amphibians and the ferns of the preceding era required moist conditions for reproduction.**

From *History of Life on Earth Charts*, Denoyer-Geppert Co., Chicago.

from which all higher vertebrates appear to have evolved. Already in the Permian, "mammal-like" reptiles, the theraspids, had appeared. They were most likely derived from the stem reptiles. The teeth of most reptiles are just rows of sharp pegs, almost all the same shape and size. The dog-toothed reptiles (cynodonts) had incisors, canines, and cheek teeth with several cusps (Fig. 6-19). These were apparently the forerunners of mammals, whose order possesses the most highly differentiated denti-

Table 6-1

Time scale and the history of life on earth

Eras (Years of duration)	Major Divisions	Periods (Years from present)	Epochs	Dominant Organisms	Events of Biological Significance	Geological and Climactic Phenomena
Cenozoic (60 million)	Quaternary		Recent	Age of Man and Herbs	Rise of civilized man.	
		2 million	Pleistocene		Extinction of great mammals and many trees.	Periodic glaciation.
	Tertiary	Late Tertiary	Pliocene		Rise of herbs; restriction of forests; appearance of man.	Climactic cooling; temperate zones appear; rise of Cascades, Andes.
			Miocene	Age of Flowering Plants, Mammals, and Birds	Culmination of mammals; retreat of polar floras; restriction of forests.	Cool and semi-arid climate; rise of Himalayas, Alps.
		Early Tertiary (60 million)	Oligocene		World-wide tropical forests; first anthropoid apes; primitive mammals disappear.	Climate warm and humid, rise of Pyrenees.
			Eocene		Modernization of flowering plants; tropical forests extensive; modern mammals and birds appear.	Climate fluctuating.
Mesozoic (125 million)	Late Mesozoic	Cretaceous (125 million)			Rise and rapid development of flowering plants; gymnosperms dominant but beginning to disappear; rise of primitive mammals.	Rise of Rockies and Andes; great continental seas in N. America, Europe climate fluctuating.
				Age of Higher Gymnosperms and Reptiles	Extinction of great reptiles.	Climate very warm.
	Early Mesozoic	Jurassic (157 million)			First known flowering plants; gymnosperms prominent but primitive ones disappear; dinosaurs and higher insects numerous; primitive birds and flying reptiles.	Great continental seas; rise of Sierras; climate warm.

Era	Subdivision	Period	Dominant Life Age	Life	Physical / Climate
		Triassic (185 million)		Gymnosperms increase; first mammals; rise of dinosaurs.	Climate warm and semiarid.
Paleozoic (368 million)	Late Paleozoic	Permian (223 million)		First modern conifers; rise of land vertebrates.	Periodic glaciation; rise of Appalachians; Urals.
		Pennsylvanian (271 million)	Age of Lycopods, Seed Ferns, and Amphibians	Primitive gymnosperms dominant; extensive coal formation in swamps.	
		Mississipian (309 million)		Lycopods, horsetails, and seed ferns dominant; some coal formation; rise of primitive reptiles and insects.	Shallow seas in N. America.
	Middle Paleozoic	Devonian (354 million)	Age of Early Land Plants and Fishes	Rise of early land plants; rise of amphibians; fishes dominant.	Shallow seas in N. America.
		Silurian (381 million)		First known land plants; algae dominant; first air-breathing animals (lungfish and scorpions).	
	Early Paleozoic	Ordovician (448 million)	Age of Algae and Higher Invertebrates	Marine algae dominant; corals, star fishes, bivalves; first vertebrates (fishes).	Shallow seas in N. America.
		Cambrian (553 million)		Algae dominant; many invertebrates.	Shallow seas in N. America.
Proterozoic (900 million)		(1500 million)	Age of Primitive Marine Invertebrates	Bacteria, algae, worms, crustaceans prominent.	Formation of Grand Canyon, Laurentians; sedimentary rocks.
Archeozoic (550+ million)		(2000 million)	Age of Unicellular Forms	No fossils; organisms probably unicellular; origin of first life.	Rock mostly igneous.
		(10 billion) ?			Beginning of present universe and the solar system.

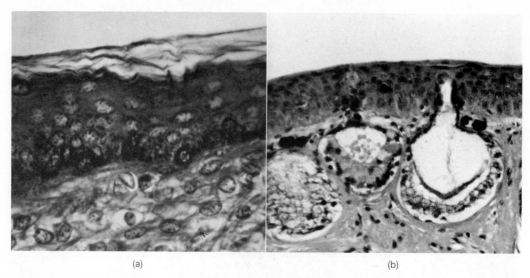

(a)   (b)

Fig. 6-20   **(a) Land vertebrate skin from reptiles through man is covered with a thick layer of flattened, dead, proteinized cells. (b) The upper skin (epidermis) of amphibians is thin with few, flattened dead cells upon the surface, which is continually moistened by mucous secretions from the glands below.**
From M. J. Zbar and M. D. Nicklanovich, *Cells and Tissues*, Saturn Scientific, Inc., Fort Lauderdale, Fla., 1971.

tion of any of the vertebrates. The best known genus of the dog-toothed reptiles, *Cynognathus*, was an 8-ft long, lizardlike creature.

The next 130 million years after the Permian period is known as the "Age of the Reptiles" (Mesozoic era). The climate, although warmer than during the preceding great ice age, was cooler and drier than that of Carboniferous times.

The reptiles were entirely free from the water. Two features allowed them to subdue the dry land: the land egg and a new skin. The dry, horny skin of reptiles is covered with scales composed of several layers of dead cells loaded with the protein keratin (Fig. 6-20a). This thick, dry, horny layer of reptilian skin is analogous to the waxy cuticle of land plants, for its function is to prevent evaporation of water from the deeper tissues. Amphibian skin is moist, covered with glandular secretions of water-binding mucus and—with the exception of the very uppermost—the majority of the surface cells are alive (Fig. 6-20b). The amphibian skin (like the catfish skin) functions as an accessory organ of respiration.

The reptilian egg can develop out of water. Since the water of amphibian times teemed with potential egg predators, it has been postulated that a reproductive advantage (natural selection) accrued to those forms which laid their eggs first on the shore and then higher and higher on the land. The shells and membranes of land eggs are porous and permit gas exchange yet retard water loss. The internal membranes are continuous with the embryo's body yet surround it (Fig. 6-21). They are called the extraembryonic ("beyond the embryo") membranes. They number four: the *yolk sac*, which grows out from the midgut of the embryo and encloses the stored food material of the yolk; the *amnion*, which encloses the embryo in a cushioning, watery developmental environment; the *allantois*, an outgrowth of the hindgut, whose cavity stores uric acid (a nitrogenous, urinary waste) and whose blood vessels vascularize the outermost respiratory membrane: the *chorion*. The yolk sac is found in sharks and most fishes ("lower vertebrates"), but the amnion and the other membranes appear in the reptiles for the first time. Therefore, the reptiles and all vertebrates above them (birds and mammals) are called "amniotes." The young that hatch from reptilian eggs

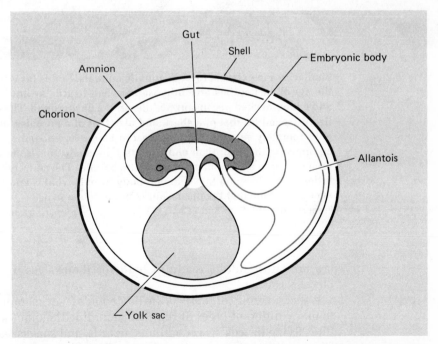

Fig. 6-21   **The land egg. The anatomy of eggs of reptiles and their bird and primitive egg-laying mammal descendants is quite the same. Note the extraembryonic membranes.**

Fig. 6-22   Similarities in external form in aquatic animals. In this Jurassic seascape from
the Age of the Reptiles, an ichthyosaur, a "fish-lizard," swims in the foreground,
and a long-necked plesiosaur paddles in the background. The body of the fish-
lizard resembles that of a shark and also that of a porpoise, although they are
only remotely related. All possess the elongate, streamlined, fishlike body,
equipped with fins or their equivalents. Obviously, this is the optimal shape in
their common environment, the aquatic habitat. The convergence of their evolutionary paths upon this common body form is what is known as convergent
evolution. Note that the plesiosaur is not unlike a seal.
From *History of Life on Earth Charts,* Denoyer-Geppert Co., Chicago.

are miniature adults ready to fend for themselves, not intermediate
larval forms.

A second group of reptiles, the *thecodonts* ("socket toothed") gave rise
to many different types of reptiles, including the dinosaurs, pterosaurs
(the "flying lizards"), crocodilians, lizards, and snakes, as well as birds.
The reptiles underwent an explosive adaptive radiation in the Mesozoic.
Besides the flying lizards (pterosaurs) in the air, there were fish-lizards
(ichthyosaurs) in the sea as well as paddle-limbed, long-necked plesi-

osaurs. The ichthyosaurs resembled fish in their external body form. This is an often-cited example of convergent evolution in which only distantly related forms come to closely resemble each other superficially, in similar environments (Fig. 6-22). The fish body form is the optimal, streamlined, external anatomy for the dense medium of water. On the land, the famed *Tyrannosaurus* rex preyed upon herbivorous dinosaurs, such as the three-horned *Triceratops* and the spade-spined *Stegosaurus* (Fig. 6-23).

The limbs of reptiles are better developed than those of their ancestral amphibians. The *pentadactyl* ("five-fingered") limb, which first made its appearance in amphibians, was "modified for various ends" in different reptiles, just as it would be in mammals and birds (Fig. 6-24). The typical

Fig. 6-23   **Late Jurassic landscape (150 million years ago). This period fell in the heart of the Age of the Reptiles. They had undergone explosive radiation, reached a peak variety, and adapted to a number of ways of life. A herbivorous dinosaur stands in the foreground. The most famous carnivorous dinosaur, *Tyrannosaurus*, stands in the background. A winged reptile, the pterosaur, flies overhead.**

From *History of Life on Earth Charts*, Denoyer-Geppert Co., Chicago.

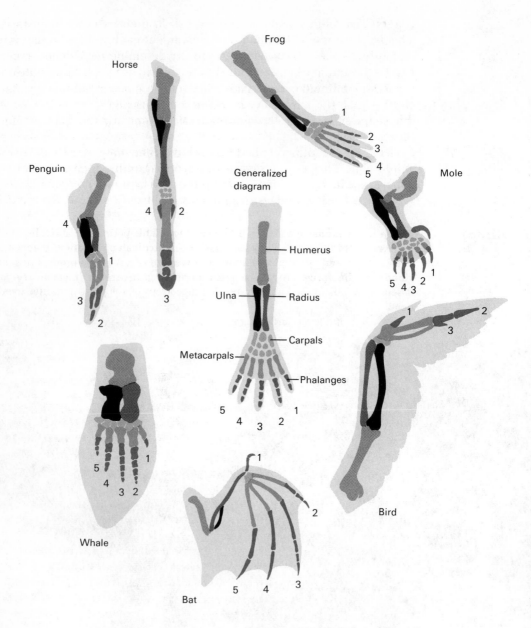

Fig. 6-24   The pentadactyl (five-fingered) limb (center) and its modification in the various vertebrates (surrounding). The generalized ancestral type has been modified for various purposes, but the basic similarities or homologies are still obvious (shown in same colors).

From N. M. Jessop, *Biosphere*, Prentice-Hall, Inc., Englewood Cliffs, N. J., 1970.

land vertebrate limb consists of one upper arm (humerus) or thigh bone (femur), two forearm (radius and ulna) or lower leg bones (tibia and fibula), several wrist (carpal) or ankle (tarsal) bones, a series of five hand (metacarpal) or foot (metatarsal) bones, and two or three finger or toe bones (phalanges) in each digit. In the flying lizards (pterosaurs), the metacarpal and phalanges of presumably the fifth finger were elongated to form support for the skin wing. In paddle or fin limbs, the bone pattern of the pentadactyl limb is still to be found although the bones are reduced in length. The same skeleton supports the flippers of sea turtles today as supported the limbs of land reptiles millions of years ago.

The snakes, a relatively recent group of reptiles, are lizards that have lost their limbs in the course of evolution. "Nature has experimented" with this "design" before. Eels and morays among fish, as well as apodan (legless) amphibians, have lost their appendages, as have the snakes. Vestiges (remnants) of the hip girdle and hind limbs persist in the perianal spurs of boas and pythons (Fig. 6-25).

Modern birds lack teeth and their tails have been reduced to a few caudal (tail) vertebrae, but the first birds possessed typical reptilian teeth and long tails with many caudal vertebrae (Fig. 6-26). The "hand" of the modern bird has been quite reduced, and representatives of no more than three fingers are found (the meatless tip of the chicken's wing). The lizard-bird (*Archaeopteryx*) had three clawed digits. It is

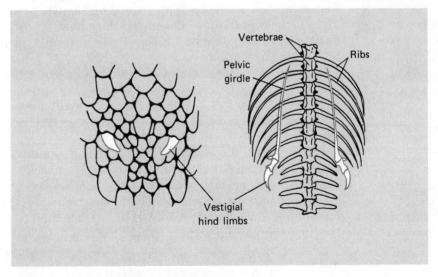

Fig. 6-25   **Vestiges of the hind limbs in the python.**
From "Guide to Reptile Gallery," British Museum, London.

Fig. 6-26   **An artist's conception of the oldest bird known from the fossil record.** *Archaeopteryx,* **or the "lizard-bird," had reptilian teeth, several clawed fingers, a long, lizardlike tail with many vertebrae (backbones) in it. Feathers are modified scales. Birds apparently evolved from a small group of running dinosaurs.**

thought that the birds were derived from a group of small, running dinosaurs, since the skull and hip anatomy of the two groups is so similar. The skin over portions of the modern bird body is scaly, as in reptiles. Feathers are composed of the same protein (keratin) as is found in the horny layer of reptilian skin.

The reptiles suffered a massive and mysterious extinction about 70 million years ago. At the height of their success, more than 16 orders were present. Today, only 4 survive. Biologists have puzzled over the causes of this great extinction for years, and several theories have been advanced.

At the close of the Mesozoic era (the Age of Reptiles), the Andes, the Rockies, the Himalayas, and the Alps were upthrust, and the regions near the poles became cooler and drier. Some have suggested that this climatic change was responsible for the spectacular reptilian extinction, but there is not sufficient reason in this explanation to account for the fact that the ancient variety of reptilian life does not remain in the tropics of South America, Africa, and Asia, or in the oceans for that matter.

More likely, perhaps, the reptiles lost out in competition with primitive mammals. Most mammals during this time were small generalized creatures about the size of rats. They are supposed to have annihilated the dinosaurs and other reptiles by eating their eggs. In contrast with mammals and birds, reptiles do not care much for their young. Most surviving reptiles take some steps at least to hide their eggs. Turtles bury their eggs in the sand before they desert them. The female alligator builds a crude "nest" of rotting vegetation for her eggs and guards the mound. Many snakes keep their eggs internally until hatching, so that the young are "born alive." Perhaps minimal egg care was a bottleneck that let the surviving reptiles through, but fossil nests of primitive reptiles have been found, and the fish-lizard brought forth its young alive. In recent times, introduced mammals have frequently destroyed reptilian *and* bird populations by eating their eggs. The appearance of mammals in the past, however, was not sudden: they evolved gradually in balanced ecosystems.

Probably mammals were more efficient herbivores than reptiles, which would have given them an energy advantage. The constant body temperature of mammals (and birds) allowed them to remain active for longer periods than the "cold-blooded" reptiles, who presumably became active by midday, retired shortly after dusk, and periodically sought the shade for periods of inactivity during the day when overheated. Thermoregulation was undoubtedly a great problem for the large reptiles with large body masses, slow to warm and slow to cool. It has been suggested that the great finlike fans on the backs of several dinosaurs were highly vascularized and could be opened and folded to radiate or retain body heat. Certainly, the ability to maintain a constant body temperature was a decisive advantage for the higher vertebrates. Mammals were active at all times, in the night, and in the dawn, when most reptiles barely stir.

Another proposed explanation for the great reptilian extinction has been the epidemic disease hypothesis, but, somehow, this too seems improbable.

Natural selection acting over long periods of time tends to make animals more and more narrowly specialized. Perhaps the ancient reptiles became overspecialized. A highly specialized animal loses its ability to adapt to changing conditions. The possibilities of the generalized animal are greater. By perfectly fitting an animal to one environment, evolution itself (natural selection) can bring a species or larger group of animals to the brink of extinction, which can follow the slightest environmental change. The generalized inherit the earth. Evolution moves from the simple to the complex, from the generalized to the specialized. The fittest make up the most brittle branches of life's tree. There is no question that the extinct reptiles included a number of the most advanced, complex, specialized creatures that the world has ever seen.

Certain evolutionary forces, such as sexual selection, tend to increase the size of species. Many of the extinct reptiles were truly ponderous.

The cycle of the adaptive radiation of a generalized form followed by massive extinctions of specialized types was to be repeated with the mammals as it had been with the reptiles and the amphibians before them. The mysteries of the great extinctions have revived the old theory of catastrophism in a more naturalistic form. The cyclic periodicity of the great terrestrial extinctions, once about every 100 million years, has led to an astronomical hypothesis of galactic revolution and a shifting of the earth's poles.

The four surviving orders of reptiles include the turtles (direct descendants of the stem reptiles), the crocodilians, the still successful snakes and lizards (more than 6000 species), and one group represented today by a single form, the New Zealand tuatara, *Sphenodon* (Fig. 6-27). Sphenodon's fossil record predates that of the dinosaurs.

The Age of the Reptiles is to the botanist the Age of the Gymnosperms, for these flowerless, fruitless, naked-seeded plants dominated the land vegetation of this era. These were the first truly successful terrestrial seed plants. Although the ferns and other lower plants produced spores that could be disseminated by the wind and were resistant to drought, the spores themselves were produced by a plant body (the sporophyte) whose conception depended upon swimming sperm following a watery route to fertilization. The seed plants have achieved reproductive independence of the water by evolving dustlike *pollen grains,* which carry the sperm nuclei and are drought resistant and wind transportable. The reproductive organs of gymnosperms are usually *cones,* both male and female. The pines, firs, and spruces are familiar trees that belong to the major gymnosperm group, the *conifers* ("conebearers"). The smaller male cones produce pollen, and seeds develop within the larger female cones (Fig. 6-28).

Seeds are roughly the equivalent of the reptilian land egg, for within the hard outer coat is a tiny plant embryo with stored food (the endo-

Fig. 6-27   **The New Zealand tuatara, the sole survivor of an order of reptiles older than the dinosaurs.**

Fig. 6-28   **Pine cones of three ages. These are seed-bearing female cones. The male or pollen cones are smaller. Pines are primarily wind pollinated.**
Courtesy of Carolina Biological Supply Co., Burlington, N. C.

sperm) (Fig. 6-29). Like the spore, the seed permits life to withstand the hard times of dry or cold weather and other unfavorable conditions, yet, unlike the spore, the seed contains a fully equipped embryonic plant already partially developed, with a "headstart on life," since it is stored with food rich in nutrients, minerals, and vitamins to nourish the growing seedling until it is big enough to make its own food and vitamins and to absorb its own minerals. For this reason more than any other, the gymnosperms spread over the land in the Mesozoic and underwent explosive adaptive radiation, like their reptilian counterparts. Many gymnosperms also resemble reptiles in that they possess an anatomy adapted for drier habitats. Recall that the Mesozoic was much drier than the preceding era.

Although the gymnosperms have declined in recent times, they are still represented by such common woody plants as the evergreen pines, firs, spruces, redwoods, sequoias, cedars, and hemlocks. The oldest living organisms are the bristle cone pines of California and Nevada. These otherwise unimpressive, small trees grow on arid mountainsides. Their ages of nearly 5000 years easily surpass those of the giant sequoias. The cypress of southern swamps is also a gymnosperm. Cycads are primitive gymnosperms whose fronds resemble those of ferns and palms, yet whose reproductive structures are cones. A few species survive in

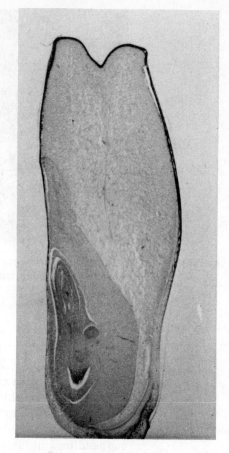

Fig. 6-29

A corn seed (kernel) is technically a thin-walled fruit; nevertheless, it is mostly seed. The small embryonic plant, lower left, is surrounded by stored food, upper right.
Courtesy of Carolina Biological Supply Co., Burlington, N. C.

Florida, the Caribbean, Asia, and Africa (Fig. 6-30). They are popular ornamental shrubs.

The exact derivation of the flowering plants, the *angiosperms*, is unknown. It seems that they arose either from early seed ferns or from very primitive gymnosperms. The angiosperms appeared in the Age of Gymnosperms and became the dominant land plants of the modern era, the Age of the Angiosperms in botanical jargon. They are the largest group of plants, containing over 300,000 species. The flowering plants adaptively radiated, filling almost every conceivable ecologic niche. Although they carried out the greatest plant conquest of the land, some of them have even gone back into the water, where they reduced many of the features that made of them an unparalleled terrestrial success. These are the "higher water plants," unrelated, of course, to the lower aquatic plants, the algae, even though some superficially resemble their remote forebears (a case of convergent evolution). Recall the reptilian invasion of the sea. Almost every time that there has been a great evolutionary explosion of a class of organisms, the "lusty" success of the group carries it even beyond its original element.

The stages in the evolution of the flower are unknown, since the geologic record has not so far given up complete fossils of the earliest flowering plants. Darwin called the angiosperms "an abominable mystery," as he puzzled over their sudden appearance and rapid rise to ascendancy.

The flower is a short stem with modified leaves bearing the reproductive organs (Fig. 6-31). The flowers of many plants carry both male and female organs; some bear the organs of just one sex. The female organ, the *pistil,* is an elongate stalk with a swollen base, the *ovary,* and sticky tip, whereupon the pollen grains are deposited. Pollen is produced in the saclike expansions of the tips of the otherwise filamentous male organs, the *stamens.* The germinating pollen grains carry the sperm nuclei down through the tissues of the pistil's stalk to the egg nuclei in the ovaries below. A *fruit* is a mature ovary, including the seeds and

Fig. 6-30   **A cycad. This fernlike plant is a nonflowering seed plant. Its reproductive structures are cones, which develop in the center of the fronds. Cycads are gymnosperms and more ancient than the pines and their relatives. They are popular garden ornamentals in the southeastern United States.**

their envelope. The flesh of the ovarian wall of the fruit is not the "meat of the seed." When fleshy fruits are eaten by animals, most seeds pass through the digestive tract unharmed and are deposited in the rich feces. In this manner, seeds can be dispersed for miles. Some fruits bear parchment wings for wind dispersal; others carry projecting hooks, spines, or bristles that adhere to the coats of animals.

Nowhere is the relationship between the evolution of plants and animals better exemplified than in the flowering plants. Although many angiosperms are wind pollinated, most flowering plants depend upon spiders, bees, wasps, butterflies, moths, flies, ants, and hummingbirds to transport pollen from one flower to the next. Flowering plants have evolved an incredible array of lures for pollinating animals. Vividly colored petals and scent chemicals mark and guide the animals to the flowers. Organs deep within the flowers, the *nectaries,* secrete sugary liquids. Insects that feed upon nectar become covered with microscopic pollen and thus transport it from one flower to the next. Many flowers possess incredible, intricate machines for powdering pollen on unwitting insects. More than 10% of all species of flowering plants are orchids, and it is among them that we find the most highly evolved pollination mechanisms. Some orchids practice a form of "sexual deception" (Fig. 6-32). Their flowers take the shape and color of female bees and wasps to attract wandering males who "pseudocopulate" with the floral structure, and, in so doing, become contaminated with the pollen of one flower at

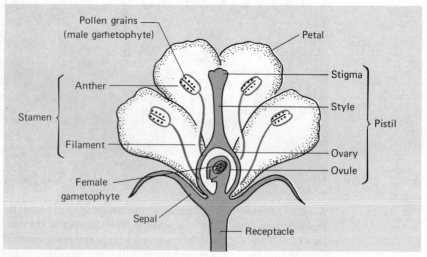

Fig. 6-31   **The anatomy of a typical flower.**

Fig. 6-32   **The flowers of this orchid (*Oncidium henekenii*) mimic female bumblebees in form.**
From M. J. Zbar and M. D. Nicklanovich, *The Biology of Reproduction*, Prentice-Hall, Inc., Englewood Cliffs, N. J., 1973.

the same time that they deposit that of others. The floral sex mannequins of these orchids are probably the most bizarre evolutionary creations. Several orchid floral fragrances are collected by euglossine bee males and used as borrowed pheromones to assist in territory marking and female attraction. (A pheromone is a chemical substance secreted by one individual to evoke a physiologic or behavioral response from another individual of the same species.) Some flowers are equipped with "slides" and bucket traps. Entrance and exit are separate, and the stigma (pistil's tip) and anthers (pollen sacs) surround the escape hatch. The fleeing bee pollinates and picks up a new batch of pollen at the same time.

The explosive rise of the flowering plants seems to have been a prelude to the great evolutionary radiation of mammals and birds that followed. As the incomparable Loren Eiseley put it,

Without the gift of flowers and the infinite diversity of their fruits, man and bird, if they had continued to exist at all, would be today unrecognizable. Archaeopterx, the lizard-bird, might still be snapping at beetles on a sequoia limb; man might still be a nocturnal insectivore gnawing

Fig. 6-33   **The duck-billed platypus is neither bird nor reptile, but a primitive egg-laying mammal.**

a roach in the dark. The weight of a petal has changed the face of the world and made it ours.[12]

To the zoologist, the "Age of Flowers" is the Age of Mammals and Birds. The climate early in this modern era was warm and moist. The earth was a hothouse, covered with jungles of flowering plants. Man's ancestors, like those of the contemporary apes, fed upon succulent fruits and shoots. Then the climate became cooler and drier.

It has been postulated that the continental upthrust cooled the world's climates. The forests retreated, the modern era became an "Age of Grass." (Grasses are flowering plants adapted to semiarid habitats.) Prairies covered the continents. The evolution of horses, antelope, buffaloes, and ostriches (all grazing animals) was correlated with the rise of the grasslands. Perhaps the recession of the forest brought man's ancestors down from the trees. It is almost impossible to separate the stories of plant and animal evolution.

Mammals and birds show a number of advances beyond reptiles. Both mammals and birds maintain high, constant body temperatures and have four-chambered, "double-pump" hearts, in which unoxygenated and oxygenated blood are kept separate. This circulatory arrangement is foreshadowed in crocodilians. The feathers of birds are believed to be modified reptilian scales. Mammals possess hair and mammary glands and, like birds, spend considerable time caring for their young. The mammals are subdivided on the basis of their mode of reproduction.

The most primitive mammals, the duck-billed platypus and the spiny anteater (Fig. 6-33), are egg-layers, and the anatomy of their eggs is nearly identical to the anatomy of the eggs of reptiles and birds. These creatures are found in Australia and adjacent islands. Their body temperatures fluctuate considerably, although nowhere nearly as much as that

---

[12] Ibid., p. 77.

of reptiles. Like birds and reptiles, the egg-laying mammals have a single opening for their digestive, urinary, and genital organs (the vent of the cloaca, the "sewer," not quite comparable to the rectum). The name of this lowest order of mammals, *Monotremata,* is derived from the Greek and means "single hole," in reference to the cloaca. The missing links between reptiles and mammals, the dog-toothed reptiles and others, are extinct but well known from the fossil record. In a sense, the monotremes are living missing links between reptiles and mammals, since they exhibit a combination of reptilian and mammalian traits. However, they are not regarded as ancestors of other mammals; they are an early and terminal branch on the mammalian evolutionary tree.

The *marsupials* derive their name from the *marsupium,* a pouch on the belly of the female, into which the "fetal" young crawl at an early stage of development, fasten to the nipples therein, and complete their development (Fig. 6-34). The *placenta* is the organ of maternal and fetal origin and physiological exchange, by means of which the embryo of higher mammals remains attached to the uterine wall until fetal development is complete. The marsupial placenta is rudimentary and short lived. Marsupial young at "birth" have the appearance of embryos in higher mammals.

The most famous and varied surviving marsupials are found in Australia. Here, they have undergone adaptive radiation, filling the roles of

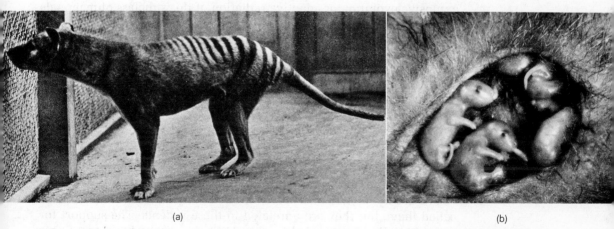

(a) (b)

Fig. 6-34 **(a) A marsupial wolf (thylacine). (b) Opossum young are shown in the pouch. Note that they are still in the fetal condition compared with the newborn of true placental mammals. Their fetal development will be completed in the pouch, where they remain attached to a nipple.**
(a) Courtesy Australian News and Information Agency. (b) Courtesy of Carolina Biological Supply Co., Burlington, N.C.

true placental mammals in other regions of the earth. In addition to the kangaroos and wallabies, there are marsupial moles, squirrels, kangaroo rats, wolves, and "bears." A few marsupials exist in southern Asia. The widest spread and most successful marsupial in the New World is the opossum. Several features of the skeletal anatomy of marsupials differ considerably from those of modern mammals (the true placental mammals). There are differences in the teeth, and most marsupials possess a pair of long, flat bones, the epipubic bones, which are embedded in the muscles of the abdominal wall and are attached to the lower front part of the pelvic girdle. Similar bones are found in monotremes and certain primitive reptiles; these bones are missing entirely from higher mammals.

The fossil record indicates that at one time (about 70 million years ago) the marsupials were widely distributed, all over the world. In fact, the oldest fossil remains of marsupials have been recently discovered in the southeastern United States, and a variety of fossil marsupials, including even great cats, have been recovered from Florida and South America. Australia put out to sea from the Asian land mass sometime after the end of the Age of the Reptiles and became a floating museum of marsupials. They survived there in isolation from competition with the more advanced placental mammals, which apparently nearly exterminated the marsupials on the other land masses.

The true placental mammals make up 95% of living mammals. The explosive evolution or adaptive radiation of this subclass of mammals was easily as great as that experienced by the reptiles in the Mesozoic. Some mammals returned to the sea, assumed a fishlike external body form, and became the present-day whales and porpoises, another case of convergent evolution. Whales are descended from land carnivores, which gradually became this mammal-fish. The hands of the arms became webbed to form paddlelike flippers; legs are absent, except for remnants of the hip girdle embedded in the flesh. Sea cows and dugongs, the sirens of mariners' tales, are also mammals that have returned to an environment to which their close relatives and even distant progenitors were not even faintly adapted (Fig. 6-35). Within a whale's fins, it is easy to make out the typical anatomy of the pentadactyl limb.

Bats are flying mammals, *Fledermaus* or "flying mice" the Germans called them, but they are unrelated to these rodents. The support for their skin wing is provided by the elongation of the hand and finger bones of the last four digits. In the soil-burrowing moles, the same basic limb anatomy has been modified to form short spades. Why does one grasp the shovel near the blade, when pulling up a shovelful of packed earth? What type of a lever is this?

Some of the best studied cases of evolution involve the development of the hoofed, grazing animals of the plains. The "dawn horse" of 60 million years ago was about 18 in long with four toes on its front feet and three on its hind feet. This dog-sized horse was a browser of soft, succulent leaves and possessed a relatively generalized (unspecialized) set of mammalian teeth. It was probably no swifter than a small dog of its size. A whole succession of fossil horses is known, in which overall body size increased, toe number was reduced, and specialized molars evolved (Fig. 6-36).

Finally, the swift modern horse ran on only one toenail or fingernail (the hoof) of each foot or hand. The bones in the axis of this middle digit (metatarsals, metacarpals, and phalanges) were enlarged and thickened. The slender "splint bones" in the lower leg (actually foot) of the horse represent the vestigial digits. This type of foot posture, unguligrade, on hooves (unguis means "nail"), is one of the most advanced and is found in some of the swiftest herbivorous animals. These features of speed are invaluable, inasmuch as the major predators of the plains are swift cats

Fig. 6-35   **A Tertiary seascape. The sea cows are one of several groups of mammals that have returned to the sea. Their front limbs have been transformed into paddles, and their hind legs have been reduced to concealed vestiges. Their oldest fossil remains show a number of resemblances to mastodons and conies. Their blood proteins are similar to those of the pig, and it is believed from these several lines of evidence that sea cows are descended from hoofed land animals. The fossil evidence strongly suggests that whales are descended from some primitive type of land carnivore.**

From *History of Life on Earth Charts*, Denoyer-Geppert Co., Chicago.

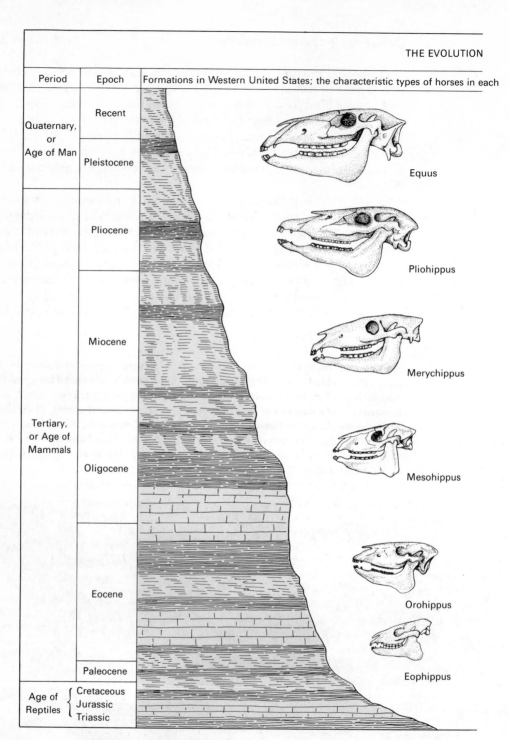

Period	Epoch	Formations in Western United States; the characteristic types of horses in each
Quaternary, or Age of Man	Recent	Equus
	Pleistocene	
Tertiary, or Age of Mammals	Pliocene	Pliohippus
	Miocene	Merychippus
	Oligocene	Mesohippus
	Eocene	Orohippus
	Paleocene	Eophippus
Age of Reptiles {	Cretaceous Jurassic Triassic	

Fig. 6-36 · **Fossil history of the horse.**
From Earl D. Hanson, *Animal Diversity,* third edition, Prentice-Hall, Inc., Englewood Cliffs, N. J., 1972.

Forefoot	Hind foot	Teeth	
One toe Splints of 2nd and 4th digits	One toe Splints of 2nd and 4th digits	Long-crowned, cement-covered	
Three toes Side toes not touching the ground	Three toes Side toes not touching the ground		
Three toes Side toes touching the ground Splint of 5th digit	Three toes Side toes touching the ground	Short-crowned, without cement	
Four toes	Three toes Splints of 1st and 5th digit		
Hypothetical ancestors with five toes on each foot and teeth like those of monkeys, etc.		The premolar teeth become more and more like true molars.	

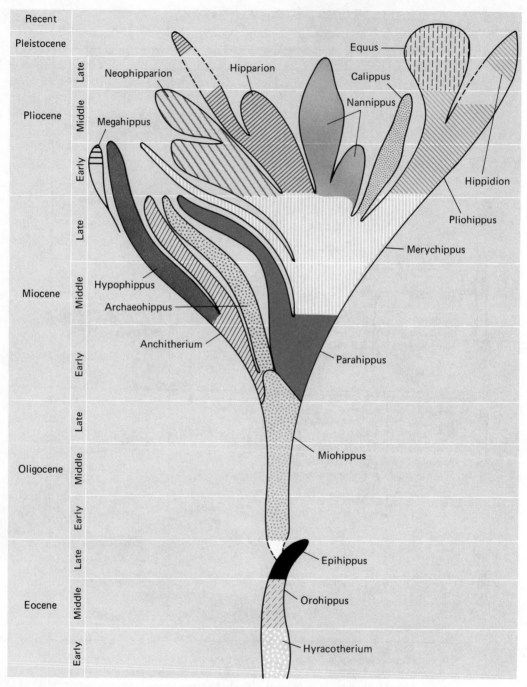

Fig. 6-37   **"Tree" of horse evolution. Note terminal branches.**
After Mayr, Linsley, Usinger, and Hanson.

and wolves. The fastest land animals are the cats that hunt in the grass. The cheetah, which hunts in the grassy savannahs of Africa, has very long legs, a greyhoundlike body, and reaches speeds of 60 mph, making it the fastest animal on the ground. It must be fast to catch the swift antelopes, whose evolution paralleled that of the horse (antelopes run on two hooves). Hooves are also defensive weapons. The horses and their relatives developed high-crowned, multiridged molars, effective grinders of the grasses with their high silica content.

The fossil record of the horse is nearly complete, and the direction of its evolution is clear. Many have used it as an example of straight line evolution, *orthogenesis,* in which an almost "supernatural hand" guides the course of evolution. Yet the map of horse evolution is a tree—with a trunk to be sure—but also with many terminal branches of extinct horses that left no descendants (Fig. 6-37). These are the experiments that failed as nature "groped."

The evolution of the elephant is also well known and is similar to that of the horse (Fig. 6-38). The "dawn elephant" of the time of primitive horses was a 2-ft high, hoglike creature, with a normal snout and 36 teeth. The second incisors were normal teeth; the molars hardly differed from those of man. In subsequent descendants, all canines disappeared, and all incisors other than the second dwindled. Great grinding cheek teeth (molars) similar to those of the horse developed, and the second incisors became the tusks. Of course, there was a tremendous increase in body size. The "dawn elephants" were shrub herbivores, whereas their descendants became grass grazers. The first elephants appeared in northern Africa about 60 million years ago. Their descendants spread over Asia and North America, following the grass. Here, mastodon and mammoth survived until recent times. The mastodon stood about 9.5 ft tall, had long, coarse hair, and inhabited Siberia, Alaska, and the northern United States. The mammoth was about the same size but was covered with a dense wool. The Columbian mammoth was 11 ft high and inhabited the southern parts of North America. Its fossils have been unearthed near Miami as well as on the Mexican plateaus, but no further south. The mammoth migrations probably never reached South America for the same reason that the buffalo failed. No proper grassland habitats extended south through Central America, an ecologic barrier to dispersal.

Many running birds evolved on the grasslands of the continents: the African ostrich, the South American rhea, and the Australian emu and cassowary. The wings of these birds are vestigial, and their breastbones are flat. The keeled breastbone of flying birds serves as an origin for the great pectoral (chest) muscles of flight (Fig. 6-39). In the ostrich, the number of toes has been reduced to two, one massive with a great toe-

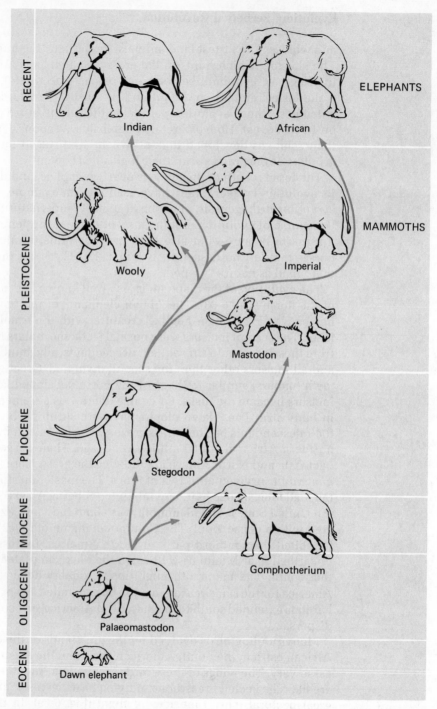

Fig. 6-38

Evolutionary history of elephants.

Adapted from Osborn, *Proboscidea,* American Museum of Natural History, New York, 1942.

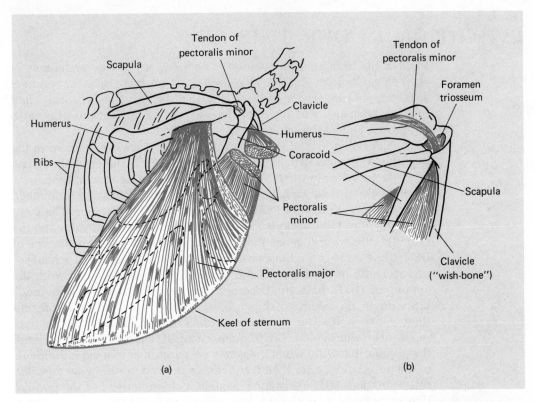

Fig. 6-39   **The anatomy of flight. (a) The great breast muscles of flight dissected from the side. The major pectoral muscle arises from the keel of the sternum and inserts upon the upper arm bone, the humerus. (b) The minor pectoral muscle (well developed here) also arises from the sternum, but it inserts upon the humerus by means of a tendon that passes through a pulley formed by the three bones. The action of the pectoralis major is to lower the wing (downbeat), and the pectoralis minor pulley raises the wing (upbeat). Neither of these muscles in man would be adequate to beat a feathered arm. The pulley arrangement in the bird's wing creates a mechanical advantage. This famous dissection was first made by Leonardo da Vinci, whose interest in flight and flying machines is well known.**
Drawing from G. B. Moment, *General Zoology*, Houghton Mifflin Company, Boston, 1967.

nail, and the other a small side balancer. If the flightless birds had been specially created, then it would be hard to explain their functionless, even comical wings. What unpoetic, natural insanity! Imagine taking a being freed of the earth, a living flying machine, a bioaviator, and making of it a bird-horse of the plain!

EVOLUTION IN OUR TIME

Most of the time, natural evolution is a gradual process, sometimes as slow as geologic evolution. Therefore, it is not surprising that no new species have appeared in the brief time (a little more than a century) that man has been aware of evolution. Except for human alteration, the earth of today is almost the same as it was 100,000 years ago.

The consistent use of DDT to control mosquitoes has resulted in the evolution of strains with higher and higher DDT resistance. Although no new species of mosquitoes have appeared, the frequencies of genes conferring DDT resistance have risen in these populations. An untouched natural population shows wide variation in DDT resistance. The commencement of DDT spraying places a premium (survival value) on the undoubtedly several genes that determine this trait. In the absence of DDT, these same genes have no survival value. They are so much meaningless static in the transmission of genetic information, but, with the coming of DDT, the static becomes a meaningful message. "Favored" individuals, possessing such genes, survive to reproduce and transmit these genes, and, after many generations of continued DDT selection, resistant strains evolve. The frequencies of DDT-toleration genes rise in the population. Only when the genes are combined in most individuals of the population, after their frequencies have risen following repeated selection, does DDT resistance become a characteristic of the population, rather than an exception. Since this "trait" is the result of the inter-

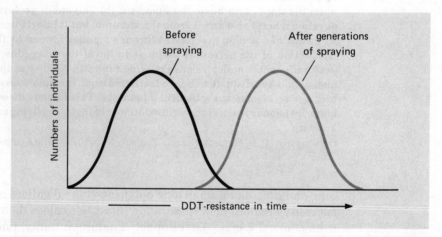

Fig. 6-40   **The evolution of DDT resistance in mosquitoes and other insects. The curve is shifted to the right.**

action of a combination of genes, the DDT resistance of the population still shows wide variation after a number of generations and describes a bell-shaped curve, the same type of distribution of DDT resistance that existed in the original, parent population. Now, however, the curve has been shifted to the right (Fig. 6-40). The new average DDT resistance is the old upper extreme. The results are more black and white when the trait is determined by a single pair of genes.

Other insects have also evolved great DDT resistance. The United States Army found to its dismay that DDT was useless in combating body lice during the Korean War, whereas it had been completely effective in World War II. It's enough to make one an evolutionary believer. It is an example of what Wallace and Srb call "man's unwitting experiments" in evolution.

Probably the most celebrated case of such a process involved pigmentation change in a type of moth. In 1845, near the industrial city of Manchester, an English butterfly catcher reported collecting a previously unrecorded dark variation of the common peppered moth, which was usually light colored. The situation is just the reverse today near the industrial centers of Britain: the dark forms are common and the light individuals rare, which was the case even in 1895.

In the rural areas of the British Isles, the light variety predominates. The moths alight upon pale lichen encrustations of tree bark, then orient themselves so that the crooked lines across their otherwise light bodies and wings blend in perfectly with the folds of the ashen lichens. So effective is their camouflage, that it is impossible to detect the still moths at a distance of a few feet. A dark form stands out against the pale background.

In the industrial areas of England, factory soot has destroyed the lichen growths and blackened the tree bark. On such trees, the light forms are very conspicuous, whereas the dark are nearly invisible. The situation is reversed. The dark coloration is due to the pigment melanin, and the dark forms are said to be "melanic." The phenomenal rise of the pigmented variation in industrial environments is called "industrial melanism." A darkened phenotype has now come to characterize 70 species of moths, many of which are not at all closely related, even belonging to different families.

Dr. H. B. Kettlewell released equal numbers of marked dark and light moths in both industrial and rural woods, recorded some of their fates photographically, and collected and counted the survivors. On blackened trunks, a greater number of light moths became the prey of birds; the reverse was true on lichen-covered, light bark.

The change of the moth population phenotype from pale to dark was an adaptive response to an altered environment (stimulus). The bird predators were the "natural selectors." The melanic variation came to have survival value, and therefore conferred reproductive advantage. The favorable variation (melanism) is heritable, and so it was passed on. Thus, the populations of the old species were modified to form "favored races" preserved in the "struggle for existence." In the beginning, melanic individuals made up 1% of the populations; at the end, they constituted 99%.

In the peppered moth, the dark trait is determined by a dominant or semidominant gene, which explains the rapidity of the change from light to dark color variants in industrial areas. If melanism were due to a recessive gene, natural selection in this case would have proceeded at a snail's pace, requiring perhaps 1000 rather than 50 years. The individuals in dark populations are almost all purebred, carrying a double dose of the gene for melanism. The melanic gene has a frequency of nearly 100%, whereas it formerly was less than 1%. The gene for melanin itself has changed from semidominant to dominant. Hybrid specimens from the last century (kept in British museums) are lighter than present-day hybrids, which are every bit as dark as their dark parents.

It is interesting to compare this case of moth evolution with an example of human evolution. Sickle cell anemia (Fig. 6-41a) is common in many African populations. The gene causing sickle cell anemia is a recessive mutant. Individuals with a double dose of the gene usually die, but hybrids (one normal and one sickle cell gene) survive. The normal gene is not really dominant to the gene for sickling, since hybrids are recognizable. Perhaps it is best to say that the normal gene is "semidominant." It seemed odd that the gene for sickle cell anemia—harmful as it was—would be tolerated at such a high frequency in the populations. It would logically seem that the gene would have been selected against rather than for, and that the frequency of the gene would have been kept low rather than high in the gene pools of the populations. Then it was discovered that hybrid individuals were immune to malaria. Populations with high frequencies of this gene occurred in the tropical equatorial regions of Africa, the "malaria belts" (Fig. 6-41b). Perhaps here we can see something of the imperfection of the process of evolution. Rome was a malaria bed, but the Romans never developed sickle cell anemia. Of course, their populations were not present very long in that particular environment. In time, it is possible that they might have developed some malaria-resisting trait, but not necessarily. Notice that a combination of two genes, one of which in double dose is fatal, is something less than a perfect plan.

Many cave creatures are blind and colorless. Albinism, the lack of pigmentation, is caused by a recessive gene that is held low in populations living above the ground and in the light, where coloration func-

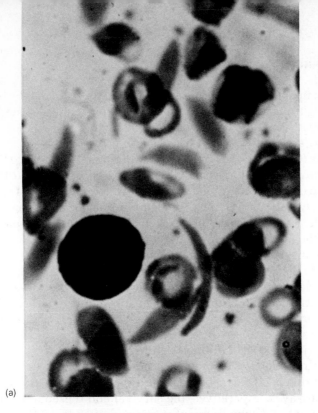

(a)

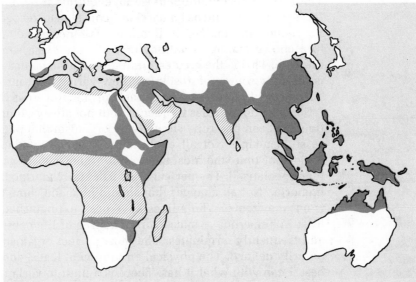

(b)

Fig. 6-41   **Malaria and sickle cell anemia. (a) Human blood, showing abnormal sickle cells
(deformed red corpuscles). (b) Distribution map, showing the coincidence of
malaria (solid colored areas) and sickling (hatched).**

(a) courtesy of Carolina Biological Supply Co., Burlington, N. C.

tions in camouflage, species recognition, etc. Albinos are blinded in the sunlight, since their eyes lack pigmentation. In the darkness of the caves, albinism is of little consequence, and the frequency of the gene for it has been "allowed" to rise. Vision is also useless in the dark cave environment. It has been proposed that easily damaged eyes are a liability. They are obviously not assets: a blindman is easily the equal of a seeing man in the dark. It is not difficult to see that defective genes for eye development would not only be tolerated but selected for, since eyeless forms are the "favored races" in the cave. Other sense organs, particularly tactile (tough) and chemical (taste and smell), have been highly developed. Cave insects frequently have extremely long, sensory antennae, their "white canes." Blind fish, salamanders, crayfish, beetles, and harvestmen (spiders) have been found in a number of caves. Most exhibit compensation in their other senses. The loss of pigmentation is not so easily explained. It is hard to see how the lack of coloration is adaptive in the dark cave. Not all cave creatures have lost their pigmentation, and many have rudimentary, although sightless, eyes.

Biologic success is only temporary. Survival is its prime criterion. Evolution is always relative. There is no such organism as "the fittest." It is always comparative: "the survival and reproduction of the fitter." The greatest disappointment is that evolution is not foresighted. A perfect evolution would be Lamarckian, consciously recognizing the "need" for change and making it. Evolution is hindsighted, always restricted to a choice among variations that already exist, some of which are obsolete. This is the greatest limitation on the choice of life. Mutation is the only source of absolutely new variety; but most mutations are harmful and recessive. The miracle of the process is that from this inferior selection life has frequently, but not always, found a way. Evolution has been painfully slow. Every organism that persists is only successful enough. Not all evolutionary change is adaptive. Over long periods of time, the most successful organisms are generalized rather than specialized. The perfect are particularly mortal; the imperfect are immortal. Not all changes that should be made come to pass. Progress is not inevitable. Whole lines have been extinguished. Failures are at least as numerous as successors. The tree of life continually branches and frequently terminates. The most perfect organisms are only temporarily defined. The physical environment leads, and life follows the best it can with what it has. There is a limit to variation, and the ultimate source of new variety—mutation—is haphazard at best. It is a dismal picture enlightened most by the mere persistence of life.

Perhaps the glory of it is that a nearly unconscious process has produced as conscious a creature as man. Herein, life redoubled its efforts to renew itself. It is as though man inherited an inescapable responsibility to life, not only his own. Life created from itself a being that overcame the shortcomings of his body with his overpowering mind, yet came to realize that the twain met so closely that they might be inseparable, despite his wishes. It was a dream fulfilled—imperfectly to be sure—as it had always been and would seemingly forever be. The gauntlet was cast down before the unknown.

DISCUSSION QUESTIONS

1. *How does the modern theory of evolution differ from classical Darwinism?*
2. *What parallels are there in the evolution of land plants and animals?*
3. *Are biochemicals formed only by living organisms?*
4. *How is evolution hindsighted? What are its limitations?*
5. *What are the possible reasons for the great and minor extinctions?*
6. *Are "missing links" or common ancestors rare in the fossil record?*
7. *What evidence do rassenkreis and clines provide for evolution?*
8. *What is the only type of species that satisfies all the requirements stipulated in the morphologic, reproductive, and genetic definitions of the term species?*

REFERENCES

DARWIN, CHARLES, *Autobiography of Charles Darwin*, W. W. Norton & Company, Inc., New York, 1958.

DARWIN, CHARLES, *The Origin of Species*, New American Library of World Literature, Inc., New York, 1958.

EISELEY, LOREN, *Darwin's Century*, Doubleday & Company, Inc., New York, 1958.

MERRELL, DAVID J., *Evolution and Genetics*, Holt, Rinehart & Winston, Inc., New York, 1962.

SAVAGE, J. M., *Evolution*, Holt, Rinehart & Winston, Inc., New York, 1968.

STEBBINS, G. L., *Process of Organic Evolution*, Prentice-Hall, Inc., Englewood Cliffs, N. J., 1966.

OBJECTIVES

1. Discuss the role of physical environmental factors such as temperature, pH, salinity, water, and aeration in determining the composition of biotic communities.

2. Diagram the carbon and nitrogen cycles.

3. List, define, and exemplify five types of organismal interactions.

4. List the four major nutritional categories of organisms in a biotic community.

5. Construct and compare the pyramids of energy, numbers, and biomass.

6. Trace several routes by which energy is lost in the transfer from one trophic level to the next.

7. Draw and interpret the phases of the growth curve of an organism in the laboratory.

8. Contrast natural and in vitro growth curves; compare the natural and the human ecosystem.

9. Express the mathematic relationship between population, biotic potential, and environmental resistance.

10. Discuss the stress theory of natural mammalian birth control.

11. Distinguish between ecologic and physiologic death.

12. List the six major land biomes, and describe their physical characteristics, climax vegetation, and several representative animals in each.

13. List, define, and exemplify three classifications of plants based upon the water conditions of their habitats.

14. Describe four adaptations of the kangaroo rat and the camel to the desert environment.

15. Analyze an Everglades and an oceanic ecosystem in terms of the physical factors, trophic levels, and representative food webs.

16. List and describe the stages of an ecological succession from pioneer to climax communities.

17. Compare the complexity, stability, and productivity of climax and replacement ecosystems.

18. Contrast biologic control and the use of pesticides.

19. List seven types of pollutants; describe their effects and recommend corrective measures or alternatives.

20. Compare the cellular, organismal, and ecosystemic steady states.

The
House of
Nature

ALMOST UNAWARE, man has entered a natural crisis, unnaturally made. Although man is a natural creation, his ways are without parallel in the natural world. Subtle environmental crises extending over millenia have resulted in the extinction of countless species. Few species have altered their environments as rapidly and to the extent that man has; rather, most species have obeyed the dictates of their surroundings, and change, when it occurred, was gradual. The conquest of habitats for which man was not biologically fit led him to believe that he was not subject to nature's ways, as long as the world was large. He came to feel that he was no longer a creature of the natural world and therefore bound by its laws. The Greek philosopher Protagoras said, "Man is the measure of all things."

The Greek word for house is "oecos," and the term *ecology* is derived from it. Ecology is the study of the relationships of organisms and their environments, both living and dead. Man is destined to be both landlord and tenant in nature for foreseeable decades—perhaps centuries—to come. His misconception of ecological freedom persists, as we can see in the writings of Eric Hoffer:

My feeling is that the humanization of billions of adolescents would be greatly facilitated by a concerted undertaking to master and domesticate the whole of the globe. One would like to see mankind spend the balance of the century in a total effort to clean up and groom the surface of the globe—wipe out the jungles, turn the deserts and swamps into arable land, terrace barren mountains, regulate rivers, eradicate pests, control the weather, and make the whole land mass a fit habitation for man. The globe should be our and not nature's home, and we no longer nature's guest.[1]

Too often, man's "conquest of environments" has involved the destruction of their occupants before he learned the wisdom of their ways. Hopefully, not too late, man has begun the study of the interrelationships of organisms and their environments. Like any science of synthesis, ecology is newborn. Man will learn the wisdom of nature's ways or perish without it. Nature will now measure man.

No living organism is an island. A cell engages in constant molecular traffic with its surroundings, and the cell that fails to maintain the steady state of its reactions declines. The organ systems of the advanced, multicellular animal have as their common goal homeostasis, "the maintenance of the constancy of the internal environment" and perhaps more, since the nervous system often integrates the internal with the external environment. Homeostatic breakdown leads inevitably to death. The concepts of the steady state and homeostasis, originally intended to apply to the cellular and organismal levels, have been expanded to include the population and species levels of organization as well. Populations are woven into functional communities. The highest level of organization is the ecosystem, which includes the interacting communities and the nonliving components of the physical environment as well. Over short periods of geologic time, homeostasis, or "a balance of nature," is the unifying theme of the constituents of an ecosystem. Ecology is the study of *ecosystems,* the structural and functional units of the *biosphere,* the thin crust of the Earth in which life arose and to which without exception all living forms are yet bound. The salvation of Goethe's Faust lay in the reclamation of land from the sea. In the future, man may be wise to "farm the sea." In a sense, man has returned to nature, for he has not yet escaped. Too long has man been guilty of the mortal sin the Greek tragedians called *hybris,* or overweening pride. The economics and technology of constructing a wholly artificial environment are completely beyond any known or nearly conceivable social organization of man. We must live in harmony with nature or perish as a species.

---

[1]Eric Hoffer, *The Temper of Our Times* (New York, N. Y.: Harper & Row, Publishers, Inc., 1967), p. 94.

THE NONLIVING ENVIRONMENT

The factors of the inorganic world determine to a great degree the kinds of life permitted. The organic evolution of life on the land has been co-ordinated with the physical evolution of the surface of the Earth. Sometimes the living world has altered the physical environment; free oxygen was liberated into the atmosphere for the first time only after the evolution of photosynthetic organisms. The humus layer of rich soils is largely formed by organic products resulting from the decay of leaves and grasses. The major media in which or upon which organisms live are the soil, the fresh water, and the ocean. Aside from climatic factors, the properties of the physical environments include acidity, salinity, aeration, and the water-holding capacities of the various soils. Temperature, sunlight, wind, and precipitation are the most important characteristics of climate.

Porosity is a crucial quality of soil. The sand and gravel "soils" of deserts and young mountains cannot hold water. Clays, rich in aluminum salts, hold water well but are poorly aerated. The richest soils, deposited by slow-moving rivers and their seasonal floods, are well aerated and rich in organic matter.

Most life forms are moderates, intolerant of environmental extremism. The few organisms, however, that dwell in radical environments offer further proof of life's plasticity.

Most organisms are adapted to live in an external environment of near neutrality, neither acidic nor basic. Salt grass, bromegrass, greasewood, and rabbit brush grow in the alkaline bottoms of dry desert lake beds. Certain cyprinid fishes survive in alkaline desert pools near Denio, Nevada, where the pH of the water is 10.0, extremely basic. A few molds

Fig. 7-1

Vinegar eels are roundworms adapted to living in the acidic environment of vinegar (acetic acid). Roundworms and insects are the two most successful classes of animals. Roundworms are found in the widest variety of extreme habitats, from glacial pools in the Alps to desert sands.

Photo courtesy of Saturn Scientific, Inc., Fort Lauderdale, Fla.

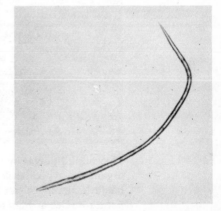

can grow in an even more alkaline range. One sulfur bacterium and a few fungi tolerate very acid media (pH range of 0 to 3). Mosses are the dominant vegetation of acidic peat bogs. Vinegar eels (roundworms) withstand the relative acidity of their medium (Fig. 7-1). The pH of sea water is about 7.8, slightly alkaline, and the pH of the internal sea of man is about 7.4, nearly neutral.

Most cells function only within a moderate and narrow temperature range, between 50 and 113°F. High temperatures tend to irreversibly alter the configurations of proteinaceous enzymes, upon whose functioning the life reactions depend, and low temperatures reduce the reaction rates so severely that a nearly suspended animation is the result. Since liquid water is the main reaction sea of biochemistry, it is not surprising that freezing halts life activities. Also, the ice crystals formed disrupt the patterned organization of membranes, upon whose mosaic surfaces the sequential reactions of life frequently take place. The majority of organisms that live only within moderate temperature ranges are called *mesophiles*. Those organisms that tolerate temperatures higher than 113°F are termed *thermophiles* (heat lovers). Some bacteria and algae live in hot springs at temperatures of 140 and 185°F. The geyser pools of Yellowstone National Park are colored by their growth, as well as by inorganic mineral deposits. Blue-green algae flourish in these hot springs at temperatures as high as 185°F (Fig. 7-2). Eggs are

Fig. 7-2   **Thermophilic algae in water of hot springs. In contrast to most organisms, which seek the middle temperature range, a few bacteria and algae can grow in environments where the temperature rises as high as 185°F, a few degrees below boiling.**
Courtesy of Carolina Biological Supply Co., Burlington, N. C.

easily cooked at these temperatures just below boiling. Egg white (albumin) coagulates at 140°F.

Cold-resistant organisms are not rare. Red snow algae color wide areas of arctic and alpine snow and ice fields pink to brilliant purplish red. Arctic and deep-sea fishes carry on their life activities at temperatures just above freezing. Some molds continue to grow below zero, even at −4 to −6.7°C. Bacteria grow slowly in ice cream at −10°C and on aging meats in refrigerator boxes; in fact, it is the bacteria and molds that tenderize aging meat by partially digesting it.

"Cold-blooded" organisms cannot maintain a constant body temperature, and their temperatures vary with that of the surrounding environment. Reptiles, amphibians, fishes, and invertebrates are such *poikilothermic* ("variable-temperature") animals. Birds and mammals keep a constant body temperature and are said to be "warm-blooded," or *homeothermic* ("same heat"). This condition liberates them from the whims of the environment. During the winter, many mammals hibernate, set their physiologic thermostats lower (at about 70°F), and sleep deeply.

The lungfishes "summer-hibernate," or *estivate*. As their water holes dry up, they burrow down in the mud, secrete a water-binding mucus about themselves, make a blowhole to the surface, and enter a state of lowered animation and metabolism for as long as 5 years, or until the water returns (Fig. 7-3). Many freshwater shrimp and microscopic, but multicellular, pond animals lay drought-resistant eggs that lie dormant during dry spells. Bacteria produce spores that are heat, drought, and cold resistant.

Most plants and animals can live only in medium that contains some salts. As a rule, they do not tolerate pure distilled water or strong salt solutions.

One way to kill most plants is to sprinkle salt upon their soil. Absorption of water from salty soils is difficult for most plants. When the concentration is high enough, water is withdrawn from the roots of these plants, causing permanent wilting and death. Those plants mentioned previously, which grow on alkali flats, face not only a very basic medium but also a salty one. Such species have become adapted to securing their water from such soils, and are known as *halophytes,* or salt-loving plants (Fig. 7-4). (Halides are chloride, bromide, iodide, and fluoride salts.) "Salt bushes" are found growing in the saline encrusted soils of Utah. Many halophytes resemble desert plants, even though they can be growing in soils with an abundance of water. The high concentration of solutes (salts) makes most of the water unavailable, so that the plants of such places are in reality exposed to desert conditions, although the total soil water content may be high. The soil is wet physically, but physiologically it is dry.

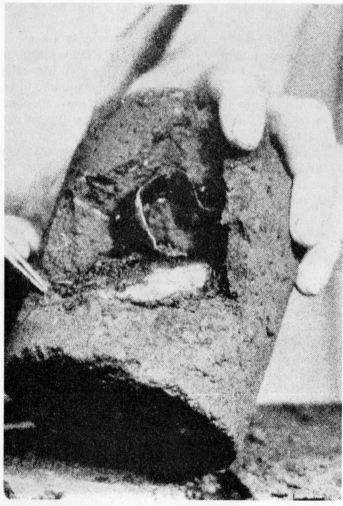

Fig. 7-3

**"Summer-hibernating" lungfish. When the waters of their habitat
evaporate, they burrow down in the mud, secrete a mucus about
themselves, make a blowhole to the surface, lower their metabolism, and await the return of the waters, a wait of sometimes
several years.**
Courtesy of CCM: General Biological, Inc., Chicago.

Salt-loving bacteria, the *halophilies,* and the brine shrimp (*Artemia*)
(Fig. 7-5) live in the waters of the Great Salt Lake, where the salt concentration is usually greater than 20%. Ordinary fresh water, sewage, and
even marine bacteria die after a few minutes in this water. (Salt was

(a)   (b)

Fig. 7-4   Halophytes, "salt plants," grow in dry lake bottoms, alkali flats, or salt flats,
                  where, although the soil is usually far from sandy and possesses good water-
                  holding capacity, it is difficult to take up water against the osmotic pull of
                  the salts. Greasewood, rabbit brush, and other halophytes have the reduced
                  leaves that also characterize other desert plants growing in sandy soils. (a)
                  Greasewood. (b) Salt grass.
                  Photos by Massoud Moezzi.

Fig. 7-5   Brine shrimp, crustaceans, are among the few organisms that
                  can live in the tremendous osmotic pressure of the Great Salt
                  Lake.

used as a preservative of cheese, fish, and hides.) The lake bacteria withstand the great salinity, and, indeed, require more than 13% for normal growth. These bacteria must represent mutants naturally selected to adapt to the increasing salt concentration of the evaporating lake, which is a remnant of a great inland sea. Several different races of brine shrimp exist in the salty waters of North America and Europe, and they have been much studied in an attempt to clarify the interplay of heredity and environment in evolution. Most saltwater organisms cannot survive in fresh water, and vice versa. Sea water has been a formidable barrier to the dispersal of many species.

The oxygen content of natural waters depends upon the quantity of photosynthetic organisms present, water temperature, and the rate of bacterial decomposition of debris. Free oxygen is the by-product of photosynthesis. One can see bubbles of it forming in masses of "water moss." The solubility of the gas in water is inversely proportional to the temperature. Heavy, deep, cold water in lakes and oceans is oxygen rich; shallow, warm water is usually poor in oxygen. Bacterial decomposition of plant and animal remains usually occurs in the bottom muck of bodies of water. Decomposition is essentially the respiration of organic matter by the microorganisms of decay, and the process therefore depletes water oxygen. Bottom muds and the water immediately above them are oxygen poor. Unmoving, or stagnant, shallow pools are usually places of foul rot and oxygen lack. A few organisms can withstand this environment. Segmented worms living in the bottom muck possess respiratory pigments such as hemoglobin that increase the oxygen-carrying capacity of their blood manyfold. These pigments also load quickly (saturate) at low oxygen tensions (concentrations). The red worm, *Tubifex,* which thrusts its head in the mud and waves its tail in the water above, can become so abundant in ponds and slow-moving streams that the bottom turns red. It is an indicator of the degree of organic pollution of streams and ponds by cattle or other sources. There is natural pollution, but the dumping of sewage and wood pulp washings into streams increases the rate of bacterial decomposition and decreases the oxygen content of the water. The fish kills that result are similar to those that occur naturally in the stagnant waters of seasonal droughts. As the water becomes shallower and warmed throughout, it barely covers the decomposing muck, and the negligible oxygen is inadequate to support higher forms of life. Trout, sunfish, and bass are among the first to die. Catfish last longer, because their scaleless skins serve as accessory organs of respiration in addition to their gills. In fact, catfish can survive out of water for several hours. Some fish have lungs.

The living community and its physical environment are undivorcably wed. As life began in the past from nonlife, so today the life processes begin with nonliving materials and end with their return to the environment. The air we breathe, the water we drink, and the salts of our internal seas have passed through countless generations of organisms.

THE BIOGEOCHEMICAL CYCLES

The vast cycles of matter moving from the inorganic world into the living and back are appropriately called *biogeochemical cycles*. The most important of them are the carbon, water, nitrogen, and phosphorus cycles.

All biochemicals are carbon compounds. Carbohydrates, proteins, fats, nucleic acids, and vitamins are synthesized by plants from the carbon dioxide of air and water and a few other elements. Plants are eaten by animals, and, after digestion and absorption, the carbon compounds are either reorganized into animal biochemicals or destructively metabolized (respired) to yield up carbon dioxide as a respiratory waste gas to be returned to the air or water. Plants, too, respire. When the bodies of dead plants and animals are decomposed (respired) by bacteria, all their carbon is returned to the environment as CO_2. The photosynthetic organisms take it up again and the cycle is repeated. This is the *carbon cycle* (Fig. 7-6).

The cycles of water and oxygen are also bound up with photosynthesis and respiration. Free oxygen was not present in the Earth's atmosphere until photosynthetic organisms evolved. Oxygen is a by-product of photosynthesis, coming from the split water molecule. Oxygen is taken from the air or water by animals and plants to serve their respiratory processes. This oxygen is combined with hydrogen and returned to the environment as the water of biological oxidation. Water rises from the seas, is precipitated on the land, and runs back again. Not all of the water absorbed by a plant is used in photosynthesis; much of it is lost to the surrounding atmosphere. The water ingested by animals becomes part of their body fluids; some of it is used to wash away their excretory wastes in urine. Water is also lost in sweating, breathing, and defecating. Much of the ground water and surface runoff makes its way back to the sea or atmosphere without ever having entered biologic side chains. The *water cycle*, the *oxygen cycle*, and the carbon cycle are all intertwined with processes of photosynthesis and respiration (Fig. 7-7).

Nitrogen is a vital element found in the bases of DNA, RNA, vitamins, phospholipids, in the amino acids of proteins, and in many polysaccharides of animal tissues. Obviously, this element is absolutely essential to

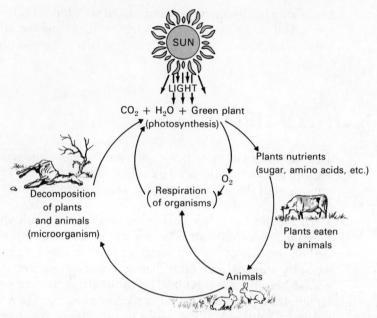

Fig. 7-6　Carbon cycle. Green plants use the energy of sunlight, carbon dioxide, and water to make all the carbon skeletons of the biochemicals. Oxygen is liberated during photosynthesis. Animals depend upon plants for food as well as oxygen. The end products of animal (and plant) respiration are water and carbon dioxide. The bacteria and fungi of decay respire the carcasses of dead plants and animals, once again producing carbon dioxide (CO_2) and water (H_2O), with which the food cycle began.

all reactions of metabolism and growth. Most nitrogen—aside from that locked in the biochemicals of organisms—is in the form of free, atmospheric nitrogen (N_2). Free nitrogen cannot be utilized by plants and animals. It is nearly biologically inert. Nitrate (NO_3^-), nitrite (NO_2^-), and ammonium (NH_4^+) salts are available to plants in water solution and determine soil fertility to a great extent. Although some ammonia is formed in the upper atmosphere along lightning and rained down, the amount is insignificant. Certain bacteria play a crucial role in determining soil fertility because of their participation in the *nitrogen cycle* (Fig. 7-8).

The *nitrogen-fixing bacteria* are able to capture atmospheric nitrogen gas by combining it with other chemical elements to form nitrogenous organic compounds, whose nitrogen eventually becomes available in usable form for plant nutrition. One group of nitrogen fixers lives free in the soil. These free-living microorganisms use carbohydrates from

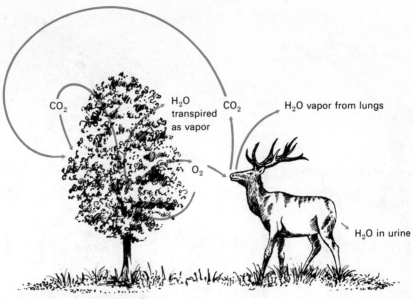

Fig. 7-7   **Interrelation of photosynthesis and respiration.**
From N. M. Jessop, *Biosphere,* Prentice-Hall, Inc., Englewood Cliffs, N. J., 1970.

decaying plant remains as an energy source and incorporate atmospheric nitrogen into their own biochemicals. When these organisms die, they are themselves decomposed by other soil microbes, and the nitrogen of their biochemicals is released in the form of ammonia (NH_3). Ammonia dissolved in soil water forms ammonium salts, such as NH_4Cl (ammonium chloride), which are briefly available to higher plants. Unfortunately, some denitrifying bacteria break down ammonia and liberate free nitrogen gas to the atmosphere. On the other hand, *nitrifying bacteria* oxidize ammonia (NH_3) to nitrites (NO_2^-) and nitrates (NO_3^-), a process called nitrification. Nitrate salts are the stablest soil nitrogen compounds and therefore the most generally available, useful form of nitrogen for plant nutrition. Nevertheless, there are denitrifying bacteria that reduce nitrates to nitrites and ammonia, finally liberating nitrogen gas.

A second group of nitrogen-fixing bacteria live in nodular swellings on the roots of certain plants, clover, peas, beans, the peanut, and the hay, alfalfa (Fig. 7-9). The *root-nodule bacteria* use the carbohydrates and other nutrients of the plant juices at the same time that they fix atmospheric nitrogen. This fixed (organic) nitrogen eventually becomes available to the plant. For decades now, farmers have rotated the pre-

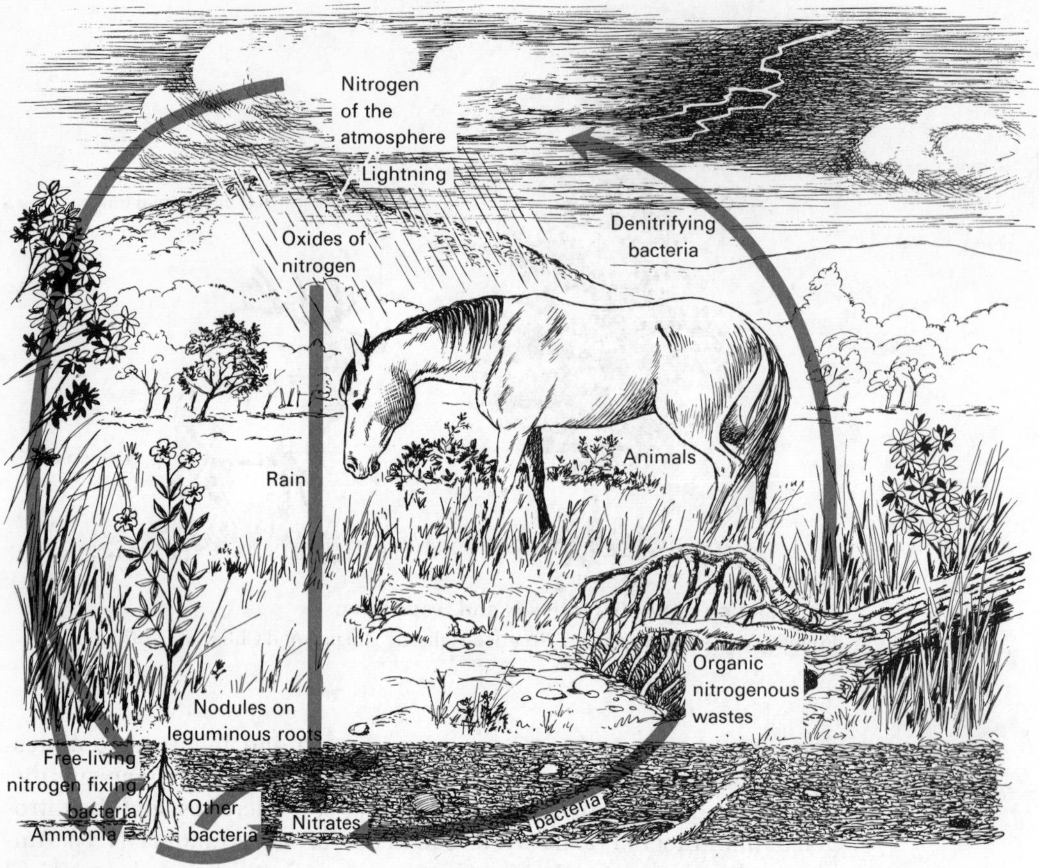
Nitrogen of the atmosphere
Lightning
Denitrifying bacteria
Oxides of nitrogen
Rain
Animals
Organic nitrogenous wastes
Nodules on leguminous roots
Free-living nitrogen fixing bacteria
Ammonia
Other bacteria
Nitrates
bacteria

Fig. 7-8 **Nitrogen cycle. See text for further description.**
From A. W. Galston, *The Green Plant,* Prentice-Hall, Inc., Englewood Cliffs, N. J., 1968.

ceding plants with soil-exhausting crops, such as cotton and tobacco, whose roots do not bear nodular nitrogen fixers. Rotational farming with peanuts restored the fertility of many southern soils depleted by successive years of cotton planting. The great American agricultural chemist and botanist, George Washington Carver, promoted this method of scientific farming in the South.

Carnivorous plants have solved the problem of growing in nitrogen-poor soils in another way (see Fig. 7-10).

In addition to crop rotation, farmers also employ mineral nitrate fertilizers. Saltpeter (potassium nitrate) is mined in several regions of the earth, notably Chile. Nitrate is also used in the preparation of gun-

Fig. 7-9   **The roots of several plants bear swellings known as root nodules, in which nitrogen-fixing bacteria reside. These bacteria possess the rare ability to fix atmospheric nitrogen directly in their biochemicals. Some of this fixed nitrogen is made available to the "host." The relationship is actually mutualistic, since neither the host plant nor the bacteria alone can fix atmospheric nitrogen.**

Courtesy of Carolina Biological Supply Co., Burlington, N. C.

(a)   (b)

Fig. 7-10   **Insectivorous plants. Venus' flytrap (a) and the sundew (b) have solved the problem of nitrogen deficiencies in the soils of their habitats by eating animals.**

Photos courtesy of Carolina Biological Supply Co., Burlington, N. C.

powder. In World War I, when Germany's nitrate supply was cut off, the German chemist Fritz Haber devised a synthesis. Nitrogen and hydrogen, under great pressure and heat in the presence of iron oxide and potassium aluminate catalysts, were combined to form ammonia (NH_3), which was then oxidized to nitrates. Living organisms have accomplished the same ends under mild conditions. How?

Phosphorus is also an indispensable element in biochemistry. Phosphate is found in DNA, RNA, ATP, and the phospholipids so important to cell membrane construction. Immediately after glucose enters a tissue cell, it is phosphorylated, and most of the series of compounds found in the anaerobic phase of respiration are phosphates. Like nitrogen, phosphorus is an essential element for reproduction, growth, and metabolism. Calcium phosphate is also found in bones, teeth, shells, and chitinous exoskeletons. The mineral phosphates deposited in Earth's rocks are not very soluble, but the small quantities of them that are washed away by streams and deposited in soils are converted into the more soluble phosphoric acid through the action of carbonic and other organic acids produced by soil microorganisms. Higher plants can take up con-

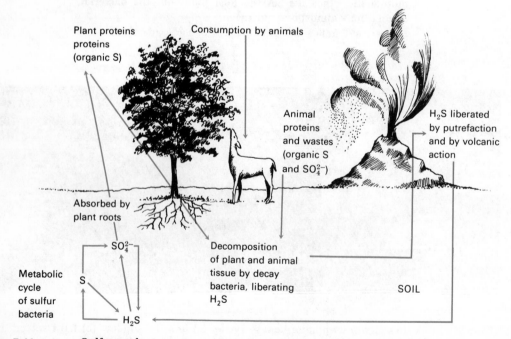

Fig. 7-11   **Sulfur cycle.**
From N. M. Jessop, *Biosphere*, Prentice-Hall, Inc., Englewood Cliffs, N. J., 1970.

siderable amounts of phosphoric acid and some phosphates. After its incorporation into the biochemicals of plants and animals, phosphate is returned to the soil through death followed by decay. Ground bone is used as a soil fertilizer. Some phosphates are carried to the sea food chains of marine fish and sea birds. The huge bird-dropping deposits (guano) of coastal islands off Peru and elsewhere are mined and used as phosphate fertilizers. The pebble phosphate rock of central Florida is also utilized for this purpose. Because phosphate is only slightly soluble, it is often the factor limiting growth in an ecosystem. Soils that have been under cultivation for a half-century or more in the Mississippi Valley have 36% less phosphate than virgin soils of the same region.

Sulfur is also an important element for life; it is found in several amino acids. The sulfur cycle is diagrammed in Fig. 7-11.

The total environment of a single organism includes not only the physical factors but the surrounding organisms as well. Although nutrition begins with compounds of the nonliving world, the biochemicals synthesized by plants travel through the dependent animals and other organisms, which are functionally related, until they are returned to the physical environment.

THE BIOLOGIC ENVIRONMENT

An organism is defined to a large extent by its total environment, both physical and biologic. As anatomy and physiology (structure and function) are frequently inseparable, so an organism is to a great degree a reflection of the demands placed upon it by its living and nonliving surroundings. Few organisms are perfect mirror images of their environments, but almost all life forms possess anatomical devices and physiological mechanisms selected by the hand of nature to allow them to cope with their environmental pressures. These *adaptations* fit organisms to their environment.

The *habitat* of a species of plant or animal is the particular environment it inhabits. Habitat is a term of spatial location and has been called the organism's "address." The species' *econiche* is the "occupation," function, or role of the organism within its environment and *not* a spatial address.

All the living organisms of an ecosystem, such as a freshwater pond, belong to a *biotic community*. No two species occupy the same econiche indefinitely; eventually, one form replaces the other. This is called the *principle of competitive exclusion*. Competition is greatest among members of the same species. Animal *intraspecific interactions* include competition for food, territory, and mates; plants compete for sunlight and water. Cooperation between individuals of the same species does exist in animal societies. The relationships between members of different species, *interspecific interactions,* are prey-predator relations, host-parasite interactions, symbiosis, antibiosis, and sometimes the competition for food.

Predators are generally large animals that feed upon smaller prey organisms. Parasites are most often minute or microscopic animals and bacteria that live in or upon, and at the expense of, their host. The most successful parasites do the least damage to their hosts, thereby perpetuating their means of livelihood (Fig. 7-12). In fact, parasitism is almost one extreme of a spectrum of intimate organismal relationships called *symbiosis* ("life together"). Parasites benefit from the relationship, whereas hosts suffer, severely or minutely.

A common type of symbiosis is *mutualism,* in which both associated organisms benefit. The relationship of leguminous plants and their nodular nitrogen-fixing bacteria is mutualistic. The roots of many plants, pines, spruces, oaks, willows, and orchids are intimately associated with soil fungi. The combination of root and soil fungus is called

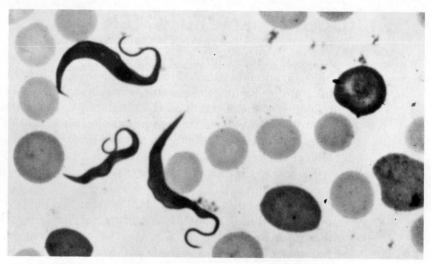

Fig. 7-12   **The sleeping sickness protozoon *Trypanosoma gambiense.* This same parasite, which fatally prostrates men and domestic cattle, regularly lives nearly harmlessly in the blood of the multitude of African antelopes, where it is almost a commensal rather than a parasite.**
Courtesy of CCM: General Biological, Inc., Chicago.

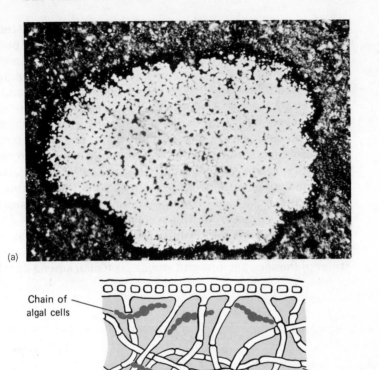

(a)

Chain of
algal cells

Fungus

(b)

Fig. 7-13   **(a) Lichens are hardy, intimate associations of an alga (primitive plant) and a
fungus; they grow on dry rocks, as shown above. The alga manufactures the
food and the fungus procures moisture from the atmosphere. (b) Diagram of
microanatomy of lichen section.**
Photo courtesy of Carolina Biological Supply Co., Burlington, N. C.

a *mycorhiza* ("fungus-root"). The fungi digest organic matter in the soil
and make nitrogen available to the plant. Perhaps fungi and their associated bacteria fix free nitrogen as well. The fungi obtain food substances
and vitamins from the roots. The relationship between fungus and root
is believed to be mutualistic. Mycorhizal plants can grow in otherwise
poor soils. The "mosses" growing upon the rocks of arid and semiarid
regions are *lichens,* consisting of an alga and a fungus (Fig. 7-13). The

algae photosynthesize food, and the fungi take moisture from the atmosphere. Both organisms benefit (even though the fungus does digest some algae). The bacteria living in the ideal environment of man's intestine synthesize vitamin K and receive a rich food supply in return. Man cannot digest cellulose, and therefore this sugar-rich compound is worthless to him as a foodstuff. Protozoa and bacteria in the guts of cattle, porcupines, beavers, and termites digest cellulose for their hosts, which gather vegetation and provide a warm, moist, and food-rich environment for the microorganisms (Fig. 7-14). This mutually beneficial process is called *symbiotic digestion*. The algae living in the tissues of flatworms, jellyfishes, and corals provide these animals with food and find an environment richer in CO_2, nitrogen compounds (animal wastes), and phosphates than is the surrounding water.

In the symbiotic relationship of *commensalism,* one species benefits and the other is unharmed. Somehow, the clown fish is immune to the stings of the sea anemone and swims protected among its tentacles, per-

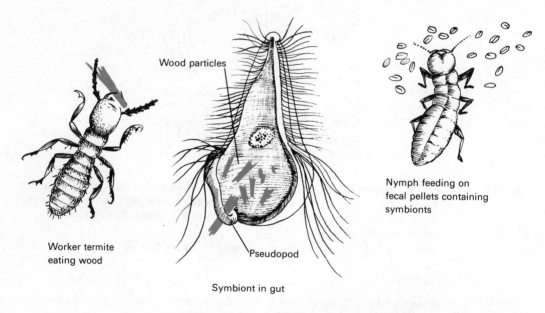

Wood particles

Nymph feeding on
fecal pellets containing
symbionts

Worker termite
eating wood

Pseudopod

Symbiont in gut

Fig. 7-14　　**Mutualism between termites and their intestinal protozoons. Several flagellates live in the gut of termites and digest the wood the insects ingest. Without these protozoan digestive symbionts, termites could not get sugar from cellulose. Young termites become "infected" with the symbionts by eating adult fecal pellets.**

From N. M. Jessop, *Biosphere,* 1971, Prentice-Hall, Inc., Englewood Cliffs, N. J., 1971.

haps decoying other fish species for the anemone. One tropical fish lives in the "rectum" of a sea cucumber.

Some organisms produce substances poisonous to other species. Certain fungi and bacteria produce these antibiotic chemicals that inhibit the growth of other species. This relationship of *antibiosis* may exist between some species of higher plants.

THE FOOD WEB

The organisms of a biotic community are conveniently classified by their nutritional roles. There are four major categories: *producers, consumers, decomposers,* and *scavengers.* Producers manufacture their own food: they are *autotrophs* (with independent nutrition). The most common and important producers are the green plants. They are the foundation of the food chain, and all other organisms depend upon them directly or indirectly. The number of producers in an ecosystem is greater than any other category of organisms; their combined biomass exceeds all others. If dried and burned, their combustion yields more energy than any other nutritional group of organisms in the ecosystem.

The second largest nutritional level in terms of numbers, biomass, and combustible energy is that of the *herbivores,* or *primary consumers,* the animals that eat the plants. These may be microscopic protozoa and crustaceans or macroscopic rabbits, antelopes, or bison.

The *secondary consumers* are *primary carnivores,* flesh eaters, and their populations, combined biomass, and total energy are sharply lower than those of the producers and herbivores. Feeding upon the first carnivores are *tertiary consumers* or *secondary carnivores. Tertiary carnivores* usually make up the highest nutritional level and are called "top carnivores." Wolves, eagles, and sharks are examples of these animals that are not the prey of any others. The numbers, biomass, and total energy of top carnivores in nature is the least. Although man is an omnivore, eating everything, he sometimes functions as a top carnivore.

Even though the top carnivores are not eaten by others, they, too, are parasitized, as are almost all plants and animals. When animals die or are slain and only partially eaten, *scavengers* pick the carcasses. Foxes, hyenas, coyotes, vultures, ravens, beetles, and fly larvae (maggots) are the primary scavengers of animal remains. Ants and termites are the most common scavengers of plant remnants. Bacteria and fungi, the *decomposers,* respire what scavengers leave to CO_2, H_2O, NH_3, and phosphates. The "food cycle" is complete; the nonliving compounds with which it began are returned to the physical environment.

The food chain is not definite. A hawk can be either a primary carnivore if it eats an herbivorous rodent, or a tertiary carnivore if it takes a snake, which has eaten a frog, which has eaten a grasshopper, which has eaten grass. A "food chain" also has side branches with scavengers and decomposers. For these reasons, ecologists prefer the term "food web" to describe the nutritional relationships of the members of an ecosystem's biotic community.

The relationships in a food chain have been portrayed as *pyramids of numbers, biomass,* and *energy* (Fig. 7-15). The pyramid of numbers has just been described: the producers at its base are the most numerous, and the top carnivores at the apex or point of the pyramid are least. Early man was primarily a top carnivore. With the agricultural revolution, man moved lower in the food chain, closer to the producer level, and therefore his population expanded. If man could devise an artificial photosynthesis, more efficient and simpler than the natural, his population would be able to rise to the maximum, short of becoming plant himself. The pyramids of biomass, numbers, and energy are scalloped on the edges, rather than smooth. The energy losses suffered in transfer from one trophic (nutritional) level to the next can be explained by several factors. Not all of a prey animal is eaten or digested. Bones, skin and fur, hard-to-clean meat, and tough tendons and ligaments are left for scavengers. (Dermestid beetles are used to clean skeletons, since their powerful jaws can cut through the literal ropes of tendons and ligaments.) Much ingested, digested, and absorbed food is respired, and the process of respiration is not perfect. More than 50% of the energy liberated from foodstuffs in respiration is lost as heat. Cellulose is indigestible to many animals, and even those that carry out symbiotic digestion do so with only a 10 to 40% efficiency. Organisms that die are deposited in the "coffins of decomposition," and their biochemicals and energy do not ascend the straight chain of a smooth pyramid.

MAXIMA, MINIMA, OPTIMA

After the appearance of a successful new species it tends to spread, but no species occurs uniformly over the entire world. Each is restricted to a definite *range,* or geographic area of distribution. The external factors regulating distribution include physical, climatic, and biologic *barriers* to dispersal. Aquatic species cannot traverse the land; freshwater spe-

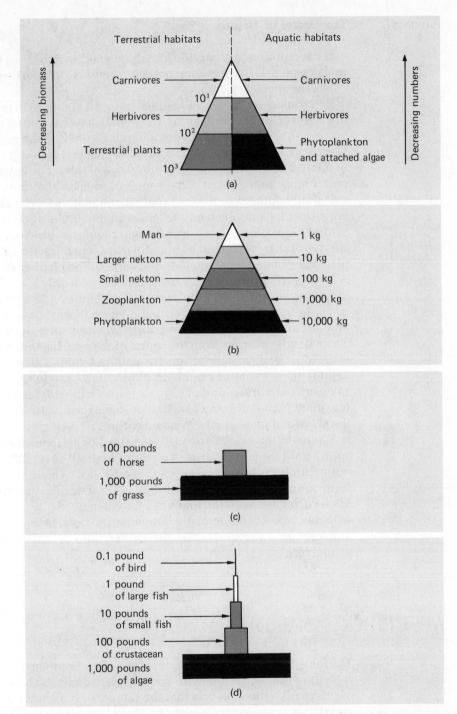

Fig. 7-15   Nutritional pyramids. (a) Pyramid of numbers. (b) Biomass pyramid. (c) A more accurate graphic depiction of the pyramid of biomass. (d) The energy pyramid; avenues of energy loss are discussed in the text.

cies cannot cross oceans. Moist-habitat species find deserts barriers to their dispersal. Competitors, predators, and regional diseases can also check the species' march.

Every species of plant and animal has a maximal and a minimal *tolerance* to each factor (biologic and physical) in its environment. The range and equilibrium level of a population is limited by the essential factor present in the least amount, or an inhibiting factor present to the point of intolerance. There is an optimum range of tolerance in which the species is most abundant; it is the center of species distribution. To either side of the optimal center, the species' populations decrease, until they are absent beyond the maximal and minimal limits of toleration.

Most plant species have definite light requirements. Some species can survive and grow in light of low intensity, as on the forest floor. These plants are said to be *shade tolerant*—the coffee plant is shade tolerant. A few species of plants grow for some time with only 5% of full sunlight. Most species are *shade intolerant,* often requiring 50% of full sunlight or more. In the feeble light reaching the rain forest floor, undergrowth may be absent. Grasses grow poorly when shaded by dense foliage. Their stands decrease from the center of the clearing toward the forest's edge. Most grasses cannot survive with less than 33% of full sunlight. Light in these examples is the limiting factor. Certain essential mineral salts, such as nitrates and phosphates, determine soil fertility and therefore plant population and distribution, but excessively salty soils inhibit the growth of most plants, as noted earlier.

A population of American oysters in Long Island Sound will not spawn until the temperature exceeds a threshold of 46°F. Polar bears are not naturally found south of certain latitudes, which represent a uniform temperature line. The rise and fall of natural populations is most often correlated with the times of year when nutrients, water, and minerals become available as the temperature rises, as well as when the temperature falls and nutrients, water, and minerals become unavailable.

MULTIPLY AND DIE

The famed British fur company, Hudson's Bay Company, kept complete records of the number of skins taken in Canada since about 1800. Study of the yearly tallies reveals that the Canadian lynx, a wildcat, reached population peaks every 10 years. Likewise, the population hit lows every 10 years. In the years between the lows and highs, the population rose and fell between the extreme points. The graphic curve of population oscillates up and down in the sinuous form of a sine wave, whose wavelength (distance between peaks) is 10 years. The chief prey of the Canadian lynx is the snowshoe hare. The hare populations rise and fall in the same fashion as those of the lynx, except that the population peaks for hares slightly precede the lynx highs (see Fig. 7-16).

At first glance, it seems to be a simple prey-predator cycle. As the number of hares increases, there is more food to support the larger lynx populations. As the predatory population of lynxes rises, the prey population of hares should decrease. The lynx population would fall as prey declines. The predators seemingly ride up and down on the prey populations. Vito Volterra and A. J. Lotka were mathematicians who first described these prey-predator cycles, which are accordingly called *Volterra-Lotka cycles*. A number of such prey-predator cycles are known: cougars and deer, coyotes and jack rabbits, snowy owls and lemmings.

Before 1907, on the Kaibab plateau of Arizona, there were about 4000 deer, preyed upon by numerous wolves and mountain lions. Bounties were placed upon the predators, and government trappers and ranchers all but wiped them out. By 1925, the deer population shot up to 100,000. Famine resulted, and the ravenous herds actually extinguished species of shrubs. Then, for two winters, 40% of the giant herds died of starvation, until the population fell to 10,000. When the natural check of predation was removed, the deer population overproduced; too many young survived. The steady state was upset by the elimination of predators. This famous case of disturbed natural balance seems to have no other explanation.

There is a notable exception to the predator-prey explanation of population dynamics. On the large island of Anticosti in the Gulf of St. Lawrence, the snowshoe hare population rises and falls just as it does on the Canadian mainland, even though there are no lynxes at all on the island! Another explanation may be required.

About 1% of the wildlife population is infected with the disease tularemia, caused by a bacterium closely related to the plague bacillus. The disease is often called "rabbit fever," or "deerfly fever," since the disease naturally occurs in rabbits and deer, but it has also been reported in ground squirrels, quail, partridges, muskrats, and beavers. The disease is often contracted by hunters dressing infected game. In the peak years of jackrabbit population, epidemics of tularemia are common; more than 80% of the population is infected. Only a few percent of the population is diseased in the years of population lows. With crowding, the communication of contagious disease is obviously increased, but this is only half the story.

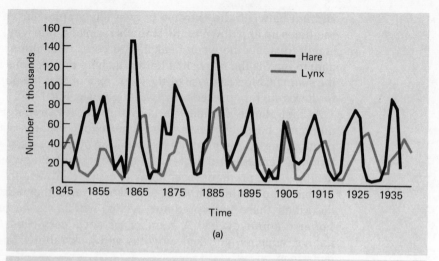

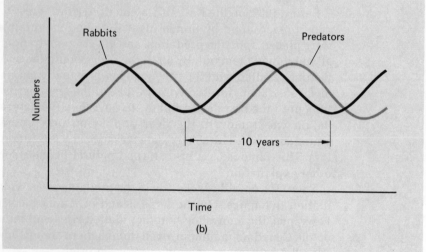

Fig. 7-16   **Population fluctuations of Canadian lynxes and snowshoe hares. (a) The actual graph of population oscillation based upon the number of pelts taken and tallied in the record books of the Hudson's Bay Company. (b) Simplification of (a); smoothed out sine wave curves, showing a 10-year wavelength for the prey and predator populations.**

(a) From D. A. MacLulich, *University of Toronto Studies, Biol. Ser.,* No. 43, 1937.

Crowding produces stress in the individual organisms as they compete for food, space, mating privilege, etc. The adrenal glands enlarge (the cortex is the actual part of the gland that expands). One category of adrenal cortical hormones, the glucocorticoids, has the effect of reducing the quantity of lymphoid, or antibody-producing tissue throughout the body. With the increased outpouring of these hormones during stress resulting from population pressure, the amount of circulating antibodies in the blood decreases. Thus, the immunity of the population falls sharply at the worst possible time—when disease communication is greatest. More rabbits die of tularemia in population peak years than are dispatched by coyotes. The hormones of stress also promote protein breakdown, and the strength of stressed organisms declines. The sperm count of sick rabbits drops to the point of sterility, but, more importantly, the reproductive cycles of females are shut off.

Sex hormones from the adrenal cortex inhibit the production and release of gonad-stimulating hormones from the pituitary and thus lead to reproductive failure. This inhibition is a negative feedback and constitutes a natural birth control. Hence, population has its legs cut out from underneath it in two directions: decreased immunity and disrupted reproduction. The population crashes. Some of the hormones of the adrenal cortex are androgens, male sex hormones, which affect the nervous system, increasing intraspecific aggression, the very behavior that produces stress. This is a positive feedback, a runaway system. The thermostat has broken down; instead of turning down the temperature, it turns on the heat, which climbs higher and higher. The Four Horsemen of the Apocalypse, pestilence, famine, war and death, ride down upon the helpless overpopulation. The hormones of stress also promote connective tissue breakdown, leading to subskin hemorrhages and easy bruising. Wound healing is retarded at the same time that susceptibility to infection is highest. A graph of average adrenal weight in a population describes the same up and down curves as that of population numbers (Fig. 7-17).

A seeming hysteria often grips an overpopulation, and the hordes begin to move in mass migrations to nowhere. Perhaps it is the species' last-gasp attempt to spread itself out and find new colonies. A migratory restlessness appears in migratory birds as their sex hormones rise; perhaps the mass migrations of overpopulations are akin. Auto traffic has been halted as seas of ground squirrels poured over the highway. The frantic migrations of lemming mobs can even take the form of attempts to swim across bodies of water, fjords. Although most of them die in exhaustion, it is not the intentional suicide portrayed by Walt Disney and others. The following is a particularly vivid description of the meaningless, insane movements of overpopulated European rabbits in Australia:

> . . . the rabbits had come in such millions that the whole ground seemed to move. Their killing-pen was filled up in a few minutes. There was no point in hitting the rabbits on the head, for they were saving the trouble

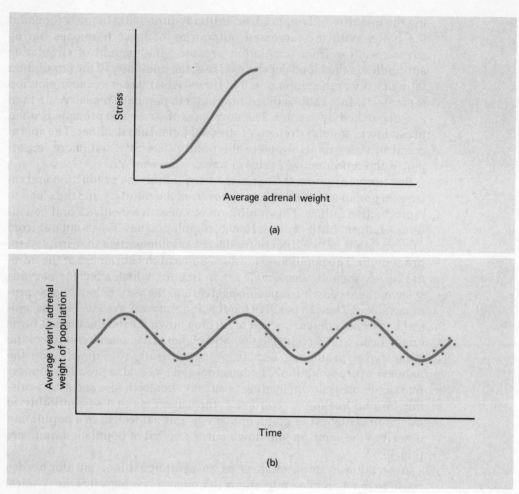

(a)

(b)

Fig. 7-17   **Adrenal weight, stress, and population. (a) Note the S-shaped curve, like a population growth curve. (b) Note the sine wave, which resembles a population cycle.**

by smothering one another. It wasn't long before the pen had been filled to the top of the netting with rabbits, the bottom ones all crushed and smothered. Then the swarm just passed on over the piles of corpses, and continued going southward.[2]

[2]Francis N. Ratcliffe, *Flying Fox and Drifting Sand,* (Sydney, Australia: Angus & Robertson, Ltd., 1947), p. 132.

See Fig. 7-18. Grasshopper overpopulations among the invertebrates also migrate as clouds of locust. Indeed, locusts have been described as a "plague."

Rats are highly social animals, and their communities have been described as "models of social virtue." There is no real fighting within the pack. The larger males allow the young to feed first, in contrast to wolves, where the highest rankers eat first. Adults even allow the half and three-quarter grown animals to take sexual precedence. Mothers put their young in a community nest, and it seems likely that each mother tends more than her own. Rat packs are usually large families, and the grandparents, uncles, and aunts are very tolerant toward the colony's young. Contact is frequent; the young crawl under the old continually. There is no peck order. The pack attacks enemy and prey in unison. Serious intraspecies fighting occurs only between members of different packs.

Fig. 7-18   **The introduction of the European grey rabbit into Australia resulted in a literal plague of the animals.**
Courtesy of the Australian News and Information Bureau.

The "domestic" or brown rat, unlike his wild cousins, reproduces all year round. Life with man is good. The estrous cycle takes 5 days, and pregnancy, 20. Six to eight newborn comprise a litter. The reproductive organs of the rat dwarf those of man in comparison. The testicles, prostates, and seminal vesicles of the male rat are almost 10 times as large as those of man when compared with total body weight. The uterus of the female rat is split into two branches, and four or more embryos can implant and develop on each side. The reproductive potential of rodents is tremendous.

When abundant food, water, and nesting materials are provided to a nesting pair, the population explodes; in nature, food is a major limiting factor. In the restricted space of the laboratory, the social mores of the group deteriorate as crowding begins. Fighting soon breaks out, and a peck order is established. Sexual excesses multiply, females cease lactating and nursing, and the nurslings are trampled, die, or are eaten. Reproduction ceases and war reigns; adults die young. The behavior of zoo-caged baboons bears many resemblances to this degenerate behavior of overcrowded laboratory rats.

The langurs, leaf-eating monkeys of India and Ceylon, show similar social disintegration, with fighting, sexual excess, and infanticide, when population density becomes too great. The langurs and baboons are primates, members of the order of man.

Every organism produces more offspring than can possibly survive. The *reproductive potential,* or biotic potential of most species is theoretically staggering. Darwin illustrated this crucial point in his theory with the slowest-breeding animal known:

The elephant is reckoned the slowest breeder of all known animals, and I have taken some pains to estimate its probable minimum rate of natural increase; it will be safest to assume that it begins breeding when thirty years old, and goes on breeding till ninety years old, bringing forth six young in the interval, and surviving until one hundred years old; if this be so, after a period of from 740 to 750 years there would be nearly nineteen million elephants alive, descended from the first pair.[3]

Actually, the elephant becomes sexually mature at 20 and stops breeding at 50. The period of elephant pregnancy is 22 months, and usually a single calf is born. Even with these additions and corrections, however, the continent of Africa would be overrun with elephants in a few centuries.

---

[3] Charles Darwin, *The Origin of Species* (New York, N. Y.: Crowell-Collier Publishing Company, 1962), p. 78.

In nature, almost all animals die *ecological deaths,* resulting from such environmental factors as disease, accident, famine, etc. An old rabbit is extremely rare. Few animals die of "old age," *physiological death.* In fact, mortality is greatest among the young, who are less experienced and weaker than the mature. The rate of population change—increase or decrease—is determined by comparing the birth rate and the death rate.

Throughout most of his history and prehistory, man died ecologically. It is only recently that he has come to physiologic death. The average life span has increased spectacularly, not so much because people live longer but because fewer infants die. It will be interesting to see if the average life span will continue to rise or if a downward "trend" will appear with increased crowding and overpopulation.

Most of the fatal diseases of man seem to have a considerable stress basis. The diseases of aging are frequently connective tissue disorders. The adrenocortical hormones are known to have a profound effect on the composition, synthesis, breakdown, and maintenance of the connective tissues. In addition to muscle, the walls of blood vessels are largely composed of connective tissues. Ulcers, whether they occur in the walls of the digestive tract or on the surface of the skin, are really wounds that fail to heal, or heal slowly. Connective tissue is the "scar tissue" of wound healing. Ulcers, high blood pressure, and atherosclerosis are frequently called "the diseases of civilization"; they develop most often in stressed individuals. The high dairy diets and low heart attack rates of Scandinavian peoples seriously question the animal fat theory of cholesterol degeneration of arterial walls. Stress seems to be more important than diet. The Mabaan tribesmen of the African Sudan are healthier at 80 than most American teenagers; the hearing and eyesight of the Mabaan are nearly perfect. Their hearts are strong, and their longevity is greater than societies with the most modern medicine. Ulcers, high blood pressure, and atherosclerosis are unknown among the Mabaan. Their old age is free of the usual degenerative diseases. Mabaan life is quiet, peaceful, and extremely moderate. Mabaan youths that go to the city of Khartoum, 650 miles away, suffer the "diseases of civilization."

In annually breeding animals in the state of nature, the population curve rises and falls in a single year. The population peak corresponds to the time of the birth of the young, usually in the spring and summer, and the population is brought low in the winter. The numbers of the breeding population of spring remain the same over a few years but rise and fall in decades. From one year to the next, the constituents of the breeding population might have changed, but their number has not. A natural equilibrium is reached, and the population fluctuates around this level—the *saturation level* or *carrying capacity* for a given environment, which can be represented by a level line drawn between the peaks and troughs of population (Fig. 7-19).

Few species attain their theoretical rate of increase, which graphically is first a curve and then a line heading straight upward. The checks on human population have been removed. No species has ever spread so successfully. The population of man has exploded. No species has or ever will sustain a population explosion indefinitely. Although man has technologically removed most of the natural checks on his population, no imaginable technological invention will allow him to continue his present growth curve course. Innumerable scientific innovations shall be required just to maintain the present level of human population in its deteriorating environments. Some authorities have already gone so far as to state that the symptoms of human social disintegration have begun to appear in crowded areas, such as the city. This author believes these prophecies. Review the symptoms of social disintegration discussed previously for a number of animals and draw your own conclusions. Zoo animals develop the diseases of civilization: obesity, hardening of the arteries, ulcers. In more natural zoos, the incidence of

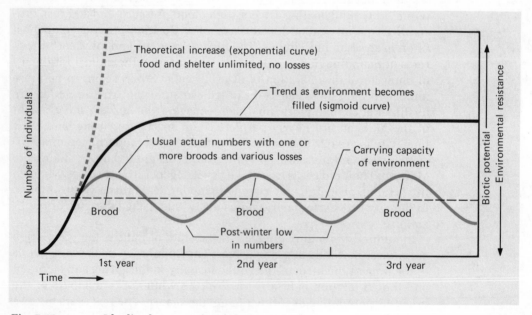

Fig. 7-19   **Idealized curves of population growth.**

From T. I. Storer and R. L. Usinger, *General Zoology,* McGraw-Hill Book Company, New York, 1965.

degenerative diseases decreases. Natural, moral behavior predominates in animals housed in modern, naturalistic zoos, whereas degenerate behavior is characteristic of animals imprisoned in the cells of archaic zoos. As Desmond Morris concludes in his best seller, *The Human Zoo,* these facts eloquently argue for the redesign of cities as well as prisons. The social and health problems of modern man probably stem as much from "people pollution" as from any other factor.

There is good reason to believe that the populations of primitive and preman followed cycles such as those of the snowshoe hare and the jackrabbit. Primitive man probably practiced infanticide, as some aborigines still do (not that this is recommended for modern man). First the Agricultural, and then the Industrial Revolution contributed to the acceleration of our growth curve. Now we have the responsibility, as conscious, knowing men, to halt it.

The potential rate of increase of most animals is checked by *environmental resistance,* in the form of competition, enemies, disease, and climate. R. N. Chapman expressed the relationship between population, environment, and biotic or reproductive potential in a simple formula:

$$\text{Population size} = \frac{\text{biotic potential}}{\text{environmental resistance}}$$

Population tends to be directly proportional to biotic potential, but it is inversely proportional to environmental resistance. When environmental resistance increases, population decreases; population grows when environmental resistance declines. Predation is a factor in the environmental resistance to a prey population. When predators are removed, as in the case of the Kaibab deer, the population explodes. Environmental resistance was decreased; the environment was opened. The population size at a given time is determined by the ratio of the two. Chapman's formula is analogous to Ohm's law in physics, which states that the quantity of electricity flowing, expressed in amperes (I), is equal to the voltage (V) or electrical pressure divided by the resistance (R) in ohms:

$$I = \frac{V}{R}$$

Voltage, the potential difference, is roughly comparable to pressure in a fluid system, and blood flow (F) depends upon the pressure (P) in the tube and the resistance (R) to the same:

$$F = \frac{P}{R}$$

Thus the population flows through time. Its pressure is reproductive ability, and its confines are the walls of the environment. The carrying capacity of the environment is finite, not infinite, and there is a definite limit to the numbers that an environment can support.

When a mating pair of fruit flies is introduced into a flask with food, or when a few bacteria are added to nutrient broth, the population is unchecked at first: the environment is open. The population increases slowly at the start, but the new individuals become part of the reproductive capital and compound the interest, adding their own share of offspring. Then the population *growth curve* swings upward, nearly to the vertical. A plateau is reached, and the population levels off. So far, the population curve describes an S-shape. When the food supply is depleted and/or the environment is fouled with wastes, the population enters the death phase and falls to zero. Kaibab deer, fruit flies, and bacteria follow essentially the same S-shaped growth curves (Fig. 7-20). Perhaps man will also.

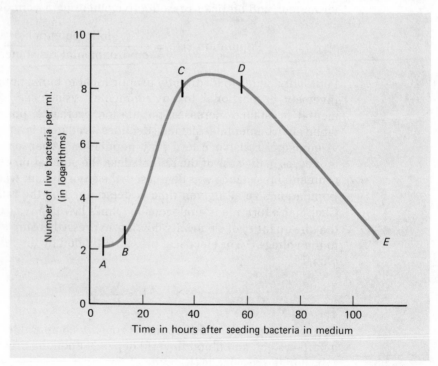

Fig. 7-20   **A laboratory growth curve for bacteria (or fruit flies) with restricted space and food supply. The lag phase is A to B; B to C is the logarithmic growth phase; C to D is the maximum stationary phase; and D to E is the death phase.**

The rate of population increase is proportional to the product of the *reproductive ability* (R) of a species and the number of individuals (N) in a population, provided that, of course, the environment is open. However, every environment has a limit, and this factor must also be considered. Let L represent the limit, the maximum possible population. The rate of change of individuals per unit time can then be expressed precisely as $RN[(L - N)/L]$. In the beginning, when N is very small, $(L - N)/L = 1$, and the population growth potential is uninhibited. The population explodes. As N increases, $(L - N)/L$ becomes less than one. Finally, $N = L$ and $(L - N)/L = 0$. At that point, population growth ceases. The slope of the growth curve varies with *fertility,* actual reproductive performance. Although the reproductive potential of the population might remain the same, its fulfillment depends upon internal as well as external factors, such as the accumulation of wastes in the medium, the change in its acidity, etc.

The laboratory growth curve is like one cycle of the population curve of a natural population, either the yearly rise and fall of an annually breeding species or the 10-year period of a Volterra-Lotka cycle. Of course, the natural system is more complicated, with a greater variety of checks on population, so that the curve hardly ever becomes so nearly vertical in the growth phase. External and internal negative feedbacks keep the natural population cycling up and down gradually about the carrying capacity of the environment, which reflects the steady state of the ecosystem. The sine waves of populations are like the oscillations of hormonal levels, and the homeostasis of a balanced ecosystem is like the internal constancy of a single advanced multicellular organism. No species has ever violated nature's steady state to the extent that man has.

Some natural population explosions have occurred in recent times (Fig. 7-21). Since about 1930, the cattle egret, a native of Asia, Africa, and Europe, has expanded its range to include Australia and the New World. It seems unlikely that a trans-Atlantic migration occurred. This bird participates in a mutualistic relationship with grazing animals, feeding upon insects stirred up by the movements of herd animals and cleaning external parasites (ticks and fleas) from the skins of these beasts. It is probable that they were inadvertently imported along with cattle. Rather than an unchecked expansion of local populations, there has been migration with the establishment of new populations in unfilled niches, and similar habitats. These white birds are to be seen scattered in among cattle and horse herds from Florida to New England in the United States, in Europe, and in South America, as well as with rhino and buffalo herds in Africa.

More mysteriously, the "crown-of-thorns" starfish (*Astrangia*) is cur-

Fig. 7-21   Cattle egrets. These birds, originally inhabitants of the Old
World, have recently invaded the Western Hemisphere and
their populations are exploding. Their way of life, or econiche,
was wide open. They are insectivores, feeding upon insects
stirred up by the grazing of the cattle with which they associate, and symbiotic or mutualistic cleaners, picking ticks and
other external parasites from the fur of the grazing animals.
Photo from Miami Herald Publishing Co.

rently undergoing a massive population explosion in the South Pacific,
and, since this species feeds on coral animals, it poses a threat to the
Great Barrier Reef off Australia, as well as to a number of Pacific coral
atolls and islands. The only hope of biologic control is a marine mollusk
predatory to this starfish. Human attempts at control have been undertaken.

The population curve of modern man is nearly vertical, like that of a
temporarily protected laboratory species. His natural checks of predation, famine, and disease have been severely reduced by his technologi-

cal innovations, the Agricultural Revolution, the Industrial Revolution, and modern medicine. Nevertheless, his environment is finite like the test tube and the culture flask, and there will be a limit. Man's technology has allowed him to "conquer" environments, but now the waste products of his technology and overpopulation have become a major part of his environment and threaten his existence as well as most of the living world. No one knows with certainty what will happen when human population reaches the plateau. Perhaps we can predict from our studies of other animal populations.

Certainly our cities have become centers of social disintegration. A nearly hysterical desire to migrate from the cities justly dominates current mass psychology, but most people are economically imprisoned there almost as effectively as animals caged by wire. The youth, with the greatest reproductive potential, continues to migrate from the rural areas to the cities, rich in "opportunity." The wealthier middle-aged and older retired people, usually postreproductive, seem to be the only ones that can afford to go the opposite way, retiring to the luxury of the countryside.

Unfortunately, in contrast to natural animals, man has his means of total destruction at hand, an arsenal for absolute suicide. The hysteria, aggression, and insanity generated in overpopulation may trigger the holocaust and bring down an ignoble curtain on the brief drama of man. The dangers of overpopulation are less obvious than the threat of nuclear war. The number one problem of concern to the twentieth-century citizen should be the population explosion. All other sources of contamination pale into insignificance alongside "people pollution;" indeed, the latter has generated most of the former. Now, while nature shrinks and withers on the vine, let us take what may be a last long look at the way things were.

BIOMES AND THE MAJOR HABITATS

The major habitats are the oceans, the fresh waters, and the land. Large areas of the Earth's surface with specific climates are dominated by characteristic biologic communities of plants and animals well adapted to the limitations of their physical environments, as well as to each other. These are the *biomes*, and each contains a number of ecosystems. None is really separate; one biome blends into another, often imperceptibly, seldom sharply. The six major land biomes are desert, grassland, tundra, rain forest, deciduous forest, and coniferous forest. Sometimes the oceans are referred to as the "marine biome."

THE DESERT

Most living organisms are about 90% water by weight. Water is the "medium of life." Some would say that life's greatest triumph was its escape from the waters, yet the fugitives took an "internal sea" with them. The desert is the environment in which the greatest threat to life on land, the lack of water, is carried to the extreme. It is quite likely that life would still be restricted to the water if desert had lain beyond the shore. Deserts rank among the most potent geographical barriers to the dispersal of species.

The deserts were not always; the land itself evolves with geologic revolutions. At times in the distant past, some of today's deserts were lush terrestrial biomes, swamps and forests. Many of the great oil deposits of the world lie beneath desert sands and represent the organic remains of Carboniferous swamp vegetation of 250 million years ago. Some modern deserts lay at the bottom of primeval seas. The Great Salt Lake is the evaporating remnant of an inland sea, and around it lie vast salt deserts. Recent borings in the Mediterranean Sea indicate that it dried up twice in the last 10 million years. The upthrust of mountain chains has not only landlocked portions of the Earth's surface, but frequently robbed them of moisture as well. Ocean currents and prevailing winds have made deserts of many shores, even though no mountains intervene. With climatic change and progressive drought, forests recede before the sea of grass, and the grass finally gives way to oceans of sand.

The wasteland is definitely the harshest land habitat. Daytime summer temperatures frequently soar to 120°F, and the surface sands commonly reach 180°F. Without the insulation of moist air, nighttime temperatures may drop 50° or more.

The daylit desert is probably the quietest place on earth, a place where one can hear the "sound of silence." The land appears lifeless, a region of death rather than life; the Latin roots from which the word "desert" is derived can be translated "an abandoned place." This image of the desert has led to its frequent use by poets as a metaphor of loneliness, as in Robert Frost's "Desert Places." The desert is an awesome place, a fitting site for the biblical temptation of Christ. The conflict between life and death is often starkly displayed here, where life is stripped to its essentials. Survival is the constant theme. The desert is an inspiring land, a favorite of poet, prophet, and biologist.

The survival of numerous forms of life in the desert seems miraculous, a triumph of nature's will, but, scientifically speaking, it is a revelation of life's plasticity. The living organisms of the desert are mostly survivors, modified descendants, wonderfully adapted to their severely selective environment. Often their ancestors were not desert creatures.

The plants that come first to mind when one thinks of desert vegetation are the cacti. Desert cacti have reduced or lost their leaves, covered themselves with thorns, and transformed their stems into water-hoarding barrels of succulent or juicy flesh. The green stems are photosynthetic. The only unconservative feature of many cacti is the extravagance of their showy flowers, which betray their origin. Plant classification is based upon leaf and stem characteristics, but, most importantly, on flower structure. Many members of the cactus family do not grow in deserts at all, but in slightly arid or even moist tropical environments. The "dawn cactus" still grows in South America; tropical and subtropical cacti have ordinary broad leaves and nonfleshy, woody stems (Fig. 7-22). Their flower anatomy allies them with the desert cacti, which apparently descended from these less familiar plants by "learning" to live with less and less water, gradually adapting over the millennia of droughts. This is a perfect example of "descent with modification," evolution. Many desert plants are not cacti, but superficially resemble them, and have converged upon them in evolutionary terms, imitating their solutions to the problem of persistent drought.

On the basis of water conditions in the habitat to which they are adapted, higher land plants can be classified as *mesophytes* and *xerophytes*. Mesophytes are the most numerous and familiar vascular plants, growing in an environment that is neither too moist nor too dry, such as temperate meadows, forests, and similar climatic regions in the tropics. Xerophytes are plants adapted to survive in habitats visited by seasonal or persistent drought. Plants rose up from the earth in competition for sunlight. Unfortunately, land plants must lose water from the photosynthetic heights to draw it up from the soil below. Leaves bear pores or stomates through which water vapor exits at the same time that other gases are exchanged (Fig. 7-23). Most mesophytes have broad, thin leaves with great surface area. Not all of the leaf surface is porous, however. Most woody plants have no stomates on their upper leaf surfaces: these pores are restricted to the underside of the leaf. Those mesophytes that are not woody have slender stems. Mesophytes are frequently mixed with xerophytes in subtropical regions with rainy and dry seasons. Surprisingly, mesophytes are the most numerous annual desert plants.

Xerophytes dominate the perennial desert vegetation. There are two types of xerophytes: the succulent and nonsucculent. Grasses are nonsucculent xerophytes, which we shall consider later. Most of the desert cacti are succulent xerophytes with expanded, fleshy, watery stems. The aptly named barrel cactus has such a succulent trunk, and has dispensed with true leaves altogether; only spines remain. Spines are considered to be reduced leaves, but they serve another function. In the desert, warfare between plants and animals has reached its peak. Most animals get their water from either vegetation or each other. The water stores of the

Fig. 7-22   **The West Indian Gooseberry is a member of the cactus family that is found in the American tropics and subtropics of Puerto Rico and Florida. It is thought to resemble the "dawn cactus" considerably. In contrast to the succulent cacti, with fleshy, water-storing stems and reduced or lost leaves, this cactus has broad leaves and a woody stem. Only its floral anatomy unites it with the other members of the cactus family. It is a typical mesophyte. It is not found in desert regions and is not xeromorphic, possessing a "drought anatomy."**

cacti are thus armed with spines to discourage animal grazing. Sometimes the pulp contains distasteful chemicals as well. The giant saguaro and organ pipe cactuses have pleated trunks that can expand in times of abundant water (Fig. 7-24). Spurges are common shrubs and herblike plants. Most are mesophytes, but one, which grows in the deserts of southeastern Africa, resembles a cactus remarkably in that it has pleated succulent stems. Another spurge "cactus" is seen in Fig. 7-25. Succulent stems and reduced leaves are *xeromorphic* ("drought anatomy") modifications. One South African xerophyte has flask-shaped, subterranean

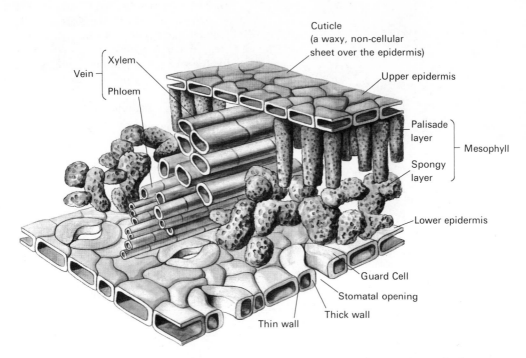

Cuticle
(a waxy, non-cellular
sheet over the epidermis)

Xylem

Vein

Phloem

Upper epidermis

Palisade
layer

Mesophyll

Spongy
layer

Lower epidermis

Guard Cell

Stomatal opening

Thick wall

Thin wall

Fig. 7-23   **Diagram of the cross-sectional anatomy of a typical (mesophytic) leaf.**
From A. W. Galston, *The Green Plant*, Prentice-Hall, Inc., Englewood Cliffs, N. J., 1968.

leaves with exposed "windows" at the ground surface. The night-blooming cereus (Mexicans call it the *reina de la noche,* or "queen of the night") puts forth extravagant blossoms in the night, during the driest season. Its wandlike stem is only an inch or so in diameter, but beneath the ground the stem is modified to form a melonlike 5- to 15-lb tuber. This is the water reservoir of this plant, and it is an alternative to succulent stems above the ground. Some desert plants have succulent leaves rather than fleshy stems. The Mexican sedum and the South African kleinia have bulbous leaves, heavily coated with waxy cuticles. The African and Asian Kalanchoe also has succulent leaves and has dispensed with the wastefulness of flowering; one of its leaves, laid on damp ground, will send out a new plant at each leaf notch. These, then, are the experiments in succulence. By storing water, succulents resist rather than endure the drought.

A poet once said of desert plants, "The folly of leaves was cast aside." The paloverde tree is not a cactus, but it has reduced leaves that can be shed in the drought. Its Spanish name means "green trunk," and its bark is photosynthetic. During the long spells of severe dryness, the ocotillo

Fig. 7-24

(a)

(a) The saguaro, the most familiar cactus of the southwestern American desert. (b) Beneath the spines lies a fleshy, water-rich stem, which is pleated to allow expansion for the storage of additional water rapidly absorbed following a desert thundershower.

Photo courtesy of Carolina Biological Supply Co., Burlington, N. C.

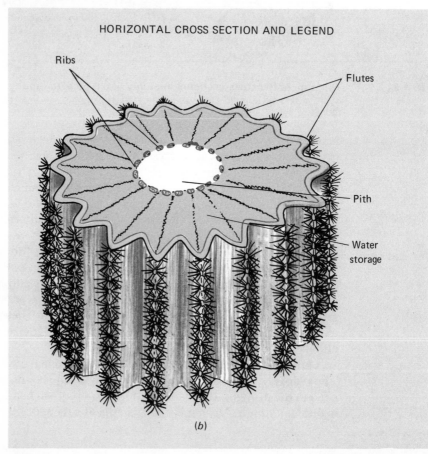

HORIZONTAL CROSS SECTION AND LEGEND

Ribs

Flutes

Pith

Water storage

(*b*)

Fig. 7-25   The candlabra cactus. This plant resembles a cactus (family Cactaceae) even though it belongs to a different family (Euphorbiaceae). This is an example of convergent evolution, where two relatively unrelated organisms come to resemble each other in form as the result of adapting to a common environment. Although the plant shown here is not a desert plant but is rather found in the Florida keys and the hammocks of the everglades, it should be borne in mind that the subtropics, too, experience quite frequently a severe, annual drought. This is one possible solution to that problem. "Cactuses" of this family are also found in the deserts of South Africa.

consists of several spined, seemingly dead wands or stems above the ground. The spines are actually the stems and midribs of modified leaves. At the first hint of drought, the rest of the leaf is shed. Some desert plants die out above the ground in the great dry spells.

One would think that all desert plants would have deep root systems, but this is true of only a few, such as the mesquite. The giant saguaro cactus has an extensive pancake root complex no more than a foot below the surface. After a heavy downpour, the efficient roots absorb gallons of water, which is quickly transferred to the stem.

Succulent xerophytes resist drought by storing water in stems, leaves, or tubers; nonsucculent desert plants drop their leaves or reduce them in the drought. The second group of desert plants (mesophytes) evades the seasons of dryness. The herbaceous annuals live quick lives in the brief wet periods, germinating, flowering, and reproducing inside a month. The chartreuse landscape of spring is quickly transformed into the brown summer hills, covered with the seeming "ashes of spring." The perennials that live more than a season die above the ground but continue to live beneath. The smoke tree grows along desert ravines, and the tough coats of its seeds must be ground away by rolling stones in the flash floods of thundershowers before they will germinate. Other desert seeds must have growth inhibitors washed away in the "wet" season before they will germinate. To insure their survival, desert annuals produce an incredible excess of seeds. In a moderately dry year, the desert annuals of Coachella Valley, California, produce about 1.5 billion seeds per acre. Although there are sparse grasses in many semiarid regions, these seeds are the real "grass of the desert," the base of the wasteland food chain. Thus, although desert vegetation is sparse and, when abundant, short lived, the tremendous seed production permits the construction of a surprisingly heavy pyramid of numbers. A square mile of Arizona desert supports 18,000 mice, 8000 kangaroo rats, 45 jackrabbits, 2 hawks, 2 owls, and 1 coyote. Obviously, rodents are the major food staple of the entire desert carnivore community; the rodent herds are the desert "cows."

It is very possible to live a whole lifetime in a desert town and never see a rattlesnake (Fig. 7-26) in the daytime, although the trails of sidewinders cover the sand. Likewise, there is hardly a rodent to be seen in the day, yet the sand reveals their trails and tail drags everywhere. The nighttime desert floor teems with kangaroo rats and deer mice. In a short jaunt along a trapline, it is not rare to encounter a rattlesnake or two, hunting rodents in the night. On Anahoe Island in the Nevadan desert Lake Pyramid, the author and his companions encountered no less than eight rattlers in a walk of no more than a hundred yards through the tumbleweed. Most biologists conducting studies there understandably arrive in the late morning and leave in the early afternoon. Most desert animals are nocturnal, restricting their activities to the cool nighttime hours. Even the daytime creatures are usually most active in the dawn and dusk. The midday desert is still; almost all animals are laid up in the shade or in their burrows beneath the burning sand.

The kangaroo rat (*Dipodomys*) is the most remarkably adapted desert animal (Fig. 7-27). His adaptations to the restrictions of his arid environment are threefold: anatomic, physiologic, and behavioral. *Dipodomys* is neither rat nor mouse, but belongs to a different family of the rodent

Fig. 7-26

Rattlesnake. Popularly known as a desert animal, rattlesnakes can swim and are spread throughout the Western Hemisphere. In the desert, these primarily rodent predators hunt after dusk. Although their eyesight is poor, they have highly developed heat-sensitive pit organs on their heads that allow them to detect prey. The rattle is not used in hunting but rather as a defensive warning device in the presence of large animals, such as man, coyotes, horses, deer, etc. A population of rattlesnakes on an island in the Bay of Cortez is without rattles.

Photo courtesy of Professor L. D. Ober.

Fig. 7-27   The kangaroo rat, the "mouse who never drinks." The adaptations of this species are primarily behavioral and physiological; the animal is nocturnal and secretes large quantities of the water-conserving hormone ADH. There is no specially constructed kidney.

order. He is biologically famous because he can live indefinitely on dry seeds without a drop to drink. Kept under desert conditions of heat and dryness for 50 days and fed only dried barley seeds, the kangaroo rat not only survived, but lost no weight by tissue dehydration; in fact, some experimental animals actually gained weight. The experiment was discontinued only because the point was made, but it could have been continued. Most desert animals get their water from vegetation or prey. In nature, the kangaroo rat does supplement his diet with green shoots in the spring.

If a live specimen of *Dipodomys* is kept in a wire cage on the sand during the day, when the surface temperature reaches 180°F, he will, like any other creature, literally "breathe his life away" in a few hours, and by the end of the day he will be nearly mummified. The kangaroo rat therefore dies almost as quickly as the rest of us at a few percent dehydration. However, even in this matter he is unique. When dry air is inspired, his evaporative loss from the lungs is only 0.54 mg water per milliliter oxygen used, as compared with 0.84 in man or 0.94 in the lab rat. The kangaroo rat is nocturnal in habit, coming out of its burrow to feed only at night. At dawn, *Dipodomys* retires to its burrow, plugs the openings with loose earth, and sleeps the day away. While the midday temperature of the surface sand may reach 180°F, just 12 in. below the surface, the temperature drops to 82°F, below body temperature. Below 10% relative humidity, respiratory water loss from the lungs, mouth, and throat begins. The relative humidity of the air above the surface sand approaches zero, but in the rat's burrow, the humidity seldom drops below 30% and often exceeds 50%. The burrow air is "half-saturated," and the respiratory water loss of the animal is negligible. The water that is lost to the internal air sometimes condenses on the cool earthen walls of the burrow, and some of it is even absorbed by seeds in the caches there. Therefore, *Dipodomys* makes its own satisfactory *microclimate*. The nocturnal habit and the burrow-plugging are behavioral adaptations, but nature's ingenuity with this animal does not end here.

Kangaroo rat urine is twice as concentrated as that of the white rat. He can concentrate his urine to the point that it has 17 times the density of blood. Man cannot concentrate his urine to more than about four times the osmotic concentration of his blood plasma. The volume of urine that the desert rat voids at one time is a small drop. For this reason, *Dipodomys* and the Mongolian desert gerbils are popular housepets. Many desert ground squirrels have kidneys with extremely elongated renal papillae, which represent the medullary region of the kidney, wherein the loops of Henle descend and the urine is "finally concentrated," has the last possible amounts of water withdrawn (Fig. 3-49).

The longer the medullary region, the greater the urine-concentrating ability of a kidney. One would be tempted to predict that the kangaroo rat kidney would have an extremely elongate renal papilla. It does not. The gross and microscopic anatomy of a *Dipodomys* kidney is almost precisely the same as that of the ordinary domestic rat. The normal blood level of ADH (the antidiuretic hormone from the brain base and pituitary, which regulates the water-saving capacity of the kidney) is much higher in the kangaroo rat than in the laboratory rat. Kangaroo rat urine is so concentrated that it is apt to solidify after it is withdrawn from the bladder. The concentration of urine in the urea reaches 23%, and that of sodium, 8.7%, as compared with 6.0 and 2.3% for those two substances in human urine. The kangaroo rat kidney can concentrate to the greatest extent of any known animal kidney. *Dipodomys* is the only land animal that can drink sea water.

Desert plants meet the water crisis by losing little, storing it, or lying low. The kangaroo rat meets the water problem with a most radical solution: he makes his own. Often, his sole source of water is the "water of metabolism," or the "water of biological oxidation," the by-product of the respiration of foodstuffs. When *Dipodomys* is fed a high-protein diet of dried soybeans, he dies in 2 to 3 weeks, after a weight loss of 34%. Death results from a decrease in the volume of body fluid, although its composition remains the same. The protein diet increases water loss in the urine as a consequence of the large quantity of urea from protein breakdown, which requires excretion. The normal kangaroo rat diet is richest in carbohydrate, which produces no such nitrogen excretion problem. Carbohydrate produces a greater amount of water of oxidation than protein. The water of oxidation of carbohydrate is 0.556 g water per gram food; protein is 0.396, and fats, 1.07. In grams of water produced per kilocalorie of heat (by-product of respiration), carbohydrate ranks highest: 0.133 to 0.092 for protein.

The kangaroo rat neither sweats nor pants. Like many desert mammals, he lacks sweat glands. He has no heat regulation problem, since the temperature of the nighttime desert air comes close to that of his cool, daytime burrow. No potential water loss is ignored. The cheek pouches of some species of kangaroo rats are lined with fur.

The large black eyes of the kangaroo rat betray the fact that his habit is nocturnal. Like most such animals, his retina contains rods almost to the exclusion of cones. Rods are dim-light photoreceptor cells. His skull has greatly enlarged hearing chambers that serve as resonating pockets, amplifying sounds in the nearly silent desert. *Dipodomys* resembles a kangaroo in that he has long, powerful legs for leaping. The scientific name for the genus, *Dipodomys,* refers to his two-legged or bipedal stance. His lengthy tufted tail is used as a counterbalance in jumping.

Rattlesnakes are among his chief predators, and the animal's erratic leaps are very effective in avoiding snake strikes. A kangaroo rat can cover 20 ft per second in 10-ft jumps.

Several widely separated desert rodents from different families have adopted the same two-legged jumping habits and essential external body form of the kangaroo rat. The African jerboa is nearly identical to the American kangaroo rat, even though the two are unrelated. The resemblance is a case of convergent evolution. In the same way, the Australian rat kangaroos have converged upon them, coming to resemble the kangaroo rat, but remaining, nevertheless, primitive nonplacental mammals.

The desert tortoise follows the way of many desert plants and stores water. The water of its vegetable diet is stored in two sacs just beneath the upper shell. In the summer heat, the tortoise is active in the dawn and dusk only, laying up in a shallow burrow during the day.

Like lungfish, the spadefoot toads of the American southwestern deserts estivate in deep burrows during the drought and emerge, mate, and feed following heavy rain. Twenty years may be required for a spadefoot toad to mature, since they do not grow or age significantly during drought hibernation.

Contrary to popular belief, the camel cannot store water. The camel's hump is a fat deposit. The food store is, of course, a potential source of "water of metabolism"; 1 lb fat yields 1.1 lb water. A camel with 100 lb of fat in its hump can produce 13 gal of water from it; however, to oxidize the fat into the water of biological oxidation, the camel must breathe, and the respiratory water loss from the lungs to the desert air easily exceeds the water of oxidation gained. The camel can go 17 days without water. How does he do it? When a man's body temperature rises only 2° above normal (from 98.5 to 100°F), he begins to sweat to cool his body by the evaporation of water from the surface, thereby maintaining a constant body temperature of 98.5°F. The body temperature is not allowed to rise except when the mechanisms of heat elimination fail. Temperatures of 100 to 110°F are disease symptoms, and death occurs above that range. Fevers of 105°F are accompanied by profuse sweating. The human body temperature does not drop below 98.5°F, except when heat production and conservation fail to equal heat loss.

The camel's thermoregulation is not so strict. This animal's normal physiological system will allow its body temperature to rise above 105°F before sweating begins. At night, the camel's body temperature drops to 93°F. Its temperature must climb 12° as the day becomes hotter before water loss through sweating commences. Whereas man tolerates a temperature fluctuation of only 2°, the camel can tolerate a 12° change, 7° rise above 98.5°F, before the cooling mechanism of evaporative water loss in the form of sweating is called into physiologic play.

In the desert heat, a man can suffer no more than a 12% decrease in body weight due to water loss before he expires in "explosive heat death." The water loss from the body fluids thickens the blood to the point that it becomes too viscous to be pumped rapidly enough to the skin, where heat loss from the blood occurs. The internal body temperature quickly rises above the fatal 112°F.

A camel can withstand more than 25% dehydration without suffering explosive heat death. The camel's blood does not thicken. Its blood volume is maintained at the expense of the other tissue fluids. No one knows how this is done, but it is. Obviously, this mechanism of physiologic response is a tremendous advantage for life in arid lands.

The nitrogenous waste of protein breakdown, urea, is less toxic than its precursor, ammonia, but it is nonetheless poisonous in high concentration. It is absolutely essential that it be washed from the body fluids by the kidneys. At the same time, water is lost in the urine. The urine can be concentrated only so far, and the camel dies of urea "autointoxication" just like a man. The camel, however, recycles urea. Urea is returned to the stomach via the bloodstream, and the urea nitrogen is combined with the breakdown products of cellulose digestion into bacterial protein in the digestive tract. In turn, the camel can digest and absorb the proteins of the bacterial cells; thus, urea is shunted out of the blood, at least temporarily. The number of bacteria in the first chamber of the camel's four-chambered stomach increases following such "fertilization."

The insulating layer of uniform fat beneath a man's skin not only prevents heat gain but also discourages heat loss. The camel concentrates fat in his hump and thus allows heat loss from the other parts of his body. Hair insulation also retards heat gain; no desert animal is naked like man, and desert nomads dress wisely in heavy wool clothing. A shaved camel produces 60% more sweat than an unshaved one.

In contrast to the kangaroo rat, the camel must drink. A camel that has lost 25% of its body weight can restore its body water 10 minutes after drinking. The donkey (like the camel a native of Asian deserts) can restore its body fluid volume only 2 minutes after drinking. It takes a man more than a day, but man was not made for the desert.

THE TUNDRA

Just south of the perpetual snow and ice of the arctic lie the vast and nearly treeless plains of the *tundra,* or the high arctic biome (Fig. 7-28). A foot or two beneath the surface, the soil remains permanently frozen.

Fig. 7-28   **Arctic tundra. Snowfall is light, and flowering plants (right foreground) bloom and go to seed quickly in the short 2-month summer.**
Courtesy of Carolina Biological Supply Co., Burlington, N. C.

The growing season is short, from a few weeks to 2 months, when the surface soil thaws out during the "60-day" summer. Surprisingly, it is not a land of great snowfall, and large areas of the tundra really resemble deserts, in terms of their annual precipitation. Strong, dry winds prevail. Although the summer days are long, the light is low, since the sun's rays come down obliquely. The barren, rock-strewn landscape was the last of the continents to be liberated from the glaciers of the recent ice age. The presence of oil in these regions indicates that they, too, were once covered by swamps when the climate sharply differed from the present.

More than 200 plants have "learned" to survive in the arctic. Most are short-lived annuals, such as the arctic poppy and saxifrage, which flower quickly during the brief summer and produce numerous seeds that can withstand the severe winter as they lie dormant. In this they resemble those desert plants that live accelerated lives following a desert rain. The leaves of arctic tundra plants bear thick waxy cuticles to resist the drying winds. For the same reason, most of them grow close to the earth. Wind pollination predominates, and many flowers open wide. Few flowering plants are found far from water on the arid plains of the tundra. Lichens grow year round on the dry, rocky surfaces. This "reindeer moss" is the staple diet of many arctic herbivores. In the spring, the light snow melts, and narrow rivulets run down into bogs and shallow lakes. True mosses grow in thick spongy masses along their banks, and many flowering plants take root in the moisture-laden carpets of moss.

Migratory waterfowl, ducks, geese, terns, etc., return to the marshes of the tundra, their ancient breeding grounds, in the spring, and the young are born and reared before the great southward migrations of early fall. Arctic insects, like the higher plants of the tundra, reproduce quickly, and their pupae lie alongside the dormant plant embryos (seeds) in the frozen ground through the long winter. The reproductive cycles of arctic rodents are also timed to take place in the short spring and summer.

The characteristic resident animals of the tundra include the great shaggy musk oxen, caribou, reindeer, lemmings, arctic hares, arctic foxes and wolves, snowy owls, and the ptarmigan ("snow grouse"). Reindeer herds have been domesticated by the Laplanders of northern Europe, and recently the domestication of the musk ox has been undertaken. The latter animals are related to the goat, and their long hair coats are a potentially valuable source of cashmere. Both reindeer and musk oxen survive the winter by browsing on lichens. Their domestication is a fine example of ecological farming, which makes greatest use of the natural productivity of an ecosystem rather than altering the landscape and destroying nature's careful design. The arctic rodents feed on the herbaceous annuals and their myriad seeds. Rodents cache seeds in their burrows for the long winter, and some of them hibernate. The populations of snowy owls and lemmings rise and fall in Volterra-Lotka cycles. When the lemming populations crash, snowy owls forage farther south; they have even been sighted in Idaho, Montana, and Nevada.

THE TAIGA

South of the tundra, in Canada, the Northern United States, and Siberia, there is an almost continuous belt of needle-leaved evergreens, pines, spruces, firs, and cedars. The Russian word for these northern forests of cone-bearing (coniferous) trees is *taiga,* and this term is generally applied to the northern coniferous forest biome (Fig. 7-29). The climates of regions where this biome is found today vary from mild to severe winters, and from heavy to light annual precipitation. The taiga extends

Fig. 7-29   **Coniferous forest. These woods are dominated by evergreens, such as the fir trees seen here.**
Courtesy of Carolina Biological Supply Co., Burlington, N. C.

southward across the face of the United States on the great mountain chains, the Cascades, the Sierra Nevada, the Rockies, and the Appalachians, to a lesser extent. Ponderous snows fall on most of these ranges, and it would seem that there would be no lack of water. Yet the conifers are xerophytes, belonging to the gymnosperms, the first truly land plants, which dominated the landscape of the Age of the Reptiles, when the continents were cool, high, and dry.

Most of the mountain ranges previously mentioned are relatively young, in the early stages of erosion. In the Sierra Nevada, pines grow in the decomposing granite (gneiss) gravels of the mountainsides. Like the desert cacti, the conifers must take water when it is available in the spring and conserve it through the dry summer, fall, and winter, when the water is bound up as ice. The southern coniferous forests of long-leaf and slash pines also stand in sandy soils, and the rainfall is seasonal. The piñon pines and junipers grow on some of the driest hillsides of the western states; in the dry season, the soil in which they root crumbles into dust underfoot. Piñons and junipers hardly seem to be trees at all, but rather large shrubs. So extensive are their "forests," however, that the piñon-juniper communities, although not considered biomes, are often described as major "biotic areas." They illustrate very well the conifers' drought adaptation.

The microscopic anatomy of the pine leaf differs from the broad leaf of mesophytes in several ways. The epidermis produces a heavy cuticle. Stomates are sunken beneath the surface. The interior photosynthetic tissue cells are tightly packed with few intercellular spaces; thus, they expose less internal surface area for evaporation. Mechanical tissues encircle and lend rigidity to the entire leaf (Fig. 7-30). Pine leaves never wilt in the drought, as is frequently the fate of broad-leaved plants.

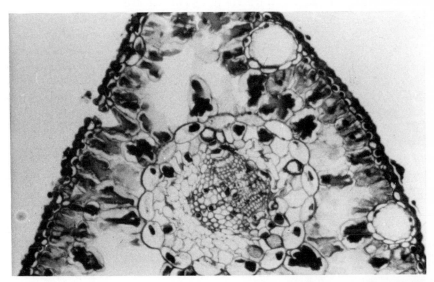

Fig. 7-30   **Xeric ("dry habitat") adaptations of pine. This photomicrograph of a cross section of a pine leaf (needle) shows a thickened cuticle, sunken stomata (epidermal pores), and low surface-to-volume ratio, all features for moisture conservation.**

Courtesy of Saturn Scientific, Inc., Fort Lauderdale, Fla.

Some conifers cast their leaves, but most are evergreen. In the Rocky Mountain and Sierra Nevadan fall, only the broad-leaved quaking aspens turn color. The yellow-gold of their autumnal stands streaks the canyon bottoms and ravines between the foothills and mountainsides, where creeks run and the soil is richer than that of the hillsides.

The characteristic mammals of the northern coniferous forest include the moose, wolverine, lynx, and snowshoe hare. The pine woods are quiet; birds are few. In the mountains of the American West, mule deer and cougars are common. The deer leave the higher elevations in the wintertime and migrate to the sheltered lower hills of sagebrush grassland, the winter feeding grounds of the herds.

THE WOODS OF FALLING LEAVES

The temperate zone is the region of the Earth with the traditional four seasons. The northeast coast of the United States, southern Canada, New England, and portions of the Appalachians are famed for their vivid falls. The hardwood, broad-leaf trees cast down their leaves, "deciduate," in autumn, which is why these forests are called *deciduous*. In the subtropics, where the seasons are not so clearly marked, several broadleaf species drop their leaves in the dry season. Winter, too, is a dry season: the water is bound in snow and ice. The deciduous habit is an adaptation to drought.

If frost killed the leaves, as is commonly thought, the leaves would wither with the first freeze. Nevertheless, their stalks would remain. A leaf breaks cleanly away across the stalk. Toward the end of the growing season, the process of abscission ("a cutting off") begins. Long before, a separation layer of cells has been formed across the leaf stalk. In the early fall, enzymes that dissolve the cement between cell walls are released. Wind or ice breaks the tissues of the separation layer, and the leaf then remains attached only by a slender bundle of vascular tissue branched from that of the stem. After the leaf falls, a protective layer of cork forms just beneath the separation layer, thus protecting the exposed stem tissues from dehydration. The fall or abscission of the leaves cuts off the primary route of water loss, evaporation from leaf pores.

The maple leaf is the Canadian national emblem, and the most common deciduous trees in the northern temperate forests are maples and beeches. Although the winters are milder in the southeastern United States, several broadleaf trees there are nonetheless deciduous (Fig. 7-31). Hickory and oak are the predominant deciduous trees in the

Fig. 7-31   **Hardwood deciduous forest, the "woods of falling leaves," a biome in which the freshwater pond is frequently found.**
Courtesy of Carolina Biological Supply Co., Burlington, N. C.

southern temperate zone; the small magnolia-oak forests of the deep South are broadleaf, evergreen, subtropical forests.

The animals of the temperate deciduous forest biome are the Virginia deer, black bear, squirrels, etc. The most studied ecosystem of this biome is the freshwater pond. Amphibians are common, and the yellow-spotted, black salamander, *Ambystoma maculatum,* is the index animal of this region.

The soils in which the deciduous forest thrives are generally rich, in the regions of worn-down mountains. The leaf drop provides a rich humus fertilizer. Rainfall is 30 to 60 inches per year. Deciduous forest communities originally covered eastern North America and central Europe. These were the forests of European folktales and of James Fenimore Cooper's *Deerslayer.* In view of their rich soils, it is not surprising that most of these areas were deforested and "put to the plow."

As the earth tilts on its axis, the temperate zones are angled toward or away from the sun, and the day lengthens and the seasons change. In the spring, the day length increases, finally reaching a summer high, the longest day of the year; in the fall, the day declines, until the shortest day of winter. There are times in the spring and autumn when the day and night are equal: the vernal and autumnal equinoxes. Just as spring is the season of the bud, so fall is the season of the fruit and the death of leaves. The rhythm of life with the passage of time in the temperate zone is poignantly marked by recurrent birth, death, and rebirth. Poets and philosophers of the temperate zones have been acutely aware of time thus measured. The pattern of peoples' lives was adjusted to the seasonal changes. There was "a time to sow and a time to reap," as well as a time to be snowbound.

BETWEEN CANCER AND CAPRICORN

Between the Tropic of Cancer and the Tropic of Capricorn lies the torrid zone, which is bisected by the equator. The temperature is uniformly very hot or torrid. The seasons do not change upon the equator; the day remains the same. The subtropics blend into the tropics of this zone. All types of biomes from desert to tropical rain forest are found in this region. The annual precipitation of the tropical rain forests in the equatorial regions often exceeds 80 in., yet a large portion of the Sahara Desert also is located in the torrid zone. The humidity and temperature of tropical rain forests are steadily high, and some rain forests ("cloud forests") are perpetually cloaked with clouds. It has been suggested that volatile compounds from these forests play a role in cloud formation. Here, indeed, would be a remarkable example of life modifying the physical environment. The rain forests of the tropics are nature's greenhouses. The rain forests' trees are mainly broadleaf evergreens.

The true tropical rain forests (Fig. 7-32) are located in the East Indies (Indonesia), Malay Asia, equatorial Africa, Central America, and the basins of the Amazon and Orinoco Rivers of South America. Portions of Caribbean islands and Puerto Rico bear what are nearly true rain forests. The subtropical ("nearly tropical") regions, north and south of the tropics, are often cited as rain forest biomes, but the monsoons of these areas alternate with droughts. These are the two seasons of a subtropic region, such as southeast Asia.

No more than several species of trees are found in the northern forests, and great stands are often composed of only one or two species. In tropical rain forests, several hundred kinds of trees are commonly intermixed. The greatest variety of species is found in the tropics.

The forest giants, ironwood, teak, and banak, tower above the others to a height of 125 ft or more. A second layer of moderately tall trees in-

Fig. 7-32   In the tropical rain forests of the Amazon and elsewhere, even the forest giants that break through the canopy carry other plants upward with them. The epiphytes ("upon plants"), or air plants, grow in the crevices of their bark and in the crotches of their branches, striving upward for the light which is limited below. The boughs of these trees forming the canopy are draped with "hanging gardens" of epiphytes.
Courtesy of Ward's Natural Science Establishment, Rochester, New York.

terweaves its branches and leaves to form a canopy. Smaller trees and a few shrubs grow in the shade below. Thick-stemmed, woody vines, "lianas," hang down from the branches above to the forest floor below. Few plants are shade tolerant enough to grow in the semidarkness beneath the canopy and in the shade of lesser trees. Ferns, lichens, mosses, orchids, members of the cactus family, and broadleaf pineapple relatives (bromeliads) grow upon the trees. They are called *epiphytes,* or "air plants." In the crotches of trees and crevices of bark, the epiphytes gather decaying vegetation about their roots; they obtain essential elements from rainwater and dust. Their higher location places them in a more favorable position to receive light.

In very few places on the forest floor can a wall of "jungle" vegetation be found. Most plants strive upward to the light. Ferns take the form of trees (Fig. 7-33) and even the grass, bamboo, is woody and grows upward, as a tree. The soil of the lush tropic forest is surprisingly poor. One might expect that it would be rich in humus, with abundant decaying vegetation, but the rate of decomposition is very rapid. The warm, moist, dark environment is ideal for bacteria. In addition, hardly any of the

Fig. 7-33   **Many plants that elsewhere grow as low shrubs take the form of trees in the rain forest. These tree ferns are found in Java.**
Photo by W. H. Hodge.

The sloths are tree-dwelling mammals found in the tropical rain forests of northern South America. The sloth was so named because of its slow movements.
San Diego Zoo photo.

trees deciduate. Even nitrate fertilization of tropical soils doesn't last long, since denitrifying bacteria quickly turn the chemical salt into atmospheric nitrogen. Neither conventional farming nor grazing is practical in the tropical rain forests. Rubber trees and jute and hemp, valued for their rope fibers, are commercially important plants from these regions. The wood of the forest giants is also highly valued.

Most rain forest animals are tree dwellers (arboreal). In British Guiana, 31 out of 59 species are arboreal (Fig. 7-34). It was in the tropics and subtropics that the evolution of man's order began, and his most primitive primate relatives, potos, galagos, and tarsiers, are still to be found here, along with the monkeys and the great apes. Tree frogs and other amphibians are numerous in the rain forest. They often breed in the water reservoirs of epiphytes (expanded leaf bases), and have abbreviated their metamorphoses. The eggs of *Hylodes,* a tree frog in the West Indies, are laid on leaves and hatch in 2 or 3 weeks directly into frogs, although a brief functionless tadpole stage does appear. Reproduction takes place during the rainy season. The largest surviving reptiles are found in the tropics and subtropics.

THE SEAS OF GRASS

Since the last ice age, a number of regions on the continents have been subjected to mild drought; the climate became drier, but far from completely arid. Forests receded before the grass; often, grass replaced fire-leveled forests. The grassland is the major land biome on the rise in recent times. The evolution of man is believed to have been somehow correlated with the ascendance of the grasslands. The various names of famous grasslands reflect their widespread distribution. The *prairie* was

Fig. 7-35   **Most of Africa is covered by savannah or veldt, a grassland with scattered, xerophytic trees. The variety of grazing animals that has evolved here is greater than that of any other grassland in the world.**
Courtesy of Dr. Douglas Schaefer.

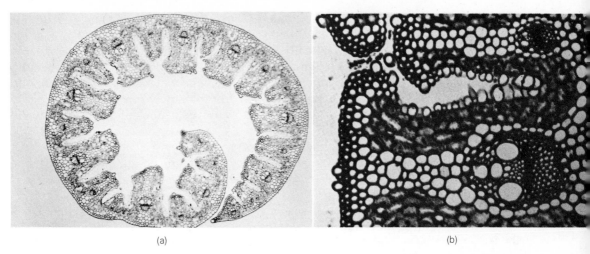

Fig. 7-36   **Drought adaptations of a grass: a cross section of a leaf at (a) low and (b) higher power. The stomates are found on the inner surface and are buried when the leaf curls up, reducing the amount of evaporative surface area exposed to the dry atmosphere.**
Courtesy of Ripon Microslides Laboratory, Ripon, Wis.

the tall grassland of midwestern North America. The *plains* were the short grasslands farther west. The *pampas* of South America are primarily found in Argentina, but the plains of the Gran Chaco extend into Paraguay and Bolivia. The South African Boers gave their grasslands a Dutch name, *veldts*. These African plains with scattered trees are also called savannahs (Fig. 7-35). The Russian *steppes* are plains.

The annual precipitation of the true grasslands is about 30 in. per year, but the rain falls erratically, or is interspersed with long, dry periods. As little as 15 in. of rain falls annually in the western United States plains. Several species of grasses grow successfully in semiarid and even desert country. Most grasses are xerophytes. Their narrow leaf blades have very thick cuticles, which retard water loss from the deep tissues. Most xerophytic grasses have the capacity to fold or curl their leaves in the drought. The stomates are located on the inside surface, and, as the leaf edges roll inward, the rate of water loss is decreased, since the stomates are buried (Fig. 7-36). Like the other nonsucculent xerophyte we studied (the pine), the grasses also have leaves with well-developed mechanical tissues, which prevent leaf collapse. This is one of the major reasons that xerophytes suffer much less than mesophytes during dry periods. The perennial grasses of arid regions grow quickly in the spring or rainy season, mature, and reproduce. As the dry season begins, the aerial shoots die. Underground parts, stems and extensive root systems, can survive for a year or more without rain. Many of the leaves of most grasses are arranged more vertically than horizontally, thus exposing less total surface area to direct sunlight. Grasses grow in thick stands and temper their own environment, breaking the wind and maintaining a higher humidity in their own vicinity. The thick thatch or mulch of fallen grasses about the bases of the living plants prevents much of the evaporation of water from the soil.

The soil of the midwestern prairie was very rich, not only because of self-fertilization but also because it lay on the floodplains of the Mississippi drainage basin, where it received fine sediments. The numerous minute grass flowers produce abundant seed. The tall grasses of the prairie waved like a sea in the wind, but they did not suffer the degree of dehydration they would have had they grown in thin stands. The short buffalo grass of the dry western plains did not grow too far off the earth and was thus protected from drying gusts. Also, the ability to flower and seed at low heights is a characteristic of positive survival value to a heavily grazed grassland, as were the plains, even before the coming of the white man.

Horses, antelopes, and wild cattle evolved in the grasslands environment. Most of the even-toed, cloven-hoofed beasts have four-chambered stomachs. The first chamber is called the rumen; it serves as a storehouse for unchewed grasses, which can be regurgitated and chewed at leisure by these "cud-chewers," or *ruminants*. The stomach and intestines of ruminants are rich in microorganisms capable of digesting cellulose with at least a 10% efficiency. What type of a relationship is this? Most of the grassland grazers have massive, high-crowned, multiridged molars, effective grinding teeth for the siliceous grasses.

Great, flightless, running birds have also evolved on the grasslands. In South America is found the rhea of the famed bolo sport of gauchos. The ostrich is native to the African savannah. Similar flightless birds, the emu and cassowary, live in the Australian biogeographic region. All these birds have long powerful legs with reduced numbers of toes and enlarged toenails. Their speed is great. In these ways, they have evolved along lines similar to the horse.

Wheat, corn, barley, rye, and oats are domestic grasses that have replaced the natural grasses of the prairies, plains, and steppes in most places. The "breadbaskets" of the world are transformed grasslands. Wisely, plant breeders have crossed native and domestic grasses to produce hardier yet desirable strains. Repeated plowing and the burning of wheat straw stalks coupled with a severe drought to produce the "dustbowl" of Oklahoma, Texas, and Nebraska in the 1930's. A moving description of this ecological tragedy is recorded in John Steinbeck's

Grapes of Wrath. The topsoil had no binder; native trees and introduced plants (mesophytes) died by the tens of thousands in Nebraska during the severe drought of 1934–1937. Now the stalks of harvested wheat are left standing in the fields until the next plowing, and shelter belts of planted trees serve as windbreaks. Nevertheless, extensive fertilization is necessary. The native grasses survive as well as ever along the fences. If unleashed, it seems logical that they could retake the land from the wheat, corn, hogs, and men, and restore the stable ecosystem of the prairie.

Overgrazing by protected livestock, cattle, and sheep produced similar results on portions of the Great Plains. The same grama grasses found on the short grass plains sometimes grow on desert rangelands. Cattle overgrazing results in an increase in mesquite, rabbit brush, and cacti, as well as in jackrabbits, which find these undesirable plants (in cattleman's language) excellent dry season forage. The sagebrush-grasslands are found in semiarid regions, not true deserts, and livestock overgrazing results in the rise of the sagebrush to dominance. Why didn't the great buffalo herds or deer and antelope bring such ecological damage to the plains?

The cattle of the Masai have likewise overgrazed areas of Africa. It has been recommended that the soundest ecologic practice in Africa would be the harvesting of the various antelopes. The African antelopes are more fit for their environment than domestic cattle, which—like man—are victims of the sleeping sickness. Although the blood of African antelopes harbors the causative organisms of sleeping sickness, the host animals are hardly harmed; the parasite is almost a commensal.

The prairie dogs and prairie chickens must be added to the list of endangered species (Fig. 7-37). Perhaps we shall have to keep them in zoos, as we have done with the buffalo.

THE MARINE BIOME

Life was born in the sea and cradled there in its infancy. A billion years passed before life forms left the stability of the waters for the insecurities of life on the land. Nine-tenths of evolution took place in the sea, even though the physical environment hardly changed. It was the living environment that changed. Nevertheless, some of the oldest, unchanged living organisms are found in the ocean. The horseshoe crab, which is not a crab at all, and the lampshell, which looks like a mollusk but is not, are sea dwellers and the oldest, living, essentially unmodified hard-

Fig. 7-37

(a)

(b)

(a) Prairie dog. (b) Prairie chicken cock in courtship display, with headdress erect and red skin throat pouches inflated. The plains Indians sometimes imitated the behavior of this bird in their dances.

From G. A. Ammann, *The Prairie Grouse of Michigan*, Department of Conservation, Mich.

shelled animals (Fig. 7-38). They have been found in fossils about 450 and 500 million years old.

Those mostly microscopic organisms that float passively with the oceanic currents are called *plankton*, which means "that which is made to wander." The unicellular microplants, or protophyta, are termed the *phytoplankton*, and the protozoans and minute crustaceans that feed upon the microplants are designated the *zooplankton*. The phytoplankton has been aptly described as the "grass of the sea," since its microalgae lie at the base of all marine food chains. Their photosynthesis produces much more oxygen than all land plants combined. The most common of the phytoplanktonic organisms are the glass-walled *diatoms* (Fig. 7-39) and the *dinoflagellates*, the "terrible flagellates,"

(a)   (b)

Fig. 7-38   Horseshoe crabs (a) and lampshells (b) are unusual organisms in that they have
persisted unchanged for more than the last 400 million years. The horseshoe
crab is closely related to spiders, and the lampshell is not a mollusk, although
it was so classified until someone bothered to open its shells and examine its
internal anatomy.
Courtesy of Saturn Scientific, Inc., Fort Lauderdale, Fla.

Fig. 7-39   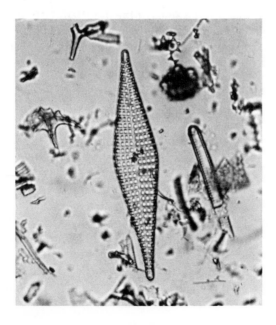   Diatoms. These one-celled
plants live in glass walls. To-
gether with the dinoflagellates, they are the "grass of
the sea," the phytoplankton.
Courtesy of Saturn Scientific,
Inc., Fort Lauderdale, Fla.

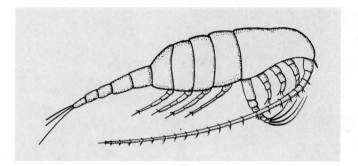

Fig. 7-40   **A copepod, an important zooplanktonic crustacean, diatom consumer.**

whose toxic blooms sometimes bloody the surface waters of the Gulf of Mexico in "red tides," killing numerous fish. The organisms that graze upon the phytoplankton include not only microscopic protozoans and crustaceans (Fig. 7-40) but small shrimp as well. Many larger and sometimes more complex animals are plankton consumers. Clams, oysters, scallops, sponges, and our own invertebrate relatives, sea squirts and lancets, are filter feeders. The baleen whales, which include the largest of creatures, strain the plankton-rich waters of the Antarctic for food. These whales swim with their jaws wide open, and the entering sea water is channeled out the sides of the mouth after first passing through fringed curtains of "whalebone" (baleen), which collect 2-in. shrimps as well as some other planktonic forms (the great schools of 2-in. shrimps are called "krill"). The toothed whales that make a meal of 5000 herrings are also dependent upon the phytoplankton, since each herring stomach can contain 7000 microcrustaceans, and each tiny crustacean gut holds more than 100,000 diatoms. Phytoplankton also lies at the base of the freshwater food chain.

The oceans are divided into a number of vertical and horizontal zones. Between the low and high tides lies the *littoral,* shore, or intertidal zone. At ebb tide, the shore is surrendered to the land with its uncertainties. The littoral is reclaimed by the stable sea at flood tide. Organisms living in the intertidal zone are alternately subjected to wetting and drying, heat and cold, wind and wave. One would think the shore between the tide lines would be a "no man's land," but it isn't. The littoral is an ecotone, where two major ecosystems meet, and, although life is tried here, its success is reflected in the variety and population densities it achieves. The sea's edge is the land's beginning, and the littoral belongs exclusively to neither but alternately to both. As such, it is a region of compromise rather than dramatic tension. A number of sea-weeds came ashore here, and the animals vary from protozoa to our own phylum, the chordates.

The substratum of the shore can vary from sand to rock. Organisms clinging to a rocky shore cannot escape exposure by burrowing. Occasionally, muddy flats are formed in the vicinity of river deltas. Sand can be either calcareous or siliceous. Quartz sand is coarsest and most resistant to erosion. Close examination of the sands of the Florida coast reveals that they are composed primarily of the ground remains and fragments of sea shells. It is staggering to contemplate the myriad of once live mollusks that concentrated all these calcium salts to the point of insolublility, taking them from a dilute sea. Then the sea grinds their remains along the shoals of shells, and the calcium is gradually returned to the water.

The tide also varies from coast to coast across the world. It averages only 2 to 3 ft in the Gulf of Mexico. Along much of the Atlantic and Pacific seaboards, the tide averages 10 ft most of the time. The surf also varies, and it is an important physical factor for life on the tidal shores (Fig. 7-41). Along the coasts of northern California, Washington, Oregon, northern Australia, Oahu, and New Brunswick, great swells of sea pound the shores.

The littoral zone itself can be further subdivided. The line of most abrupt change is at the high tide mark. Highest is the dry beach or *dune* area. Many land plants have adapted themselves to the seashore environment, where they grow on the dunes. In some ways, shores resemble deserts, and the dune plants frequently show characteristics of plants in arid regions. Some have deep roots and small, leathery, hairy, or waxy leaves. Several are succulents, capable of storing water in thickened stems and leaves. Sea figs, the seaside daisy, and the seaside goldenrod are succulent herbs. Sand verbena grows in deserts as well as along the shores. Members of the grass and sedge families have been most successful in the dune areas; sea oats and sandbur are common shore grasses. All dune plants are water conservationists and are adapted to resist drought, wind, and salt spray.

Some sand hoppers or beach fleas, which are crustaceans rather than insects, live on the dry beach areas. Sand and ghost crabs also live in dry beach sand.

Although the *uppermost beach* lies above the high tide mark, like the dunes area, its sands are occasionally moistened by the highest tides, storm waves, and sea spray. Fiddler crabs (Fig. 7-42) burrow in the sand of this relatively dry belt.

Organisms in the *high tide zone* or upper beach are exposed 60 to 90% of the time. It is flooded by tides twice a day, but its life forms spend more time on the land and in the air than under the water. The last sea-weed, the channeled wrack, is found in the lower high tide zone. It is the seaweed that ascends to the greatest height on the sea cliffs. Some bar-nacles are found on the rocks of the upper littoral. Periwinkles are snails that inhabit this zone; they attach their spiral shells to the rocks with a mucous glue, withdraw into their shells, and close the openings with a plate, the operculum.

Between the particles of sand, the semiterrestrial beach crustaceans inhabit a *microhabitat* (Fig. 7-43) with its own *microclimate*. Although the temperature at the surface of the sand may be 100°F with a relative humidity of 70%, only a few grains beneath the surface the temperature is 86°F and the relative humidity is 98%. Under these conditions, there is little danger that these animals will lose body water to the sur-rounding air through their respiratory surfaces or gills.

The *midtide zone* is covered with water most of the time. In this broad

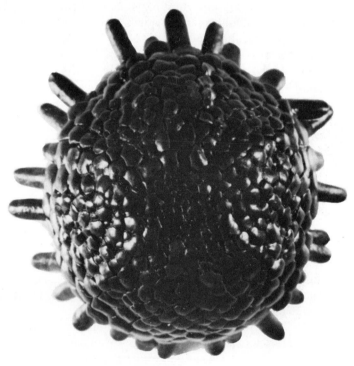

Fig. 7-41   This sea urchin (*Podophora*) is adapted to life in the rocky pools along seashores where heavy surf pounds. Its architec-ture includes armor plates (above) and thick spines like pil-ings (below).

Fig. 7-42   **The fiddler crab is a creature of the intertidal zone that spends considerable time out of the water. It lives high on the sand. The male has one enlarged pincer, which is used in territorial combats.**

belt, the organisms lowest on the rocks are exposed for less than 20% of the day, whereas the highest are uncovered more than half the time. The knotted wrack, the predominant variety of seaweed, provides some cover for the animals underneath its long stringy fronds. *Fucus,* the brown rockweed (Fig. 7-44), also grows in the midtide zone, and is exposed at low tide. Its broader, flattened strands provide excellent cover for intertidal animals. The rockweeds and the wracks are leathery and flexible, adapted to withstand the pounding surf. Animals of the midtide zone include barnacles, mussels, and limpets. The conical "shells" of rock barnacles are cemented to the stone below and are well adapted to heavy surf. Barnacles are extremely modified crustaceans that "stand on their heads and kick food into their mouths" with their hairy legs (Fig. 7-45). When exposed at low tide, they withdraw their legs into the cone and await the water's return. Mussels are two-shelled mollusks related to the clam and the oyster. They secrete a threadlike mucus that hardens and anchors them to the rocky surface. At low tide, they draw in their fleshy siphons and close the lips of their shells tightly. Limpets (Fig. 7-46) are snail relatives with shells like "Chinese coolie hats." They clamp down on the rocky surface with a force of 70 pounds per square inch and release a corrosive secretion that erodes circular grooves in the stone

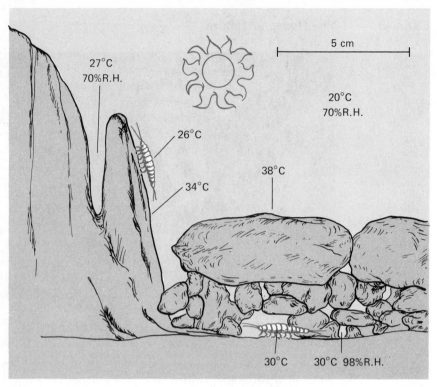

Fig. 7-43   **The microhabitats of semiterrestrial beach crustaceans. Compare the temperature and relative humidity above and below the sand.**
After E. B. Edney, from G. B. Moment, *General Zoology,* Houghton Mifflin Company, Boston, 1967.

Fig. 7-44   **The brown alga *Fucus* is also known as the rockweed, since these seaweeds are frequently seen on rocky shores and are exposed at low tide.**
Courtesy of Ward's Natural Science Establishment, Inc., Rochester, New York.

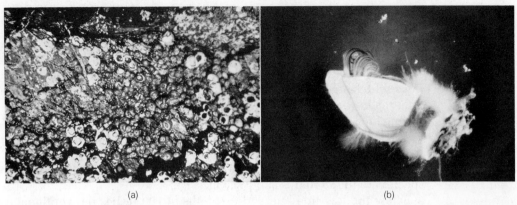

(a)   (b)

Fig. 7-45   **Barnacles. (a) Barnacles are organisms often found growing on rocks in the tidal zone. They are crustaceans, not molluscs. Because of their attached way of life, their heads are degenerate. (b) Barnacles have been described as animals that "stand on their heads and kick food into their mouth." Their curled, jointed, hairy legs are seen projecting here.**

(a) courtesy of Ward's Natural Science Establishment, Rochester, New York. (b) courtesy of Saturn Scientific, Inc., Fort Lauderdale, Fla.

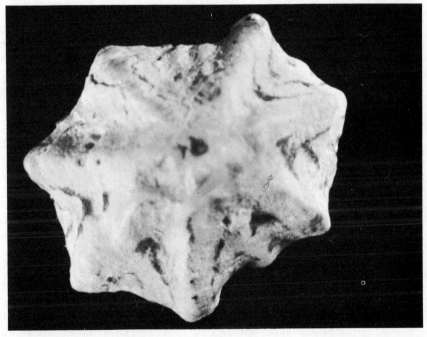

Fig. 7-46   **The limpet is a tidal zone relative of the snail. Its conical shell deflects the impact of surf.**

Courtesy of Saturn Scientific, Inc., Fort Lauderdale, Fla.

face, fitting the limpet shells perfectly. The intertidal rock pools of the midtidal zone are refuges for saltwater organisms when the tide is out. Rain dilution or sun evaporation of the pool water can result in osmotic death of the contained organisms.

The *lower beach* is exposed only twice per month during the highest tides, which are also the lowest, the *spring tides;* they occur at or shortly after the new and full moon. *Neap tides* are the smallest monthly tides, having the least difference between high and low tide.

The largest seaweeds, the kelps, are anchored to the rocky surface of the *low tide zone* by nubby holdfasts, and their great stemlike stipes and leaflike blades float in the water. The greatest number of littoral animal species are found in this subdivision.

Below the low tide line and to depths of 600 ft, lies the continental shelf or the *neritic zone.* Neritic waters are rich in nutrients and therefore life. Wave action, currents, upwellings, and river inflows fertilize these waters with mineral nutrients in particular. Many of the world's greatest fisheries are located over the continental shelves.

Beyond the continental outcroppings, the waters of the oceanic zone are divided into several vertical zones. The surface waters are rich in plankton, but since light does not penetrate appreciably beyond 600 ft, photosynthetic plankton are limited to the surface waters, the *photic zone.* Between 600 and 6000 ft, there is little or no light and permanent animal populations are few. Schools of squid and shrimp are found at these depths during the day, but they rise at night. This *bathyal zone* lies between the photic zone above and the abyssal zone below 6000 feet.

In Greek, *abyssos* means "without bottom," and it referred to the primeval great deep or chaos of myth. The black water of the *abyssal zone* is cold, only a few degrees above freezing. It is oxygen rich, since cold waters sink at the poles and creep over the bottom toward the equator. Most of the few fish of the abyssal depths are known only by their scientific names. They are small flesh eaters with great jaws and teeth often exceeding the dimensions of the rest of their slender bodies (Fig. 7-47a). Populations are sparse, since they are so far removed from the base of the food pyramid, the phytoplankton above. These deep-sea fish—in contrast to blind cave animals—have great eyes with retinas filled with rod photoreceptor cells for dim vision. The only light in the depths comes from the intermittent flashes of these fishes' bioluminescent organs. It is believed that the light patterns produce play roles in hunting and species' recognition. This is a stark example of evolution's two-edged sword. The same light devices that allow the sexes of the proper species to find each other to reproduce also give one's position away to potential predators. One species (*Melanocetus johnsoni*) has a luminescent lure on a stalk above its jaws. The deep-sea angler fish female car-

Fig. 7-47   **Most deep-sea fish have small populations and are carnivorous (a). (b) The
deep-sea angler female solves the problem of finding a mate by carrying the
male around with her, permanently attached as an external parasite.**
From N. M. Jessop, *Biosphere,* Prentice-Hall, Inc., Englewood Cliffs, N, J., 1970.

ries a miniature male around permanently attached to a fleshy protuber-
ance of her head (Fig. 7-47b). This "sexual parasitism" solves the
problem of finding the opposite sex in sparsely populated waters.

The bottom forms of the abyss include tube-dwelling worms, sea
anemones, sea lilies, glass sponges, starfish, sea urchins, etc., not unlike
those found on the bottoms of the continental shelves. Stalked sea lilies
(starfish relatives) and glass sponges, however, are found only in deep
waters. Dead diatoms and dinoflagellates drift down to the bottom
where they once again form the basis of the food chain, even in the black
abyss, as the microdebris is eaten by bottom-dwelling filter feeders.

FRESH WATER

The freshwater habitat is less stable than the sea and more variable in
its composition. There are two major categories: running streams or
brooks, and lakes or slow-moving rivers.

The shallow, splashing, descending creeks of mountainsides are well
oxygenated. The life forms within such waters are streamlined for life
in the rapids. Such animals often have flattened bodies, and some forms
have hooks and suckers to aid in clinging to rock surfaces (Fig. 7-48). The
green algae ("moss") growing upon the rock surfaces are the base of the
food chain. Most insect larvae feed on the algae of the rocks, and the

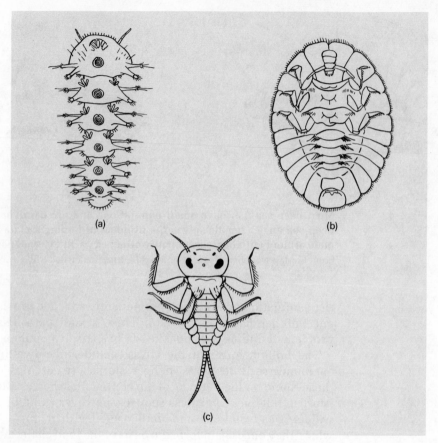

(a)

(b)

(c)

Fig. 7-48   **These insect larval forms are adapted to life in swiftly flowing water. Their bodies are flattened, and they cling to the underside of rocks with legs or specialized suckers.**
Redrawn from E. P. Odum.

carnivorous trout wait, facing upstream, for any unfortunate dislodged from its perch by the rapid currents.

Freshwater lakes of moderaté to great depth are divided into several zones similar to those of the marine habitat. In the littoral zone, rooted aquatic plants grow up from the bottom. The lighted waters of the photic zone overlie the dark depths of the deep lakes. In deep lakes, the waters of the depths are as black as the lower bathyal and abyssal zones of the sea. In ponds and even moderately deep lakes, some light penetrates to the bottom. In the surface waters of the *limnetic zone,* photosynthesis predominates. Respiration is dominant on the bottom, as the bacteria of

decomposition respire the detritus of sunken plant remains. At a point
between the waters of the limnetic zone and the depths, the rate of photosynthesis equals the rate of respiration. This line is called the light compensation level. Below are the waters of the *profundal zone*, where respiration exceeds photosynthesis, even though light still penetrates (Fig.
7-49).

The students of lakes (limnologists) distinguish three types of lakes.
Eutrophic ("good nutrition") lakes are relatively shallow, and therefore
the zone of decomposition is close to the photosynthetic layer. Algae
and pond weeds are abundant, as are crustaceans, insect larvae, small
fish, and large fish, such as bass. Most ponds are eutrophic. *Oligotrophic*
("scarce nutrition") lakes are deep, poor in nutrients, but rich in oxygen.
Lake Michigan, Lake Tahoe, and the Finger Lakes of New York are oligotrophic. Here we find the great lake trout, "lunkers." *Dystrophic* lakes,
or bogs, are rich in acids but poor in life.

The pond weeds and water weeds are higher plants (land plants) that
have returned to the water and undergone a kind of "regressive evolution." These higher aquatic plants, *hydrophytes,* have reduced their

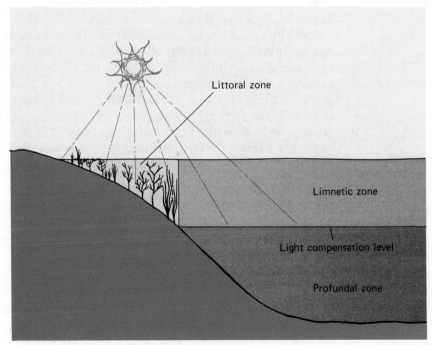

Fig. 7-49   **Life zones of a freshwater lake.**

quantities of vascular and supportive tissues. Why? Their leaves are often thin with reduced cuticles and epidermises, and their stems have air chambers. Some lack roots. Most have reduced roots and float wholly or partially submerged. Some floating plants have stomates on the upper surfaces of their leaves. Water lilies and water hyacinths are common hydrophytes.

THE EVERGLADES

The Everglades of south Florida is a unique freshwater biologic area (Fig. 7-50). The temperate zone meets the subtropics here, and a rich blend of plants and animals from both climatic regions results. The Everglades have been classified as marsh-grasslands, and, although similar swamps occur along the Mississippi drainage basin, nowhere is such an extensive development of a related biologic area to be found. Thousands of square miles of land were included in the Everglades at one time. To say that the Everglades is a threatened ecosystem is an understatement; it has declined, perhaps beyond the point of reclamation. Let us consider the Everglades as they were.

South Florida is hardly above sea level. There are no hills, not to mention mountains; the surface profile of the land of the Everglades has been aptly compared with a shallow saucer. On the eastern coast of the peninsula, a limestone ridge rises highest on the rim of this saucer. There is only a slight elevation of land. The rim is highest to the north and missing in the southwest, along the Gulf of Mexico. The shallow and incomplete bowl is tilted to the south and west. The seaward slope is no more than a few inches to the mile, but it is enough. This is the drainage basin of the Everglades, more than 100 mi from north to south and east to west. The lay of the land and the climate were the two major factors of the physical environment dictating the biologic possibilities of the ecosystem.

Rainfall is about 60 in. per year, and, like other subtropical parts of the world, there is a rainy season and an annual drought. The plains of the glades were often flooded by the torrential rains of late summer and early fall, then nearly bone dry in the winter-spring drought. The natural drainage of most of central Florida flows into the great Lake Okeechobee, the second largest freshwater body in the United States, located on the northern rim of the Everglades. In flood seasons, the waters spill over the lake's low south shore and creep southward and seaward across the basin of south Florida, which bears the Everglades.

Fig. 7-50   **The Everglades. This photograph of a sawgrass plain was taken during the rainy season, at which time in the past (before the construction of the drainage canal system) such flats were flooded with fresh water and only the tips of the sawgrass protruded. Islands of broadleaf vegetation dot the horizon.**
From M. D. Nicklanovich and M. J. Zbar, *Everglades Ecosystem,* Saturn Scientific, Inc., Fort Lauderdale, Fla., 1971.

A Florida poet called the Everglades "a river of grass." The vast expanse of water flows slowly seaward between the blades of sawgrass, which is really a sedge. The peat deposits of sawgrass and other marsh plants have enriched the poor sandy soil of eroding limestone.

A typical aquatic food chain in the Everglades begins with microscopic freshwater algae. Protozoans, small crustaceans, and insect larvae (Fig. 7-51) feed on the algae and, in turn, are eaten by mosquito fish (Fig. 7-52) and bream. These species are eaten by larger, carnivorous fish, such as the spotted gar (Fig. 7-53), whose elongate, spike-toothed jaws betray his profession. Garfish are the alligator's staple food, and the alligator is a top carnivore. Another carnivore unique to the Everglades is the long-necked anhinga, or snake bird, which dives and impales fish on its spearlike bill (Fig. 7-54).

During the dry season, much of the sawgrass plain dries out, and water

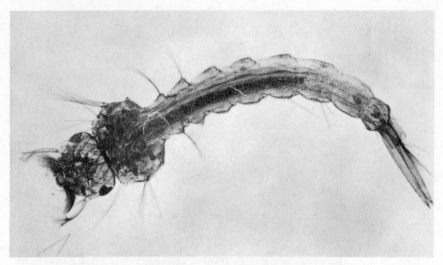

Fig. 7-51   **Mosquito pupa. The early stages in the life cycle of the mosquito are aquatic, and the larvae feed upon the freshwater algae and serve as food for small fish.**
Courtesy of Carolina Biological Supply Co., Burlington, N. C.

Fig. 7-52   **Mosquito fish. These small fish are the secondary consumers or primary carnivores in many freshwater ecosystems. They consume insect larvae, mosquito larvae in particular, and serve as food for the higher carnivores.**
From M. D. Nicklanovich and M. J. Zbar, *Everglades Ecosystem,* Saturn Scientific, Inc., Fort Lauderdale, Fla., 1971.

Fig. 7-53   **The garfish. The extremely prolonged jaws bearing simple, sharp teeth are
adaptations betraying the gar's carnivorous way of life.**
From M. D. Nicklanovich and M. J. Zbar, *Everglades Ecosystem,* Saturn Scientific, Inc., Fort
Lauderdale, Fla., 1971.

Fig. 7-54   **The anhinga, which is also known as the water turkey or the
snake bird, is a spear fisherman of the Everglade sloughs and
ponds, as its lancelike beak reflects. Unlike most waterfowl,
the anhinga lacks oil glands, and its feathers are not water-
proofed, indicated by this wing-drying pose.**
Courtesy of Colourpicture Pub., Inc., Boston. Photo by F. Truslow.

is retained only in ponds and deeper drainage channels, the sloughs or slough runs. Alligator wallows form some of the deepest pools in the Everglades and are among the last places to dry during the drought. As such, the "gator holes" along with sloughs and ponds form a refuge for fish and other wildlife during the dry season. The sheltered populations of these more permanent waters repopulate the glades following the wet season.

During the periodic visitation of the drought, the pond waters become stagnant. The foul water is filled with dead vegetation, whose bacterial decay consumes much of the dissolved oxygen. Shallow ponds are also thoroughly warmed, and this heating diminishes the quantity of oxygen in the water, since the solubility of a gas in water is inversely proportional to the temperature; oxygen decreases with rising temperature. The movement of many fish in small quarters also roils the water, stirs up sediments, and the mud settles on fish gills, making respiration that much more difficult. Under these conditions, sunfish and bass die quickly.

The garfish and the mudfish of the glades are particularly well equipped or adapted to survive the drought and stagnant water. This fact is implied in the name "mudfish." Most modern bony fish possess a hollow, white bag called the swim bladder, which is a hydrostatic or buoyancy device. By varying the quantities of gas in the interior, the fish can float at various depths. It operates something like the ballast tanks of a submarine. The swim bladders of many fish do not open into the throat or pharynx. Instead of swim bladders, the gar and the mudfish have blood-rich, spongy lungs (Plate I) that extend almost the entire length of their bodies and open into their throats. Microscopic studies of sections of the gar's lung reveal a rich capillary circulation in the walls of the respiratory pockets and great masses of striated muscle in the interwoven meshes of tissue. The gar is said to surface once every 7 min in stagnant water to gulp air. Aside from mammals, no vertebrates possess a muscular diaphragm for breathing. The contraction and relaxation of the striated muscle in the gar's lung brings about gas exchange. The anatomy of the mudfish lung is quite the same as that of the gar. In "fresh water," these fish rely wholly upon gill respiration.

The gar and the mudfish belong to a group of fish much older than the modern bony fish, and the fish lung is believed to predate the swim bladder. It seems that the fish descendants of the first vertebrates early evolved a lung to cope with the dangers of freshwater life. Marine descendants had no such problem in the stable sea, and the lung was apparently transformed into a swim bladder. Although the mudfish and the gar are not lungfishes in the strict sense of the word, the anatomy of their lungs is essentially the same as that of the true lungfishes. The peri-

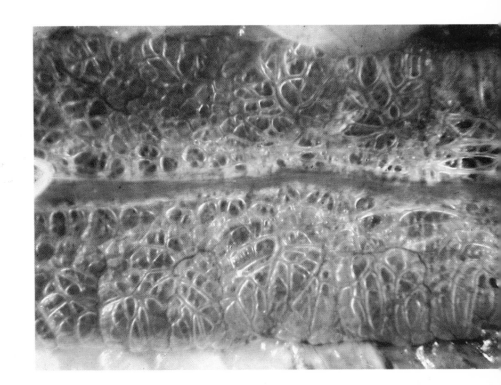

PLATE I

Garfish
Lung

(a)

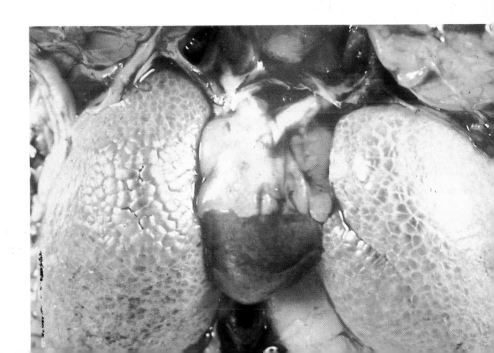

(b)

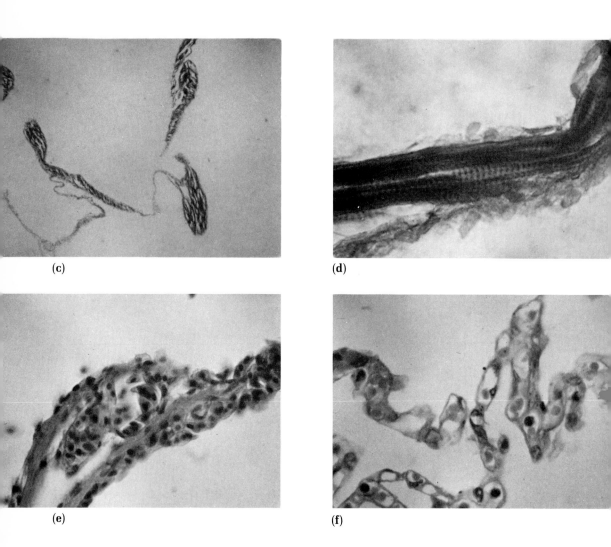

(c)

(d)

(e)

(f)

PLATE I.   Garfish lung. (a) and (b) overleaf. (a) Closeup photograph. Note the rich blood supply and the spongy nature of the organ. Compare with amphibian lung (b). In a low-power photomicrograph (c), the muscular beams and thin walls of the respiratory air chambers stand out. Higher-power photomicrography (d) shows that the muscle in the tissue beams is cross-banded, striated or "skeletal" muscle, the contractions and relaxations of which are involved in the inhalations and expirations of air. The thin walls of the air chambers (e) bear numerous capillaries and are nearly identical to the walls of alveoli, the microscopic air chambers of the lungs of higher vertebrates (f). In (e), the nucleated red blood cells can be seen coursing single file in capillaries over both walls of this partition.

From M. D. Nicklanovich and M. J. Zbar, *Everglades Ecosystem*, Saturn Scientific, Inc., Fort Lauderdale, Fla., 1971.

odic drought that visits the Everglades and the adaptations of the gar and the mudfish to it are reminiscent of pre-Devonian times, before nature made amphibians from "lunged fish."

The Everglades is mostly, but not all, sawgrass savannah. In elevated places, pinewoods and small forests of broadleaf trees are to be found. These are the homes of numerous mammals, deer, cougars, racoons, etc.

At the southernmost tip of Florida, the sawgrass gives way to mangrove swamps, where the fresh water meets the salt water of the Gulf of Mexico. Mangroves can withstand the salt water and are spread throughout the Caribbean. The ocean is no barrier to their dispersal.

The brackish water of the mangrove swamps and inlets is a mixture of salt and fresh waters, as are the estuaries of most rivers. The meeting of the sawgrass and the mangroves (Fig. 7-55) is an ecotone, a place where one ecosystem is transformed into another and vice versa—a transition zone. The mingling of the sea and the brackish waters of the mangrove swamps is also an ecotone, and, like most ecotones, this one is filled with life forms. The offshore waters are the breeding grounds of numerous crabs and shrimps.

Primarily for agricultural and real estate purposes, a large number of

Fig. 7-55   **Sawgrass-mangrove ecotone. Near the tip of Florida, the sawgrass gives way to the mangrove, as the fresh water mixes with the salt water.**
From M. D. Nicklanovich and M. J. Zbar, *Everglades Ecosystem*, Saturn Scientific, Inc., Fort Lauderdale, Fla., 1971.

canals have been dug in south Florida to "reclaim" land from the Everglades by draining the "surplus" waters into the sea and preventing their southward flow. At the same time, the Everglades National Park at the tip of Florida is generally in a state of drought. The park is only a mere fraction of the former Everglades; many of the other glade areas are privately owned and scheduled for drainage and "development." Conservationists recently halted the construction of a 39-sq-mi jetport in the heart of Big Cypress Swamp. The ridge of Miami has long since been occupied, and the suburbs of the city are presently moving out into the glades that have already gone through the transition stage of agricultural "development." Realtors and land development firms, considerable political pressure groups, are hungrily eyeing the remaining glades and openly favor the continuation of the great jetport construction. Undoubtedly, more water will be allotted to the small national park, but it seems likely that "people pollution" will destroy the rest of the Everglades. The author and others believe that the federal government should halt all commercial development of the Everglades, repurchase the land from present owners, and expand the confines of the present park. At this writing, it is also felt that a majority of citizens would favor this action.

SUCCESSION

The principal types of terrestrial biomes, forest, grassland, desert, etc., represent the most stable and enduring ecosystems. They contain the vegetation and animals best adapted to the environmental conditions of the areas. These ecosystems have *climax* communities, which remain unchanged unless the physical environment is altered. If the climax community is destroyed by man, then left unmolested for years, a series of communities will succeed one another until the climax is attained once again. This is an *ecologic succession*. Forests destroyed by fire are replaced by grass, but, given enough time, and provided that the climate remains the same, the woods may well return. Natural successions follow climatic changes. Grasslands have replaced many forests in dry, modern times. The tundra, although it is at present a climax ecosystem, represents a frozen stage of succession. If the climate were to become progressively warmer and wetter, the present tundra would be transformed eventually into a coniferous forest, and perhaps later even into a mesophytic deciduous forest.

In the high arctic biome, succession on the barren rock surfaces resulting from glaciation began with lichen *pioneers,* the first plants established. These plants grow when moisture is available and survive when it is not. Their chemical action on the underlying rocks loosened particles, which, together with the decaying lichens, formed a thin, primitive soil. Mosses succeeded lichens and produced great mats of water-holding peat. The third stage of succession saw annual herbs growing in moss beds. On the higher elevations of mountainsides in temperate climates, grasses as well as herbs succeed lichens and contribute to the developing soil; finally, pines take root and the forest begins. Succession can proceed no further than climate allows. On the tundra, succession has hardly proceeded beyond the annual herb stage.

Shore lines advance into ponds, as submerged, floating, and swamp hydrophytes contribute their organic remains to silt deposits and raise the old pond bottom. When the bottom surfaces and becomes dry enough, grasses and sedges grow on it. These are succeeded by shrubs, and finally trees. Pines may be succeeded by deciduous trees or vice versa. The successive plant communities modify the environment for their successors until the climax is reached.

The mature ecosystem (climax) is complex, with a large number of highly specialized species. Food webs are intricate, and there is a very efficient use of solar energy input by the community. Population sizes remain about the same; the member species of a mature ecosystem enjoy environmental stability. The stability and efficiency of climax communities explain their evolution. The steady state, or homeostasis, a dynamic stability, is the major theme of a climax. Mature communities can buffer the effects of normal, minor environmental fluctuations, but, since its species are so specialized and their relationships intricate, such communities are particularly susceptible to abnormal, extreme environmental change. The destruction of the climax results in the ascension of an immature, replacement ecosystem, which is composed of only a few unspecialized species with high reproductive rates. Replacement ecosystems are unstable and inefficient. Decades and even centuries are usually required for the comeback of the climax.

If the environmental change is severe enough, the natural climax may never be reestablished. Only one agent produces such extreme environmental alteration in short periods of time. He flatteringly calls himself *Homo sapiens,* "wise man." Hopefully, not too late, his ecologic ignorance has been revealed. Even the stable ecosystem is fragile. The elimination of a single species by exceeding the limits of tolerance can disrupt the "balance," or steady state, of an entire mature ecosystem. The species must be a key one, pivotal in the food chain. The productivity of replacement ecosystems is extremely low compared with that of the climax.

PESTICIDES, POLLUTION

Hardly any biology book has ever been written that did not contain the prophetic statement, "Insects are in contest with man for supremacy on the land." Insects are the most successful class of higher animals, containing the greatest variety of species, with the largest populations, and occupying the most habitats. Nature has equipped them well. They have been evolving for almost 300 million years; man has been in existence for less than 20 million.

Insects are the vectors, or carriers, of many human and domesticated plant and animal diseases. Mosquitoes spread the virus of yellow fever and the malarial protozoan; the tsetse fly carries the flagellate that causes the African sleeping sickness. The common flea transmits the plague bacillus, houseflies spread typhoid, the sandfly transmits the disease kala azar of the Orient, the human louse spreads typhus, and ticks carry Rocky Mountain Spotted Fever and Texas Cattle Fever. In medicine, the discovery that insects transmit disease organisms ranks second only to the germ theory of disease itself. The biblical "pestilence" referred not just to disease, but also to the descent of the "plague of locust" upon the crops. The Mormons, tilling the deserts of Utah, considered their salvation from the grasshopper hordes by sea gulls miraculous. The discovery of the insecticidal properties of DDT on the eve of World War II was likewise hailed as a man-made miracle. Like so many of man's technological armaments, dichloro-diphenyl-tricholorethane (DDT) might well turn out to be a two-edged sword.

DDT threatens the land birds of prey and many fishing birds. Even the robin shows signs of DDT poisoning, although its species is not immediately threatened. Perhaps, it will be, one of these years, as the late Rachel Carson warned in her historic best seller, a "silent spring."

DDT and its related *halogenated hydrocarbons* are very resistant to bacterial decomposition. They are not *biodegradable,* which is to say that they persist in the environment for long periods of time. DDT and its relatives accumulate in the fatty tissues of animals, and, more seriously, they are concentrated by those organisms high in the food chain, the nutritional level of man.

The diatoms (phytoplankton) are the "grass of the sea," at the base of most oceanic food chains, and they store food in the form of oil droplets. It is here that DDT, worked into the coastal waters by rivers draining sprayed tidewater farmlands, first appears in much less than 1 part per million (ppm). The crustaceans that eat phytoplankton concentrate the poison in their fatty tissues. In shrimp, the concentration may be 0.5 ppm. The flesh and livers of fish are rich in oils and they eat many crustaceans. The tissues of their prey are digested and respired or assimilated, but DDT accumulates in their fat depots to the extent of several parts per million (DDT and the halogenated hydrocarbon pesticides are fat soluble). Here man and the birds of prey enter the food chain as top carnivores. By this time, through "biological amplification," the amounts of tissue DDT have reached more than 5 ppm. The accumulation of pesticides can kill adult birds directly, as has been observed in robins and others in areas sprayed with DDT. Lower levels of pesticide accumulation cause reproductive failure in birds. DDT appears to interfere with calcium metabolism and egg shell formation. Affected brown pelicans produce defective shells for their eggs, and thus their attempts at reproduction are abortive. For the same reason, peregrine falcons, ospreys ("fish hawks"), eagles, and owls are in real danger of extinction, and their former ranges of distribution are rapidly decreasing.

The weed-killer 2,4-D (dichlorophenoxyacetic acid), a herbicide, has plant hormonelike effects, and plants sprayed with it have increased sugar contents, making them more palatable to livestock. The effects on animals are less well established, although the chemical has caused human pathologies.

Other commonly used pesticides (chlorinated hydrocarbons) are chlordane, dieldrin, aldrin, and endrin. Massive kills of millions of fish in the Mississippi have apparently resulted from endrin contamination, just as heavy DDT spraying for spruce budworms killed young salmon in the Miramichee River of New Brunswick, Canada. The chlorinated hydrocarbons other than DDT are increasingly more toxic to wildlife and man.

The other major group of insecticides, the *organic phosphates,* have been widely employed, since they are more biodegradable than the halogenated hydrocarbons. The insecticidal properties of organic phosphates were discovered by a German chemist just before World War II. The German government declared work on these compounds secret, and from them came the most deadly nerve gases. These chemicals block the enzyme cholinesterase, which breaks down the chemical transmitter substance, acetylcholine, which is released at the nerve endings of one cell to initiate a new impulse in the next nerve cell in the chain. The transmitter substance must be destroyed, so that meaningful, coordinated messages can be relayed again. If acetylcholine remains at the junction, impulses continue to be generated. Uncoordinated, massive muscular spasms and convulsions result in death.

The major organic phosphates are parathion and malathion. Many cases of human and animal poisoning have already resulted from their use. Unlike the long-lived chlorinated hydrocarbons (DDT *et al.*), the organic phosphates are quickly broken down. This advantage could be offset, however, by the addition of phosphate to ecosystems, upsetting the phosphorus balance.

The balance of nature is easily destroyed, but nature is nevertheless difficult to control. Just as antibiotic-resistant strains of bacteria are on the rise, so many populations of mosquitoes are becoming increasingly immune to higher and higher dosages of DDT. It is not unlikely that strains of absolutely DDT-resistant mosquitoes will soon dominate the earth. It may well be that certain regions of the globe are unfit for human habitation. Perhaps they should be left to the insects. Nor should it be forgotten that many noxious insect larvae, such as those of the mosquito, lie very close to the base of freshwater food chains, and their elimination threatens the entire ecosystem. The small fish that feed on mosquito larvae are the prey of larger carnivorous and game fish.

The alternative to the use of poisonous insecticides is natural or *biological control*. The screwworm fly produces maggots that attack the flesh of cattle and deer. The female lays eggs in open wounds. The screwworm was a serious problem in the southeastern and southwestern United States. From 1957 to 1959, X-ray sterilized males were released over Georgia, Florida, and Alabama. The sterilized males competed with the normal wild males, and the females began to deposit unfertilized eggs. Within two years, the screwworm was wiped out of the Southeast. The British intend to try the method on the tsetse fly, and it is hoped that the technique will lead to the control of corn borer, fruit flies, sugar cane borer, and even mosquitoes. Malaria and viral sleeping sickness are still spread by mosquitoes on the continental United States.

The analysis and synthesis of insect *sex attractants* is another opportunity for nearly natural control. Female insects normally secrete these substances to attract males. Poison traps baited with commercial sex attractants are certainly preferable to widespread DDT spraying. This method of eradication is being tested on the gypsy moth, an introduced species that has considerably damaged the forests of New England. Cockroaches have a similar sex attractant.

Germ warfare has been practiced with some success in controlling rampant undesirable animal populations. The myxomatosis virus was introduced in Australia to control the great European rabbit plague there. Mosquitoes were man's ally in this case, for they spread the virus from infected to healthy rabbits. After thousands of birds, other wildlife, and even livestock were killed by heavy aldrin (DDT relative) spraying in an effort to control the Japanese beetle in the Midwest, and after introduced wasp predators of the beetle yielded less than satisfactory results, bacterial warfare all but eradicated the beetle in the United States. Viruses sprayed on young alfalfa fields in California effectively controlled the alfalfa caterpillar. All these germs are quite specific and are harmless to man.

The juvenile hormone of insects prevents the metamorphosis of insect larvae, thus holding them in the nonreproductive stage. A chemical with similar properties has been traced to the wood pulp of the balsam fir. Do the trees produce this "hormone" as a defense against insects? This may be the ideal insecticide.

The use of pesticides has grown alarmingly in the last few years, and finally the public outcry has aroused some action in legislative halls. It is only by law that the growing pesticide menace can be halted, and the multimillion dollar chemical industries have powerful lobbies. DDT and its stable relatives, continue to accumulate in our soils, streams, oceans, and even our own tissues.

The *organic pollution* of streams and bays is hardly less of an ecologic problem than pesticide poisoning. Untreated sewage carries disease bacteria and is therefore, obviously, an extreme health hazard. Even treated sewage, however, often contains food materials for the omni-present bacteria of decomposition, and these—thus fed—respire, repro-duce, and deplete the natural waters of their oxygen content, destroying fish in the process. Fresh waters, fertilized with semitreated sewage out-spills and washings from wood pulp mills and breweries, experience cycles of putrification and algal blooms that culminate in the destruction of the mature freshwater ecosystem. Nor are the salt waters of bays im-mune to organic pollution. An outflow of sewage from duck farms in New York made its way by tributary streams into the Great South Bay of Long Island, New York, and caused cycles of algal blooms there that eventually destroyed the oyster beds. This story has been repeated a thousand times along the coasts of the country.

The rivers of life have become the "rivers of death," with low-grade unproductive ecosystems, in which algal blooms alternate with decom-position. Although the waters move, they are essentially like the stag-nant, foul ponds of droughts. This is particularly true of the great, slow-moving rivers.

Salts of copper have been used to control algal blooms in water reser-voirs, decorative ponds, and fountains. Similarly, the water effluents of smelters have destroyed the stable ecosystems of innumerable streams and lakes all over the world. How? (Don't forget that algae lie at the base of the food chain.)

The heavy metal mercury has been used in many industrial processes. It was thought that since it was heavy and sank to stream bottoms there would be no repercussions. However, certain bacteria combine mercury with organic compounds. Organic mercury becomes incorporated into food chains and concentrated as it ascends the biologic ladder. Many freshwater fish are no longer safe to eat, and even marine fish show contamination. In one incident, 50 Japanese died of mercury poisoning after eating fish taken from bays contaminated by industrial wastes. In 1970, swordfish was called off the market after tests revealed high mercury content. Certain brands of tuna were also taken from the grocers' shelves. Recent studies indicate that an amount of mercury "contamination" is natural; in the future, however, almost all fish placed on the market will be routinely analyzed. Recently, Pacific seal liver pills were removed from the health food market because of their mercury content. (Seals eat fish.) Pregnant women have been cautioned about fish in their diet. Mercurial fungicides are routinely used to treat wheat seed grain. Even if all the mercury pollution were halted today, it has been estimated that more than 100 years will pass before there is a significant decrease in this long-lived pollutant. Not only does mercury remain in the environment, but it also stays in the tissues; it cannot be cleared from the tissue fluids by the kidneys. Mercury poisoning causes a horrible deterioration of the human nervous system. A research unit at the Mayo Clinic is attempting to piece together the symptoms of mild organic mercury poisoning, from which untold numbers may be suffering. There is no cure for mercury poisoning.

Thermal pollution is another ecologic danger. To cool the nuclear reactors of electricity-generating plants, many companies divert and recycle river and bay waters. Even if the temperature of the water is changed by just a few degrees, the effects may be disastrous to fish and microorganisms of importance in the food chain. Many studies of thermal pollution are under way at the present time. The burning of coal and petroleum during the last century has doubled atmospheric carbon dioxide. Atmospheric carbon dioxide tends to reduce heat loss from the earth to space. During the past 50 years, the average temperature of the earth has warmed about 1°C, and some glaciers are melting. Prophets of doom have predicted that if the trend continues, continental coastal plains will be flooded. The temperature fluctuation may be just part of natural climatic changes. The particulate matter of smoke reflects incoming solar radiation and would thus tend to offset the CO_2 insulating effect.

More seriously, the tars by-product to petroleum burning are known to be carcinogenic. The carbon monoxide (CO) produced by incomplete combustion is poisonous, since it combines irreversibly with the blood respiratory pigment hemoglobin. During thermal inversions, in which the warm air fails to rise above the city for dispersal and stagnates instead, many people die of carbon monoxide poisoning.

JUDGMENT

As long as the human population was small and the earth large, man's ecologic ignorance was inconsequential, but, now that the earth is small and the human population large, the unconscious subversion of nature may well be fatal. Almost all pollution stems from the population explosion of man, made possible by technologic innovation (the Industrial Revolution). The ultimate cause of environmental contamination is thus "people pollution." No imaginable technologic invention can sustain the present rate of population growth; it must be curbed. Paul Sears has called ecology "the subversive science." The principles of ecology, newborn in human consciousness, could well bring about the overthrow of man. The ecologic crisis must be brought to an end; we cannot continue to live in ecologic sin. Peter Farb described the present predicament of man as "being a ruler over the earth without knowing the rules."

In the past, conservation meant the economic usage of nature for practical purposes, the scientific farming of forest and field. Modern man's relation with the land has been strictly economic, but there are signs of the development of an "ecologic conscience" and a "land ethic" that includes more than an economic valuation of nature. Aesthetic appreciation of the dwindling wildlife is on the rise. Since the science of ecology is yet young and so vital, there is a very practical reason for preserving the undisturbed wilderness for study. Much remains to be learned about the stable ecosystem. The principles of ecology must be built into the deteriorating environment of man.

Man cannot return to the "state of nature," but he must return to the steady state. Although man does not live in natural surroundings, he cannot escape the natural consequences of unnatural behavior. In the final analysis, man never really escaped from nature. Not even man can violate natural law for long. Francis Bacon said, "We cannot command nature except by obeying her."

If most of life should be destroyed in nuclear war, it is not unlikely that the few surviving forms, appropriately the simplest, would begin that vast trek through time again, changing form and function, ecosystem succeeding ecosystem, until the climax of higher communities is reached. Perhaps man would not reappear.

DISCUSSION QUESTIONS

1. Why do no two species—even closely related ones—occupy the same econiche?

2. How does nitrate and phosphate fertilization of farm ponds increase bass production?

3. What is the primary cause of pollution?

4. What is meant by the phrase "land ethic"? by biological amplification?

5. Could the present world population be supported without the use of DDT or other pesticides?

6. Why do natural populations seldom show the terminal curves exhibited by organisms in the laboratory?

7. Has man escaped from nature?

8. What factors should be considered in determining the fate of a tract of undeveloped land?

REFERENCES

BATES, MARSTON, *The Forest and the Sea*, Random House, New York, 1960.

CARSON, RACHEL, *The Sea Around Us*, Oxford University Press, Inc., New York, 1951.

CARSON, RACHEL, *Under the Sea Wind*, Oxford University Press, Inc., New York, 1952.

CARSON, RACHEL, *The Edge of the Sea*, Houghton Mifflin Company, Boston, 1955.

CARSON, RACHEL, *The Sense of Wonder*, Harper & Row, Publishers, New York, 1965.

EHRENFELD, D. W., *Biological Conservation*, Holt, Rinehart & Winston, Inc., New York, 1970.

EHRLICH, PAUL R., *The Population Bomb*, Ballantine Books, Inc., New York, 1968.

HARDIN, GARRET, *Population, Evolution, and Birth Control*, W. H. Freeman & Co., Publishers, San Francisco, 1969.

KRUTCH, J. W., *The Voice of the Desert*, William Sloane Associates, New York, 1957.

LOVE, G. A., and R. M. LOVE, eds., *Ecological Crisis*, Harcourt, Brace & Jovanovich, Inc., New York, 1970.

ODUM, E. P., *Fundamentals of Ecology*, W. B. Saunders Co., Philadelphia, 1953.

SHEPARD, PAUL, and DANIEL McKINLEY, eds., *The Subversive Science*, Houghton Mifflin Company, Boston, 1969.

OBJECTIVES

1. Define instinct and drive and give several examples of each.

2. Compare and contrast reflexes, instincts, and learned behavior.

3. List, define, and exemplify five major types of learning.

4. Outline the steps in classical conditioning.

5. Enumerate several roles or functions of territory in animal life.

6. Compare the components of courtship and territorial behavior in several species.

7. Contrast vertebrate and insect societies.

8. Explain the adaptiveness of herding, schooling, and dominance hierarchies.

9. Trace the occurrence of instinct and learning in the branches of the animal kingdom.

10. Define and exemplify arena behavior.

11. Summarize Darwin's discussion of instinct.

The
Ethics of
Nature

IRONICALLY BUT NECESSARILY, the struggle for existence is fiercest among members of the same species, since the individuals of a population play the same role in their ecosystems, have the same needs. They compete for food, space, and sex. Yet nature is not a perennial battlefield. There are behavioral as well as anatomic and physiologic adaptations that conserve the energy as well as the lives and limbs of the species. Indeed, it seems probable that most behavior has an anatomic basis. Anthropologists agree that human culture had its beginnings in biology; sociality has obvious natural advantages. Society is more ancient than human: nature explored the horizons of social organization long before man raised himself from his knuckles or was even a creature deserving the title of man.

The study of comparative animal behavior, *ethology,* is nearly newborn. In search of the principles of the highest type of behavior, learning, psychologists ran laboratory rats through mazes for decades before anyone set out to study the wild rat and his society. The term ethology is derived from the Greek root *ethos,* which is still used to refer to the attitudes or habits of a group and is related to the word ethics, or morality. The idea that the behavior of animals in the state of nature is ethical or moral is also quite recent and stands in opposition to the common misconception of a ruthless nature, "red in tooth and claw." The opposite of ethos is pathos, which literally means "disease." Pathology is the branch of medicine dealing with sickness. Psychiatry began with the study of diseased personality before normal human behavior was adequately analyzed. Likewise, many of the first studies of animal behavior were conducted in zoos, where the activities of the "beasts" were more pathologic than normal. Nature is highly moral, and the new science of ethology is therefore appropriately named.

Man is the most difficult animal to study behaviorally, and it is almost impossible to observe him in the state of nature. His history is short, even though it is the greatest of any animal. How did preman behave? The anatomical evolution of man from animals is quite well established. His behavior left little in the way of relics and fossils, aside from his braincase, weapons or tools, and quite recent ceremonial burials. The ways of man's near relatives may reveal something of the social behavior of preman. It is almost as difficult to define modern man behaviorally as it was when the Greek Heraclitus advised, "Know thyself." The great American playwright, Robert Ardrey, began his *African Genesis* with the statement, ". . . if man in a time of need seeks deeper knowledge concerning himself, then he must explore those animal horizons from which we have made our quick little march." [1]

We are vertebrates, animals with backbones, along with fishes, amphibians, reptiles, birds, and our other mammalian relatives. Many of the greatest discoveries in animal behavior were made by birdwatchers.

TERRITORY IN BIRD LIFE

In the early spring, migrant birds return to the temperate zone. The males arrive first. The males of resident species leave their mixed flocks of winter, where a tolerant—if not social—behavior had held sway. Definitely antisocial activities begin. The males space themselves out and defend their territories against intruders of the same species. The defenders fly from perch to perch marking the boundaries of their territories, proclaiming their dominion with song. Neighbors build invisible walls between themselves. It seems as though unseen leashes, radiating outward from the center of the territories, restrain the cocks. Occasionally, a duel erupts in midair, and then the sound of shrill, scolding chatter, accompanied by a frenzy of wings, replaces the usual sedate song. Almost no invasions succeed.

Territorial behavior is aggressive behavior, but the aggressions seldom lead to actual fighting. It is mainly matters of threat and retreat,

---

[1] Robert Ardrey, *African Genesis* (New York, N. Y.: Atheneum Publishers, 1961), p. 3.

bluffs uncalled. The intruder is always severely psychologically handicapped. The defender seems invincible, armed with his confidence of possession. As Voltaire said, "The spirit of property doubles a man's strength."

The combatants approach each other stiff legged with their feathers held out from their bodies. Thus they present the most impressive appearance of size and strength. Sometimes they posture broadside, displaying their full length as well. Threat failing, the defender will move to the attack with beak pointing and wings flapping. The territorial trespasser invariably yields and cools the wrath of the aroused occupant by making some submissive or appeasing gesture before he retreats or renews his invasion attempt (Fig. 8-1). The subdued intruder may crouch or turn away, exposing the vulnerable back of his head. The defender halts his attack. At times, at the boundaries of neighboring territories, both combatants will simultaneously disengage (Fig. 8-2). The displays of threat and submission are rituals, like bird song. Sometimes at the

Fig. 8-1   **Rockhopper penguins, showing threat by bird on right followed by withdrawal or submission of bird on left.**
San Diego Zoo photo.

point of greatest tension in the confrontation, the males will resort to some seemingly meaningless activity, such as preening, rather than engaging in combat. Herring gulls will furiously tear tufts of grass. This behavior would be appropriate for nest building. Such activity is said to be *displaced;* in territorial "fights," nervous energy is often released harmlessly in this fashion. Bloodshed is avoided; the species is conserved. Combats are mock rather than actual. Violence is rare; only feathers are ruffled.

The sexes in birds are most often differently marked. The female is usually drably colored, even camouflaged, since incubation is her responsibility. The male is more vividly colored, and his bullfighter's dress serves as an additional marker of territory (Fig. 8-3). In Wilson's phalarope, a wading bird resembling a sandpiper, sexual roles have been reversed. The female is brightly colored and establishes and defends the territory, while the drab male incubates the eggs. Colored feathers are sometimes the anatomic tools of territorial duels. The English robin's red breast is a *sign stimulus* that calls forth threatening behavior from other males. An immature robin with brown breast feathers is rarely

Fig. 8-2   **"Facing away" display of the black-headed gull. This appeasement posture serves as a means of reducing tension in aggressive encounters between males, but also as a way of reducing aggression between prospective mates. The facial mask is worn only during breeding. At that time it serves as an additional territorial device, beside the postures, gestures, and calls by means of which the birds space themselves.**
Courtesy of Dr. N. Tinbergen.

Fig. 8-3

Pheasants. Sexual dimorphism is common in birds. Usually, the male is the more vividly colored of the two (at least in breeding season), and the female is arrayed in drab camouflaging dress. The pheasant rooster at right is perhaps the most gorgeous of birds, whereas the homely hen (left) blends into the grass and earth of the fields.
Courtesy of General Biological, Inc., Chicago.

attacked, but a mature cock robin will threateningly posture at even a tuft of red feathers (Fig. 8-4).

The bird's song is no more a frill of nature than feather finery. It is a key part of territorial behavior. In many species, the song is wholly predetermined in the nervous system and appears in precise form in males reared in isolation. A basic song is elaborated by learning in other species. In a few, such as the mockingbird, many songs can be learned.

COURTSHIP IN BIRDS

Logically, territorial behavior precedes mating. Females arrive after the males have established their dominance over a piece of "ground." It is comical that their first attempts to trespass are met with the same aggressive territorial displays by the male. The females make submissive gestures, retreat, and return repeatedly until the male finally accepts the female on his territory and a pair bond is formed. Frequently, it is difficult to separate territorial and courtship behavior (Fig. 8-5). Some birds perform elaborate courtship dances prior to copulation. Whooping cranes leap into the air, spread their wings, and weave back and forth. Storks clap their bills together. Many species "ritual feed"; one partner

Red tuft

(a)   (b)

Fig. 8-4   **The robin's red breast acts as a releaser of aggressive behavior. A tuft of red feathers will call forth threat displays and attacks, whereas a stuffed immature robin lacking the red breast of adult males is ignored.**
From N. M. Jessop, *Biosphere*, Prentice-Hall, Inc., Englewood Cliffs, N. J., 1970.

crouches and gapes like a nestling while the other thrusts its beak into the open mouth. The nervous stimulation of courtship rituals apparently coordinates the mating pair. All birds except ostriches, ducks, geese, and swans lack penises; the lips of their vents are apposed. Precise synchronization is essential.

In some birds, geese, lovebirds, and crows, the pair bond lasts for life; in most, it is just a season. In a few, the male and female remain together only long enough to copulate and then go their separate ways. The female raises the young by herself. This is the most promiscuous relationship of all. In these last species, territorial behavior has been modified most bizarrely.

ARENA BEHAVIOR

Ninety-nine percent of the types of birds establish a pair bond of at least long enough duration to share the work of raising the young. Only 1% of the bird species has no true pair bond. Here, the sexes remain apart for most of the year. In the breeding season, the male clans spread themselves out in typical territorial fashion by singing, displaying plumage, charging, and fighting the mostly ritualistic combats. Females wander into an arena of males, displaying their wares upon individual territories or courts, and choose their mates or copulatory partners. These territories are mating stations only; the females exit after intercourse.

The bowerbirds of New Guinea build elaborate "bowers" upon their territories (Fig. 8-6). Although the bowers sometimes resemble nests, they are used solely as display and copulatory sites. Nevertheless, it has

(a)

(b)

(c)

Fig. 8-5   **Mutual courtship displays in the herring gull: (a) mew calling; (b) choking; (c) facing away. Synchronized performance of these acts seems to overcome fear and aggression and allow pair bond formation. These ritualized activities resemble several mutual appeasement gestures that males make following aggressive encounters.**
After N. Tinbergen, *Behavior*, 15, 1959. From N. M. Jessop, *Biosphere*, Prentice-Hall, Inc., Englewood Cliffs, N. J., 1970.

Fig. 8-6

A bowerbird, display court, and mating site, the "bower." This sexual mansion is constructed solely for the most promiscuous affairs: it will never serve as a nest. This most complex of bowers, decorated with piles of berries, colored pebbles, and even flower petals, was built by one of the dullest bowerbirds, suggesting, perhaps, that sexual selection values have been transferred to external objects.

From Knaur, *Tierreich in Farben.* Photo by John Barhan.

been postulated that their construction is a holdover of the nest-building impulse; nesting behavior might have been displaced. The bowers are quite specific, and their complexity is greatest in the drabbest bowerbirds. The most vividly plumed species build the simplest. Some bowers have the forms of tepees; others are like matting-lined avenues. Some species decorate the mossy floor of their bowers with white pebbles, colorful insect skeletons, bright berries, and fresh flowers. The satin bowerbird even paints its bower with earth pigments; it is one of the two tool-using birds. The bowerbirds are the most complex arena species.

Sexual selection has operated in the arena birds, often building over the generations a fabulously abundant and vivid plumage, as in the birds of paradise and others. Since these colorful and therefore conspicuous birds frequently display on the ground, they are obvious to predators (Fig. 8-7). Thus sexual selection has moved them in a direction opposite to the course natural selection would dictate. The substitution of external display objects for body modifications is an alternative. Many species of bowerbirds have sacrificed their sexual finery in the course of evolution. One species has lost a pink crest at the nape of its neck, but still turns its head repeatedly to display the missing crest to the female. Perhaps colored berries have been substituted for the crest. This is known as the transferral effect. Sexual plumage becomes unimportant, and natural selection moves the species in the direction of protective coloration.

Sage grouse and prairie chickens of the American West are also arena species. Prairie chicken cocks have elegant headdresses and great red throat skin pouches, which they inflate and display (Fig. 8-8). Studies of

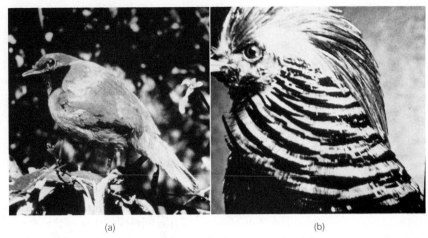

(a)   (b)

Fig. 8-7   **(a) The blue cotinga, a vivid blue member of the Cotinga family. (b) The head of the Golden pheasant. Such feathered finery is the result of sexual selection. Decorative plumage is carried to the extreme in the birds of paradise, which, along with some pheasants and cotingas, are arena birds.**
Courtesy of CCM: General Biological, Inc., Chicago.

Fig. 8-8   **Sage grouse cock in strutting display.**
Courtesy of CCM: General Biological, Inc., Chicago.

sage grouse arenas have revealed that the arrangement of cocks on the ground determines breeding privileges and therefore reflects a mating hierarchy. In one grouse arena of 400, 74% of the matings went to 4 master cocks; 13% of them were done by "submasters." The remaining matings were scattered among the peripheral cocks, after the masters tired. Position and status on the arena were determined by territorial jousts long before the hens arrived.

The cocks of the European ruff spread out collars of colorful neck feathers in display when a hen enters the clan's arena. The female pecks the collar of her chosen mate, and the entire ground of rejected males collapses, as if in faint.

Although arena behavior is a bizarre modification of territorial behavior, it has arisen independently a number of times in otherwise unrelated families of birds. It has been suggested that it is a mechanism for speeding up evolution.

TERRITORY AND AGGRESSION IN OTHER VERTEBRATES

Territorial behavior and ritualized aggression are well developed in several fish. A fight always draws a crowd, and the most familiar fish gladiator is the Siamese fighting fish (*Betta*). The males are usually dull in color, but during the breeding season they assume a brilliant blue pigmentation and fiercely defend a nesting territory. In close quarters, the aroused males bite and batter each other to ribbons. In large enough spaces, the famous death duels of the fighting fish do not take place. The oriental fans of fish fights used to place two males in separate, small jars, placed side-to-side, to excite the combatants before they were introduced into the ring. Males and females must be kept apart, but at the appropriate time the male will wrap himself about the female's body, squeeze the eggs from her, take them in his mouth, and spit them into the bubble nest he has prepared. The fighting fish has air pouches in its throat and expels mucus-encased bubbles in with the eggs for aeration.

In many perchlike and other bony fishes, opponents swim straight at each other only to halt nearly mouth-to-mouth, where they blow out gill covers, enlarging their contours. Sometimes they snap at each others' mouths. Primitive or injurious fighting, in which fish ram each other in the unprotected flank, is less frequent than ritualized combat.

Cichlids are aggressive, small, freshwater fish. All gradations of intraspecific fighting, from genuine attack to harmless ritual, are encountered in them. In some species, antagonists swim at each other from their respective territories. Before contact, one or the other (or both) turns broadside and erects his spiny, sawtooth dorsal fin, thus presenting the most imposing dimensions, but also the vulnerable flank. If the bluff fails, his side is attacked. When two cichlids are placed in a small aquarium, one will claim the entire tank as his domain and relentlessly attack the other, hour after hour. The gaudy, but passive goldfish placed in a cichlid tank will be reduced to a tattered wreck in a day. One cichlid species safely mouth-wrestles. The rivals swim toward each other, display broadside, then lock specially thickened lips, and push and pull until one surrenders and swims away.

Another freshwater fish, the European stickleback, has been a favorite of animal behaviorists. At mating time, the male stickleback builds a nest of sand and moss threads and thereafter defends his territory. His back turns bluish-white and his belly becomes yellowish-red. When two males meet at their territorial borders, a series of attacks and retreats begins. Each fish takes the offensive in his own territory. The "fight" goes back and forth across the border, but neither fish touches the other. Under crowded conditions, when territories are too small and fighting intensifies, the two males will stand on their heads with their sides to each other and dig in the sand. This maneuver is the initial part of nest-building behavior. It is obviously a displacement activity, as it is in birds. Experiments have proved that the vermilion belly of males serves as a sign stimulus or *releaser* of territorial defense similar to the robin's red breast (Fig. 8-9). Likewise, the most territorial of marine fish are the "poster-colored" (reef species of tropical waters). When a colorless, swollen stickleback female approaches a territorial male, a different sequence of behavior is released. The male performs a zigzag dance and leads the female to the nest (Fig. 8-10).

Amphibians have either lost territorial behavior in the course of evolution or evolved from a group of fish that were not territorial. Salamander courtship ceremonies will be discussed with instincts.

Territorial behavior is pronounced in reptiles, the first land vertebrates from which mammals and birds evolved. The male of the common lizard or "chameleon" of the southeastern United States and the Caribbean, *Anolis*, was frequently sold by traveling circuses, since it could change its color from bright green to brown, depending on the color of the background. The anole's throat skin is drawn out into a brilliant orange dewlap. When challenging another male, the dewlap is suddenly thrown out while the animal stands broadside in stiff-legged display with its body raised up off the earth (Fig. 8-11). If the threat display fails, males will resort to biting and something akin to mouth-wrestling.

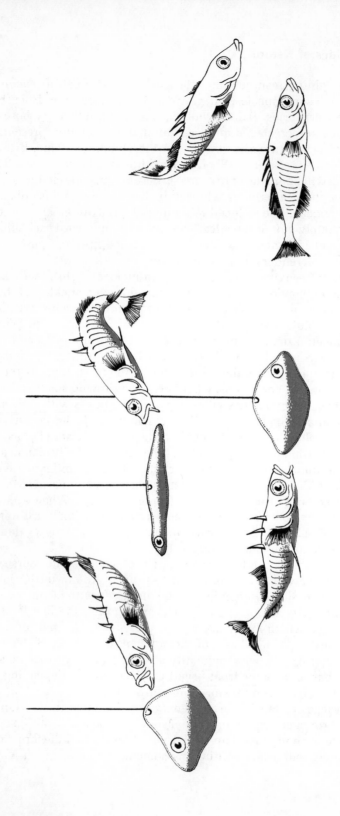

Fig. 8-9   **These experiments demonstrate that a red underbelly is the key sign-stimulus releasing aggressive behavior in the male three-spined stickleback. Note that the realistic model (right) lacking the red underside is ignored.**
From N. M. Jessop, *Biosphere*, Prentice-Hall, Inc., Englewood Cliffs, N. J., 1970.

Fig. 8-10   **Courtship and reproduction in the three-spined stickleback. The male establishes a territory and builds a nest. He defends the site against all intruders, but
halts his attack upon recognizing a swollen female. He performs a zigzag dance
to the nest, pokes his head in, withdraws it, taps the female's tail once she is
in the nest (releasing egg-laying), then swims over the laid eggs and sheds semen
(milt) over them.**
After N. Tinbergen. From N. M. Jessop, *Biosphere*, Prentice-Hall, Inc., Englewood Cliffs,
N. J., 1970.

Fig. 8-11   **The anolid lizards and several other reptiles possess a throat pouch, the dewlap, which is often colored. This device is extended in broadside displays with head-bobbing.**
Photo courtesy of Professor Lewis D. Ober.

In submission, the body is lowered to the ground. The same display is used in female courtship.

The bull alligator roars to attract females as well as to warn off other males. He will also respond to the challenge of gunshot or thunder and to a French horn played in B flat two octaves below middle C.

In the San Diego Zoo, male rattlesnakes were observed in a type of "Indian wrestling" (Fig. 8-12a). Both males raised about one-third of their body length off the ground, faced each other while weaving, then made side-to-side contact and attempted to pin each other. The winner pins the loser and crawls over him.

The male marine iguana of the Galapagos Islands defends a small territory of several square yards of rock near the shore. This large, seaweed-eating lizard is adorned with a crest of buckskin running down the midline of his neck, back, and tail. (Many iguanas, as well as the tuatara of New Zealand, bear such fringe or spines along their backs.) The forehead of the marine iguana is covered with hornlike scales. When an-

(a)

(b)

(c)

Fig. 8-12   **Ritualizing aggression. Members of a dangerous species seldom injure their own kind, owing to the ritualization of combat. (a) Threat display of male iguana, involving dewlap extension, head-bobbing, and the erection of the buckskin fringe. (b) Rattlesnake combat dance. The contest ends when one animal is pinned. (c) Rhinos butting heads; horn-clashing is relatively harmless, compared with the goring of sides or underbellies.**

(a) courtesy of CCM: General Biological, Inc., Chicago. (b) and (c) courtesy of San Diego Zoo photos.

other male approaches his territory, the defender opens his mouth, bobs his head, presents his broadside to his opponent, rears up on his legs as far as possible, erects his crest, and goose steps back and forth before the intruder, who likewise postures. If this threat display fails to intimidate the intruder, the defender attacks with lowered head. The two reptilian bulls lock horned heads and engage in a pushing match. If neither gives ground, they break contact, display, and reengage. Finally, the intruder usually assumes a submissive posture, dropping to his belly and flattening his crest before retreating. Biting seldom occurs, except when a repulsed invader wanders across another territorial border unaware. Anoles also head-bob in their territorial and courtship displays. The dewlap of the anole is the counterpart of the iguana's buckskin display crest. Several iguanas also possess dewlaps (Fig. 8-12b).

Mammals arose from reptilian ancestry 200 million years ago, and perhaps they carried territorial behavior with them, for ceremonial fighting and courtship are sharply defined in a number of mammals (Fig. 8-12c). This type of behavior has been most extensively studied in various deer and antelope. Their beautiful horns and antlers are most often the tools of ritualized combat rather than the weapons of violent defense.

The rapier horns of the African oryx antelope are used in ceremonial fencing. The ritual begins with two bulls displaying broadside to each other. Then they charge past one another, clashing horns in passage. After a pause, the bulls butt foreheads and beat their horns together with rapid head-shaking movements. Never are the horns used to inflict stab wounds in intraspecies fights. The stylized combat of European fallow deer begins with bucks displaying broadside while they goose step, nod their heads, and thus wave their antlers up and down. As if by command, they halt, turn face-to-face, lower their heads, crash antlers together, lock them, and swing their heads to and fro. Once again, strength is tested. Successful bucks defend harems of does. on the African plain, it is not unusual to see a bachelor herd peacefully grazing alongside a buck with does. Thus, these rival fights function in sexual selection.

The narwhale of arctic waters is also known as the sea unicorn, since the male has a long, spiral tusk extending from the upper jaw. Although the behavior of the narwhale remains unstudied, speculation that the great tusk is a ceremonial tool of courtship ritual seems probable.

Strictly territorial behavior is not so conspicuous in mammals, but aggression is common. We have seen that courtship behavior in several mammals at least strikingly resembles the territorial rituals of lower vertebrates. Combat in the kangaroo rat involves confrontation, leaping blows with the great hindlegs, and the kicking of sand in an opponent's eyes. The behavior of monkeys and apes has attracted a great deal of attention, since they are members of the order of man, Primates. Territorial behavior is rare in the great apes. In most primate societies, polygamous groups with dominance hierarchies are the rule of social order. Sometimes the group defends a "moving territory." The howl of the howler monkey (Fig. 8-13) clans of Central America has the same function as song in bird life.

Among the great apes, the gibbon alone is a strict territorialist, using the monogamous family as the unit of social organization. He will defend 30 acres of forest. The gibbon makes 11 distinct vocalizations, 9 of which are employed in territorial behavior. His studies of howler monkeys and gibbons led C. R. Carpenter, the greatest scholar of primate behavior to propose, "It would seem that possession and defence of ter-

Fig. 8-13   **Male red howler monkey howling. He produces loud yells with his large voice box; they announce the location of one troop to another, thus avoiding territorial trespassing and aggressive engagements.**
San Diego Zoo photo.

ritory which is found so widely among the vertebrates, including both the human and subhuman primates, may be a fundamental biologic need. Certain it is that this possession of territory motivates much primate behavior."[2]

THE FUNCTION OF TERRITORY AND AGGRESSION

Since the establishment of territory involves intraspecific aggression, both real and ritualized, there are those who would not classify territoriality as a form of social behavior, but instead categorize it as antisocial. They contend that truly social behavior begins with cooperation rather than aggression. Yet the aggressions that lead to territory perform extremely social functions.

Aggressive behavior spaces the members of a species, and the territories so formed often function as feeding grounds. The territory of a bird of prey is a hunting reserve, and that of the eagle can cover almost 40 sq mi. Nests are built upon territories and the young are raised here; thus, territories serve as breeding grounds. The spacing out of a population lowers the frequency of the communication of contagious disease. Spaced species have less stress. The dispossessed or unterritoried individuals form an insuring reproductive surplus, but also a spreading force for the species. As many historians have argued, it is the outcasts of society that take to the frontiers as pioneers.

The harmless ritualization of violent combat has likewise led to a preservation of species' members. Submissive animals are rarely attacked. This constitutes a high morality, as Konrad Lorenz, the founding father of animal behavior, has noted. The pair bonds formed by strongly aggressive and territorial species endure longer. The length of time expended in the care and training of the young increases. The lavish care of the young is an advanced feature of birds and mammals, accounting for much of their success.

Comparison of human and animal behavior is an irresistible temptation. The poet of animal behavior, Robert Ardrey, introduced the public to ethology with his best sellers *African Genesis* (1961), *The Territorial Imperative* (1966), and *The Social Contract* (1970). Ardrey describes himself as a "scientific amateur." He received his bachelor's degree in

---

[2] Ibid., p. 39.

natural science from the University of Chicago in 1932 and made his living as a playwright during the Great Depression. After eight stage-plays, two novels, and several screenplays, Ardrey returned to science with his pen. His books caused an academic uproar. It was not consid-ered scientific to make such bold and sweeping comparisons between human and animal behavior. Yet most of his themes were not original; as he noted in his first book, most of the information was buried in scien-tific journals, obscure to the public.

About his landmark book *On Aggression,* which did not appear in English translation until 1966, Konrad Lorenz stated, "I am honestly convinced that in the near future very many men—indeed perhaps the majority of mankind—will regard as obvious and banal truth all I have written in this book about intraspecific aggression and the dangers which its perversions entail for humanity." Raymond Dart, the South African anatomist, had spoken of the "killer instinct" of the South Afri-can ape-man. New Zealand biologist Robert Bigelow and others have hypothesized that the rapid growth of the human brain was catalyzed by cooperative warfare.

In his writings, Robert Ardrey has made a case for the supposition that territoriality has played a major role in the evolution of such facets of human behavior as nationality, domicile, and free enterprise. Indeed, his works have led to a recrudescence of social Darwinism on a natural behavioral basis. Is private property a biologic necessity?

For decades, psychologists have maintained that human aggression is the result of the frustration of normal drives, the sex drive in particular. In contrast, animal behaviorists have proposed that aggression is nearly instinctive, arising within the organism, often as the result of the effect of certain hormone levels upon built-in nervous patterns. Konrad Lorenz warns that human intellect is a two-edged sword. Rather than relying upon "well-adapted instincts" of ritualized intraspecific aggres-sion like most animals, man has conceived long-range weapons and im-personal warfare. In nature, intimate combat is usually less than fatal. Man's preeminent occupation with weaponry is pathological by natural standards. Yet these sophisticated weapons have been most often de-ployed by "base" desires. War is a kind of harmless-seeming ritual at long range.

New Guinea is the home of primitive tribes as well as bowerbirds and the birds of paradise. War parties of neighboring tribes meet in brief skirmishes along their territorial borders; few are injured in the 10-minute fights. The combatants retreat just out of short-bow range and about face. Then fists are raised and threats tossed back and forth. The wind ruffles the gaudy plumes of the birds of paradise thrust into the warriors' hair. With redoubled irony, a feathered arrow protrudes from the buttocks of a youth. The scene is quite removed from two modern nations casually poised with nuclear-warheaded intercontinental ballistic missiles surrounding their territories.

Robert Ardrey emphasized the role of "killer instinct" in human evolution. The first human "tool" seems to have been a weapon. In *African Genesis*, Ardrey wrote the following:

Our ancestry is firmly rooted in the animal world, and to its subtle, antique ways our hearts are yet pledged. Children of all animal kind, we inherited many a social nicety as well as the predator's way. But most significant of all our gifts, as things turned out, was the legacy bequeathed us by those killer apes, our immediate forebears. Even in the first long days of our beginnings we held in our hand the weapon, an instrument somewhat older than ourselves.[3]

Anthropologists, studying the habits of *Australopithecus*, the hunting ape-man, have proposed that this ancestor of modern man bequeathed mankind a dangerous "carnivorous mentality," since protomen broke with their ape relatives' herbivorous way of life. Ardrey's prose—although poetic—goes against a basic rule of animal behavior; predators' teeth are seldom red with the blood of their own species. In fact, lion and wolf societies are remarkably tolerant and moral in construction. The south African ape-man was more of a carnivore than any other contemporary ape, but he was still an omnivore. Lorenz suggests that man is an imperfect carnivore without natural weapons or sufficiently reliable inhibitions to prevent species' self-destruction. He proposes that curiosity and its outgrowth, rational responsibility, will counterbalance the sophisticated fratricide that has been developed by the application of conceptual human thought to aggression.

Human war is said to reveal the animal side of man's nature, but there are no comparable intraspecific animal wars in nature. Massive, fatal civil strife is unknown in the wild. The misconception is the result of a naïve view of nature, the melodramatization of the "struggle for existence," as "nature red in tooth and claw." The greatest competition lies between members of the same species, but nature, the great compromiser, has found harmless solutions to the potential problems of intraspecific aggression. Predation is not to be confused with fratricide, species suicide.

The role of territory in human behavior is still the subject of hot debate. The hypothesis is very attractive. Many questions remain unanswered. Was territoriality universal in vertebrates? We have seen that

---

[3] Ibid., p. 3.

amphibians are unterritorial and that territoriality is reduced in mammals, rare in the great apes, in fact. Few invertebrates exhibit territorial behavior. The solitary wasps are an exception. Is human social behavior a modification of basic territoriality?

OTHER ANIMAL ORGANIZATIONS

A *herd* or a *flock* is an aggregation of relatively anonymous individuals. In strongly territorial species, the individual is most distinct and independent. Many fish swim in *schools* (Fig. 8-14), which are quite like herds and flocks. The several advantages of such behavior are not very obvious at first glance. A school of small fish creates enough of a disturbance and casts a combined shadow sufficiently large to be mistaken

Fig. 8-14   **A school of fish. A school acts almost as a single organism, presenting the impression of size to, and confusing, the potential predator. The disturbance created by an attack on one flank of the school can cause the whole unit to turn.**
Courtesy of Saturn Scientific, Inc., Fort Lauderdale, Fla.

often for a sizable creature by potential predators. A small school of fish can almost be waved back and forth by a diver's hand. The lateral line systems of the fish are extremely sensitive to waves in the water, and the school almost responds like a single organism. A commotion on even a large school's flank (such as predator attack) is soon transmitted throughout the school much as a ripple follows a splash; the school changes course. An attacking predator also encounters a confusion of targets as he closes with a school; the advantage is obvious. On the land, startled flock or herd animals spread the alarm to the rest by calls. Small birds and even fish will often attack a potential predator, such as a hawk or carnivorous fish, en masse. This *mobbing* is a curious sight.

Lorenz has defined the *group* as a social organization in which the individual member is well defined by the reactions of his fellows. A *dominance hierarchy* is frequently the prevailing theme of group organization. Members are ranked according to sex, age, experience, and strength. Peck orders were first described in chickens. The highest-ranking hen is pecked by none and is first at the feeding trough; the lowest is pecked by all and is last at the feeding station. The best-studied dominance hierarchy of primates is that of the baboon.

Baboons are monkeys on the ground. Since they are members of the order of man and terrestrial, their behavior has been given a great deal of attention, as it might well reflect that of man's ancestors, descended from the trees. Baboons travel in groups of 25 to 100. Older males lead the troop. In the artificial conditions of the zoo, a male bully is "king of the hill" and defends his spoils, the harem, but in the state of nature a baboon band is led by a "senate" of several old males who maintain their authority collectively against the younger and stronger males. Females and infants travel in the center of the column, whose flanks are protected by the powerful young males. In dangerous situations, the prime defenders will encircle the whole troop; nevertheless, the most experienced lead the band.

There is no question that the members of the baboon group are strongly bonded to each other. They will defend the corpse of the fallen. A war party of several young males will fall out from the troop to attack a leopard (the baboon's prime enemy) in a struggle to the death if necessary. Thus, the baboon group is a soldier society with Spartan values, befitting a monkey on the ground. There is "strength in numbers" and efficiency in military organization. Still there is democracy in sex. Baboons are promiscuous: no sexual privileges are denied the lower-ranking younger males. The monogamous family seems to be the exception rather than the rule of primate social organization, yet most groups are composed of siblings and cousins. Thus, the band is a family in a sense. Although baboons are not strictly territorial, the range of one troop is respected by another; bands do not intermingle.

In the insect societies of bees, ants, and termites, a more rigid society has evolved (Fig. 8-15). Here there is an actual anatomic basis for the division of labor. The specialized castes can often be recognized by the naked eye. Honeybee workers develop from fertilized eggs but are sterile females. Only the queen is a reproductive female. The drones, or males, develop from unfertilized eggs. In termite colonies, we find soldiers with large jaws. Instinctive behavior is most elaborate in the invertebrates, whereas learning plays a more important role in the life of many vertebrates.

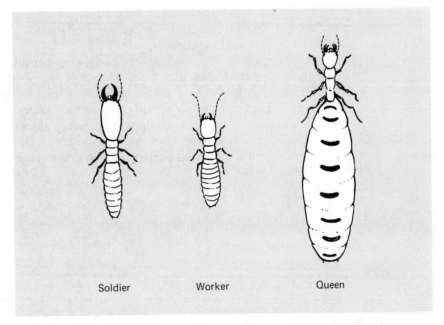

Soldier   Worker   Queen

Fig. 8-15   **Termites. Among the termites and many other social insects, there is an anatomic caste system. The individuals specialized as soldiers possess powerful, enlarged jaws. The huge egg-laden abdomen of the queen completely immobilizes her. She is wholly dependent upon her "attendants," who feed her, remove the thousands of eggs she lays to the nursery, and move her in emergencies. The young queen has a slender abdomen and is winged like the king (not shown). Following the nuptial flight and with the hatching of the first workers, her abdomen becomes very distended.**

INSTINCT

The dictionary defines instinct as "an inborn tendency to behave in a way characteristic of a species; a natural, unacquired mode of response to stimuli." The prophet of animal behavior, Konrad Lorenz, defines instincts as "unlearned, species-specific motor patterns."

Some psychologists have challenged the very existence of instincts. The loose usage of the term for any habitual or unconscious human act is unfortunate. The mere mention of even a vestige of instinctive behavior in man is academic dynamite—indeed, heresy in some schools of psychology. Hatred of the term instinct has led to the recommendation that it be dropped from the language. Instincts (or whatever we are going to call them) are widespread in the animal kingdom. Most *human* psychologists consider instinct nonexistant in *human* behavior, and resent the common usage of the term for human aptitudes, talents, "knacks," or "second sense." Animal instinct is very real to zoologists and a growing number of psychologists.

In Darwin's time, the mind and the body were considered more separate than they are today, and the existence of complex instinctive behavior was put forth as an argument against the theory of evolution. Chapter VIII of the *Origin of Species* begins, "Many instincts are so wonderful that their development will probably appear to the reader a difficulty sufficient to overthrow my whole theory." Darwin's definition of instinct should satisfy all parties to the controversy:

I will not attempt any definition of instinct. It would be very easy to show that several distinct mental actions are commonly embraced by this term; but every one understands what is meant, when it is said that instinct impels the cuckoo to migrate and to lay her eggs in other birds' nests. An action which we ourselves require experience to enable us to perform, when performed by an animal, more specifically by a very young one, without experience, and when performed by many individuals in the same way, without their knowing for what purpose it is performed, is usually said to be instinctive. But I could show that none of these characters are universal. A little dose of judgment or reason, as Pierre Huber expresses it, often comes into play, even with animals low in the scale of nature. [4]

Although Darwin "declined" to define instinct, he did anyway as you can see, and his definition remains unsurpassed.

---

[4] Charles Darwin, *The Origin of Species* (New York: Crowell-Collier Publishing Company, 1962), pp. 243–244.

The example of instinctive behavior he cites is called *nest parasitism*.
The English cuckoo lays its eggs in the nests of other birds. The baby
cuckoo immediately proceeds to eject the unhatched eggs and fellow
nestlings from the nest. This behavior is obviously unlearned, and must
therefore be hereditary, genetically determined. The adult cuckoo will
again lay her eggs in another bird's nest; no English cuckoo ever builds
its own nest. Other species of cuckoos are not nest parasites. The American cowbird is also a nest parasite with the same instinctive behavior;
thus, instinctive behavior is species specific, characteristic of a particular species. The series of movements by means of which the cuckoo
nestling throws its nestmates out is nearly always the same. So the behavior is said to be stereotyped, or definitely patterned, blueprinted, or
cast from a mold, apparently—unimaginably—the genetic code! The
head of the cuckoo and its gape are larger than those of the nestlings of
the species it parasitizes. The wider gape of the cuckoo hatchling is a
stronger stimulus to the feeding foster parent, who will feed the cuckoo
in preference to its own young (Fig. 8-16). Also the parents will not reclaim ejected nestlings or feed them outside the nest. Thus, it is seen that

Fig. 8-16   **Nest parasitism. The European cuckoo lays its eggs in the
nests of other birds. When the young cuckoo hatches, it instinctively ejects all other eggs or occupants of the nest. Even
if some other nestling should remain, its parents will feed the
cuckoo often in preference to it, since the gape of the cuckoo
is usually wider, an irresistible, supernormal stimulus to
brooding adult birds.**
Photo by Karoly Koffan.

the instinctive behavior of the cuckoo is immediately adaptive, insuring the survival of the species.

In summary, instincts are species specific, hereditary, unlearned, adaptive activities. Instinctive behavior is more complex than the simple reflex.

A nineteenth-century philosopher proposed that instincts were compounded reflexes. Reflexes usually consist of a single movement; instinctive behavior most often involves a series of movements. Logically, the number of nerve cells and circuits comprising instinctive behavior must be much greater and more complex than they are in reflex. Both instinct and reflex are unlearned and inborn, but the reflex is not species specific. Withdrawal reflexes are essentially the same in the frog and man. Many of the stimuli that result in reflex responses are very elemental and harmful—heat, pressure, imbalance—whereas the stimuli evoking instinctive behavior are often very subtle and harmless. A corner where two twigs meet is the stimulus that calls forth the highly stereotyped cocoon-building response of the caterpillar of the *Cecropia* moth. Stimuli that result in reflex arise from changes in the external or internal environment. Likewise, instinctive behavior may be evoked by changes in hormone levels resulting from alterations in the external environment. A single individual is involved in reflexive behavior, but instinctive behavior often takes place in a social setting: more than one individual is involved. An internal readiness is frequently generated within an individual to respond to the stimuli of another. After maturity, and sometimes even before, the organism is always capable of reflex response, whereas instinctive behavior is usually limited to a certain time of life or season.

Instinct and reflex have been confused and on occasion used nearly interchangeably. The suckling response of the newborn mammal has been called both instinct and reflex, yet a baby will not always suck when presented the breast—only when it is hungry. The response itself is uncomplicated, but it does involve head-turning as well as mouth and respiratory movements.

A *drive* is the urge or impulse to perform an activity or behave in a certain fashion. It is a state of stimulation, readiness. "Drive" is a psychologist's word, and it fills a need. All drives arise from body conditions, particularly internal stimuli that signal the physiological necessities of hunger, thirst, sex, etc. The word "drive" is not without considerable biologic meaning. Drives arouse general activity in animals. When the sex hormonal levels rise in the bloodstream of a migratory bird in its southern range (or drop in its northern range), a "migratory restlessness" appears. Several days of heightened activity follow, and finally the flock takes wing. A female rat will run a mile a day in a revolving cage, except when she's in heat, when she will run 15 miles.

Copulation is the mechanism of response involving a sequence of behavior that reduces the sex drive. In higher animals, such as birds and mammals, the essential preconditions of instinctive behavioral responses are commonly hormonal or nutritional. Academic subject matters themselves are unfortunately too often territorial battlegrounds. Psychologists should accept the existence of instinct, and animal biologists have a definite need for the term "drive." Robert Ardrey colorfully describes the foolish debate of psychologist and zoologist (taking the zoologist's side):

We tend, in our contemporary vocabulary of human motivation, to refer to "drives." This word is a bastard child of a common-law marriage between our rejection of the concept of instinct and a necessary acceptance of certain facts of life. It is a euphemism, as were those Victorian words and phrases referring in most genteel terms to a variety of undeniable human activities revolving about sexual intercourse. But euphemism has no lasting place in the sciences. As a psychological cynic, Professor Cyril Burt, once commented, a drive is an instinct under a new name. "Flung out at the front door, the old instincts are allowed in at the back after assuming an alias and a slight disguise." It is no help to the student of man, groping for an understanding of his fellow human being, that psychology has arrayed the open instinct, a form of innate behavior exhibited by man and all higher animals, in a crepe beard and well-placed rouge, has termed it drive, and has expected of its objective study any superbly significant conclusions. [5]

As is so often the case, an enlargement of vocabulary is all that is necessary to conclude the argument. Instinctive behavior is prefaced by an essential precondition, or drive. "Instinct" and "drive" are often used synonymously, while the phrase "instinctive behavior" definitely describes the act of fulfilling a drive. Ardrey is correct, however, in accusing human-oriented psychologists of confusing the issue, since they have been too critical of the term instinct, too man-centered, and have attempted to replace the term for patterned behavior with its precondition.

To the list of the psychologist's drives (hunger, thirst, sex, etc.), the animal behaviorist would add aggression and perhaps territory, the end or goal of most animal aggression. The greatest question is this: Is human aggression pathological or natural, the product of frustration, or an inwardly generated necessity, a vestige of our animal ancestry? Does instinct play any role in human behavior?

It is often said, with more than a grain of truth, that the lower animals

[5] Robert Ardrey, *The Territorial Imperative* (New York: Atheneum Publishers, 1966), p. 27.

are locked in reflexive and instinctive behavior. There are those who propose that learning, the dominant form of human behavior, results from the unlocking of instinct. Yet learning is encountered very low on the animal scale, as Darwin noted. Instinct has two great advantages over learning. There is no time required for trial-and-error learning in instinctive behavior: it is automatic. The time and energy of the species are also conserved, in that parental teaching and social tradition are unnecessary for the transmission of instinct, which is hereditary. The major disadvantage of reflexive and instinctive behavior is the restriction of the repertory of behavior to a few modes of response, not necessarily the best; innumerable generations must pass away before new mechanisms of response can evolve. Learning is the type of behavior that permits the maximal behavioral adaptations in a time so brief as an individual life span. Let us examine the variety of instinctive behavior before we move on to the subject of learning.

INVERTEBRATE INSTINCTS

Instincts predominate in the behavior of the lower animals. The two most successful animal groups on the land are the insects and the vertebrates. The vertebrates exhibit a greater degree of learned behavior than unlearned. At least part of the success of insects can be attributed to their highly developed instinctive behavior patterns. The evolution of human society is largely a cultural story; the development of insect societies has had a more obvious physical or biologic basis. Darwin chose the slave-making of certain ants and comb construction in honeybees as further examples of evolved instinctive behavior. Let us follow the master's reasoning, as fresh today as yesterday.

One species of European (Swiss) social ant (*Formica rufescens*) is completely dependent upon its slaves (*Formica fusca*) and without them would become extinct. As in many other insect societies, males and fertile females do no work; here, however, the "workers" or sterile females do hardly any work either! The only task the workers perform is the capturing of slaves (Fig. 8-17). Otherwise, they are incapable of making their own nests or feeding their own larvae. In the migration of a colony from the old to a new nest, the slaves carry their masters in their jaws, as well as determine the course of the migration and the new nest site.

A Swiss biologist shut up 30 masters without a single slave, but with plenty of food and their own pupae and larvae to stimulate them to work. They did nothing. They could not even feed themselves, and many died of hunger. The introduction of a single slave saved the survivors. She immediately set to work, fed the adults, made several cells, and

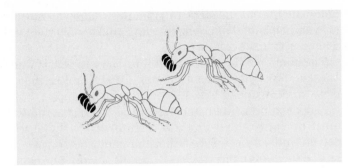

Fig. 8-17   **"Slavemaster" ants raid the nests of other, slave species to carry away larvae, which will mature to serve as workers in their masters' nest. Some slavemaking ant species are so dependent upon their captive workers that they can do no more than raid. In the movement of normal species of ants, the larvae are carried in the jaws of workers. Larvae are also frequently the spoils of insect war, in which case they are often carried off as food.**
From N. M. Jessop, *Biosphere*, Prentice-Hall, Inc., Englewood Cliffs, N. J., 1970.

tended the larvae. The story of this insect makes interesting reading, but Darwin chose it to exemplify the possible mechanism of the evolution of behavior patterns, or instincts.

Another slave-making species (*Formica sanguinea*), a large red ant, keeps small black ants of the species *Formica fusca* in bondage. This red, slave-maker has been studied in southern England and Switzerland. Here, the master workers determine where and when a new nest shall be constructed. During the migration, the masters carry their slaves. The slaves have the exclusive care of the larvae. Masters alone go on slave-making raids, wherein the pupae of the slave species are carried off. In England, the red workers usually do most of the other work. They leave the nest alone to collect building materials and food, such as aphid "milk," for themselves, their slaves, and their larvae. In Switzerland, slaves and masters share the work of collecting nest materials and food, but it is mainly the black slaves that search for aphid milk.

In ants which are not slave-makers, the workers of the species perform all tasks, including care of the young. Non-slave-holding ants carry off the pupae of other species as food. Such pupae might develop in food caches, and the foreign ants so accidently reared could follow their instincts and work. Darwin speculated that natural selection might have strengthened the habit of collecting foreign pupae until it became subverted to slave-making as a rigid instinct. Although his hypothesis is questionable, all the steps or gradations between normal, independent ants and absolutely dependent slave-makers overwhelmingly argue for an evolutionary sequence.

The other instinct whose evolution Darwin considered was that of cell construction in the hive bee. It has been computed that 12 to 15 lb of dry sugar are consumed by a hive of bees to produce 1 lb of wax. Nectar is the sugar solution of flowers that attracts insects. Almost 500 lb of nectar must be gathered to produce 1 lb of wax. Obviously, it is advantageous—indeed necessary—that the comb be constructed with the utmost efficiency, using the least amount of wax. The cells should be built in the best shape to hold the greatest possible quantity of honey. All these specifications are met in the honeycomb of the hive bee. Yet all construction is carried out by a leaderless crowd of bees in the darkness of the hive!

The comb consists of a double layer of cells. Each cell is a hexagonal prism, something like a cylinder with six flat sides rather than a circular wall (Fig. 8-18). The basal edges of the six sides are slanted to form an inverted pyramid. The sides of one cell are also the sides of adjacent cells. There is no duplication of wall construction, which would waste wax. The side walls and basal plates are less than $\frac{1}{100}$ in thick. Although

Fig. 8-18   **Honeybees upon the hexagonal cells of their combs.**
Courtesy of CCM: General Biological, Inc., Chicago.

the thickness varies, it is more uniform than can be yielded by most freehand craftsmen, and would draw the admiration of a machinist, used to thinking in terms of tolerances of hundredths of an inch. The mathematician would admire the nearly perfect geometry of the comb's architecture. Such are the works of instinct.

Here, also, Darwin saw evolution and its mechanism, natural selection, at work, and pointed once again to the great principle of gradation. At one end of the series he placed the bumblebees, the "humble bees." They are primitive in that they construct rough combs, and at the end of the summer the workers, males, and old queens die. Young queens hibernate in the winter and begin new colonies in the spring. A bumblebee in flight in the early spring is a young queen and the founder of a future colony. When she finds the appropriate nest site, she builds two crude, spherical cells with the wax that exudes from her abdomen. In one cell she places nectar and pollen gathered in her trips; in the second she lays eggs, about eight. When the eggs hatch, the mother feeds the larvae on the stored nectar and pollen. A few days later, the larvae spin their cocoons from which they shall emerge as workers. They shortly go to work, gathering nectar and pollen to care for their successive broods of younger sisters. The founder of the colony, the mother, in the meantime constructs additional cells of nectar-pollen and eggs. After the emergence of the first batch of workers, the aging queen devotes all her time to laying new broods. The workers enlarge the nest into a rough comb and forage for food, nectar, and pollen, "bee bread." This food is for the succeeding generations of a summer. The workers produce wax for the irregular comb and honey as a food store in cells for the geometrically increasing brood. Finally, the population reaches hundreds, or even 1000, as the "socially fertilized" process unfolds. The honeycomb cells of the colony are irregularly rounded and therefore waste the wax of the group as well as the energy of the community (Fig. 8-19a).

Darwin did not know of the existence of "subsocial" or solitary species, which build their nests in soil, wood crevices, or heaps of plant material, lay their eggs in the caves so made, and stock them with honey or pollen before leaving the young to fend for themselves subsequently. The behavior of the solitary species of bee would have expanded the spectrum at its lower end. Nonetheless, his recognition of the gradation existing between the irregular, rounded and separate cells of the "humble bee" and the nearly geometrically perfect cells of the "hive bee" vividly illustrates the course of evolution.

Between the perfection of the cells of the "hive bee" and the crudeness of those of the "humble bee" lie the intermediate cells of the Mexican bee, *Meliposa domestica*. Here we can see the "intelligence of nature" at work. The comb of this bee is composed of cylindrical, nearly circular, almost equally sized cells fused in an irregular mass. Where the circular walls would intersect, the Mexican bees have erected flat walls (Fig. 8-19c). Thus, each cell consists of an outer rounded portion and two or three flat surfaces where the cell joins two or three other cells.

A comb of cylindrical cells fitted together would not only have greater surface area than a set of fused hexagonal prisms (Fig. 8-19d) but would also have numerous intercellular spaces, and would therefore waste space, labor, and wax in addition to making for a weaker comb. It is easy to see that natural selection has favored those hive bees constructing the best cells with the greatest economy of labor, the least waste of energy in the secretion of wax. It is the accumulation of vast stores of honey in extensive combs that allows the honeybee colony to survive the winter. The "hive bee" colonies number 30,000 workers at times. We have noted

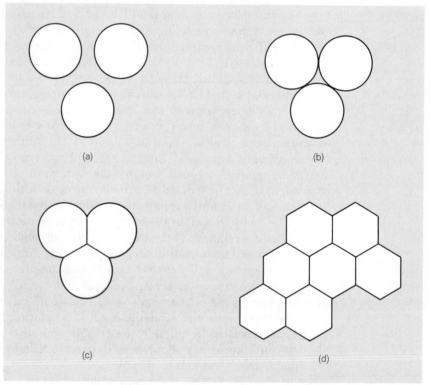

(a)   (b)

(c)   (d)

Fig. 8-19   **Stages in the evolution of comb construction. (a) Separate cylindrical cells (bumblebee). (b) Partially fused cylindrical cells. (c) Fused cylinders with flattened intersections. (d) Hexagons with common walls (honeybee).**

that the lower social bees that make less perfect and extensive combs seldom have colonies of more than a few hundred. Only their young queens survive the winter by hibernation. Bumblebees represent a primitive stage of socialization, and the honeybees exemplify the highest.

THE DANCE OF THE BEES

We have glimpsed the vast quantities of energy required to support the hive and the evolution of one type of instinctive behavior, cell construction, which in its perfected form conserves labor. It would obviously be a waste of precious energy if workers circled at random seeking flowers. It is amazing but logical that honeybees have developed a "language" to communicate the direction and distance of flower sources from the hive. If the food supply is less than 100 yards away from the hive, returning workers perform a round dance, tracing out a figure eight of one continuous line with its two loops superimposed upon each other (Fig. 8-20a), which simply means that food is in the near vicinity of the hive. The bee's highly developed sense of smell is in this case enough of a guide. The scent of a forager or scout can serve as an adequate director to the nearby source. For distances greater than 100 yards, however, the worker performs a wag-tail dance that signals not only the direction but also the distance to the food. During the wag-tail dance, the performer also traces out a figure eight, but in this case the loops are not so superimposed and are separated by a long or short straight line (Fig. 8-20b). In the tracing of this straight line, the worker wags her whole body from side-to-side. If the straight part of the dance is directed upward on the comb, it means "fly in the direction of the sun"; if directed downward, it signifies "fly away from the sun." A run of 60° to the left of vertical directs workers to fly 60° to the left of the sun; a run 120° to the right of vertical informs others to fly 120° to the right of the sun. The workers use the sun as a compass. As it changes position during the day, the workers change the angle of their runs accordingly (Fig. 8-20c). The number of turns in a dance indicates the distance from the hive. If food is 100 yards away, there are 10 complete cycles of the dance in 15 seconds; at 200 yards, there are 7; at 3.7 mi, only 2. The counts vary with wind conditions. Thus, the number of turns actually indicates flight time, which is more accurate.

Other insects, such as the blowfly, perform an irregular dance after eating. Probably the language of bees evolved from such precursors.

More has been written about the behavior of social insects—ants,

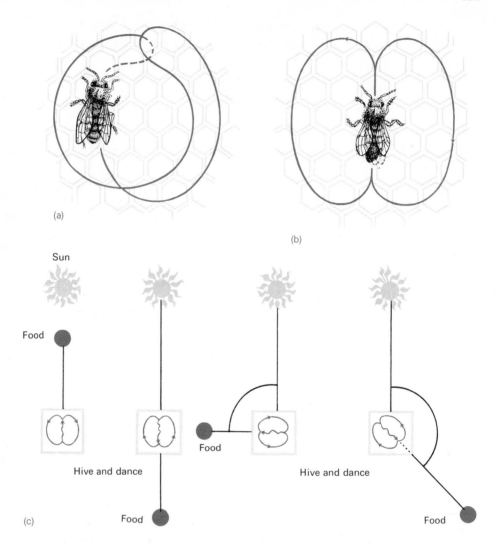

Fig. 8-20   **Honeybee communication. (a) Round dance alerts workers to food source near hive. (b) Wag dance gives directional information on the location of a food source more than 100 m from the hive. (c) Relationship of wag run to position of sun, hive, and food source.**
From N. M. Jessop, *Biosphere,* Prentice-Hall, Inc., Englewood Cliffs, N. J., 1970.

bees, and termites—than that of any other invertebrate animal. Their complex behavior is fascinating, but perhaps, more importantly, it illustrates the heights to which instinct can lead. Although ants and their relatives can learn to run mazes with difficulty, the social insects are, nevertheless, primarily slaves of instinct, and their strict differentiation of body forms accompanied by a rigidly defined behavior stands in marked contrast to the society of man. In Greek myth, the population of the island of Aegina, destroyed by pestilence, was replaced by ants transformed into men, the myrmidons of unparalleled industry and ferocity: they were the soldiers of the nearly immortal Achilles at Troy. Let us hope that people never become psychologically transformed into specialized ants in the hives of the city.

Let us examine some other invertebrate instincts. Some kinds of spiders are as easily identified by the structure of their webs as by the anatomy of their bodies. Territory, although not widespread, is not absent from the invertebrate world. So far, we have only examined territory in vertebrates.

The fiddler crab inhabits the shore and defends a territory of 1 or 2 yards about his burrow. The defense is ritualized and instinctive. One of the claws or pincers of the male is greatly enlarged and brilliantly colored. During the breeding season, each male stands by his burrow, waving the colored claw and rising on his toes. In the nonbreeding season, the claw is used in territorial defence. Two males meet at the boundary of their areas and clasp claws. Then they engage in an unusual arm-wrestling match, whose object is to break off the other crab's claw. The cicada-killer wasp male also defends a territory of several square feet. He is unusual among insects. The occurrence of territorial behavior in the two major branches of the animal kingdom must indicate that the pressures leading to its establishment are common.

VERTEBRATE INSTINCTS

In the vertebrate line of the animal world, especially the warm-blooded vertebrates, learning plays a predominant or at least more important role than it does in the arthropod line. Nevertheless, instinctive behavior permeates the lives of vertebrates, too. The territorial behavior encountered in fishes, lizards, birds, and some mammals seems to be largely instinctive.

Among amphibians, some salamanders perform rather elaborate, stereotyped courtship rituals, even though there is no copulation. In the common yellow-brown pond salamander (*Diemictylus viridescens*) of the northeastern United States, the male grasps the female from above with his hind legs around her shoulders. He then bends down and rubs each of his cheeks in turn against the nose of the female. Following this ritual, the male walks forward on the pond floor and deposits cloudy packets of spermatozoa on the bottom. The female follows and picks up the mucinous packets with the lips of her cloaca. For several months thereafter the female will lay internally fertilized eggs. The courtship rituals of other species of salamanders differ. The dances are species specific. In overview, however, they have movements in common with the mating behavior of their relatives the frogs. The male frog mounts the female from above and clasps her behind the shoulders. As the female sheds her eggs, the male deposits sperm upon them.

The behavior of reptiles is intermediate in complexity between that of amphibians and that of birds. The roaring of the male alligator and the territorial behavior of anoles and iguanas are most likely instinctive. Sea turtles come up on beaches, untaught, to lay and bury their eggs in the sand. The newly hatched turtles head immediately for the ocean, provided that they can see the sky. Green turtles migrate 1400 miles from their feeding grounds off Brazil to their breeding beaches on the shores of the mid-Atlantic Ascension Island. This navigational feat is hardly less remarkable than the stunning migrations of eels from the rivers of Europe to the Sargasso Sea near Bermuda off the coast of America. All such behavior must belong to the realm of instinct.

Recent experiments indicate that migratory birds navigate celestially, by means of the relative positions of the stars, moon, and sun. The feat of learning the patterns of the constellations seems beyond bird mentality. What role does instinct play here? Instinctive behavior plays substantial parts in bird life. Weaverbirds, hand-raised for generations, will build typical weaverbird nests when provided with the proper materials. Obviously, this behavior is instinctive. Bird song, that most distinctive aspect of territorial behavior, is not easily classified as simply instinctive or wholly learned. Most birds can be recognized by their special songs as handily as by their appearance. One would think that such species-specific behavior would be inherited precisely. Surprisingly, there is a great deal of variation in the method of song establishment. The male reed bunting, raised in isolation without opportunity to hear adult song, or even reared with other species with different songs, will nevertheless sing only its own song. The tree pipit sings its complex song as instinctively as the reed bunting. The linnet must learn; raised with other birds, it never learns the linnet song. A young skylark reared under similar conditions will learn to sing the complete songs of the chaffinch, goldfinch, or yellow bunting; nonetheless, it sings a few notes and phrases that are particularly "skylark." The English chaffinch partly learns and partly "knows" its own song instinctively. It has been suggested that in birds, rigid instincts have been partially opened. Instincts are more widespread in birds than in mammals.

The question of the extent of instincts in mammalian behavior is surely one of the greatest of our time. In fact, it may be crucial to the analysis of human behavior. Are there vestigial instincts in man? Certainly, learning plays the most important role in the development of mammalian and particularly human behavior.

LEARNING

Learning is a relatively permanent change in behavior due to past experience recorded in the nervous system. The pattern of learned behavior is established during the life of the individual organism. Unlike instinct, learning is not physically inherited. Intelligence, the ability to learn, is most likely inherited. The drives that lead to learning are often the same in those which are satisfied by instinctive behaviors. Compared with instinct, learning is in many respects an insecure way of life. Appropriate responses are not defined beforehand, as is the case in reflex and instinct. Costly trial and error is often involved in the process of learning. The solutions to the problems of satisfying drives are not readily at hand; rather, they must be established. Learning has the advantage, however, of providing a greater variety of possible responses than instinct alone can supply. The learning organism evolves during its own lifetime, and new behavior does not require the passage of innumerable generations necessary to establish, increment by increment, an elaborate instinct.

Nature has valued the way of learning too much to allow many animals to lock themselves absolutely in reflex and instinctive behavior. A degree of plasticity is desirable even in low animals. A robot-animal could never be completely programmed to deal with every situation. The difference between a robot and an animal, between the computer and man, lies in the ability of the living organisms to learn something new. A computer can solve a problem only if it is programmed with the necessary procedure. It cannot devise an entirely new method of solution. Most animals can learn. Only protozoans, sponges, and perhaps coelenterates cannot. There is a graded series of learning capacity in the animal kingdom. Man is the greatest learner. He alone among animals is bound upon a perpetual adventure of self-evolution.

Some have said that man's quest for security is in reality an unconscious expression of his desire to return to reflex and instinct. At the same time, it was the striving for security that led to the evolution of most of our culture. The object of life is the maintenance of the steady state, a status quo. Progress in nature is often the by-product of survival. Life has found not one way, but many ways, some better than others. Intelligence is the best. There are several kinds of learning.

TYPES OF LEARNING

The scientific study of the learning process began with the famous experiments of the Russian biologist, Ivan Pavlov. Pavlov presented dogs with food following the ringing of a bell. Dogs normally salivate at the sight and smell of food. In these experiments, they came to associate the stimulus of the bell with the reward of food, and so salivated subsequently at the ringing of the bell alone, a wholly unnatural response. The dogs had been *conditioned* to respond to the stimulus of the bell with salivation. Food is the *unconditioned stimulus* for the response of salivation. The ringing bell was a *conditioning stimulus* to which the dogs were conditioned to respond by salivating. This experiment is *the* classic of learning studies, and this type of learning is called *classic conditioning*.

Psychologists have conditioned rats and other animals to pull levers, press buttons, climb ladders, etc., for the reward of food. They have called this training *instrumental learning*. The animal learns by trial and error, but eventually operates the device. This type of learning is therefore also called *operant learning*. Maze-running is a similar process. Perceptual, or *insight learning* is said to have occurred when the animal responds correctly without trial and error. Perhaps the animal makes his trials and errors mentally. Scientists as a group strongly believe that insight has led to the greatest discoveries and reckon it as the most gifted and crucial type of human learning.

Our whole theory of education up until now has relied heavily upon the principles of conditioning. A student must be encouraged to learn, "motivated," provided with rewards for exhibiting learned behavior. In animal experiments, the conditioned response (learned behavior) soon becomes extinct if the learned behavior goes very long unrewarded.

Rats will learn the passages of a maze without reward, reinforcements, or punishments. Such exploratory behavior borders on pure curiosity and unimitative play. The learning that results from it has been tentatively labeled *exposure learning*. At first glance, it seems to be an abstract, impractical type of learning, but it is quite practical for an animal to explore a new environment and learn to "find his way around." Many psychologists and educators are excited by the "discovery" of this type of learning. Some are inclined to believe that "learning for learning's sake" might become a reality. With the technological revolution, we may be approaching the day when students may have to learn without the rewards of good grades and jobs. Perhaps the day will come when a learned man will be valued for his wisdom, but a man is defined by what he does, and learning is a mechanism of response.

Another type of learning takes place in very young animals only during a short critical period. Konrad Lorenz was followed by newly hatched goslings in preference to their own mother, provided that they saw him first in the few hours after hatching. Goslings have even followed wooden boxes! This type of early learning is called *imprinting*

Fig. 8-21   **Imprinting. These goslings were exposed to the famous animal behaviorist Konrad Lorenz shortly after hatching, and thereafter followed him in preference to the mother goose.**
Nina Leen. Courtesy of *Life Magazine,* Time, Inc., 1955.

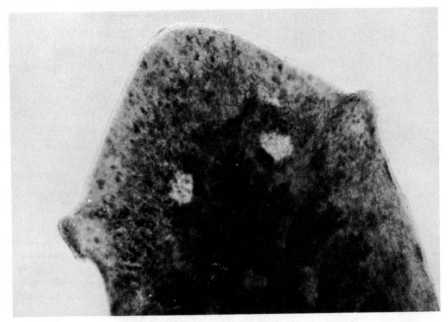

Fig. 8-22   **Planaria. The world of scientists was surprised to find that these lowly flatworms were capable not only of learning responses but of "unlearning" them as well.**

(Fig. 8-21). It has been suggested that the taste of the streams of their birth is imprinted by salmon fry and later accounts for the successful return of adults from the sea.

Just higher than sponges and jellyfish are the lowly flatworms. The common freshwater flatworms, the planarians (Fig. 8-22), have been classically conditioned. The worms contract when a mild electric shock is applied to water. Light was turned on them simultaneously with the shock treatment. The planarians came to contract when only the light was shown upon them. Incredibly, worms will remember what they have been taught after their heads have been cut off and new ones regenerated. Flatworms fed on trained worms learned faster than those fed on untrained.

Planaria have also been operantly conditioned to run a Y-shaped maze. Wells were located at the ends of the tunnels of the Y. The water was drained from the well in which a worm lay. The flatworm would then move up into the tunnel of a limb of the Y. When he reached the crossroads, a light would be shined down one tunnel; the other remained dark. By trial and error, the worm could learn to select the light or dark tunnel. If he chose the correct tunnel, it was immediately flooded with the water he was seeking. The first day's trials ended 50-50. By the third day, their success reached a peak of 80%. Then, for some as yet unexplainable reason, their score fell off quickly to 30%. This means that 70% of the time they chose the wrong way, which is significantly lower than the untrained worms' score of 50%. The planaria had not only "unlearned," but had begun seemingly to make the wrong response by preference!

Planarians usually avoid light, but they could be trained to tolerate it, in fact, to choose it over the greater danger of drought; however, not even the punishment of drought could restore their high scores. Investigators tried adding the planarian delight of liver juice to the proper wells, but even this added attraction failed to restore the conditioned response. Enlarging the wells at the end of the correct tunnel temporarily restored the "learning" of the planarians, but, finally, this inducement also failed to maintain the high level of correct responses. No rewards had been withdrawn; in fact, they had been trebled. Why do trained planaria not only unlearn, but seemingly balk at the correct response? Once again our experience fails us. We have conditioned ourselves to view learning as classical conditioning. Insight is needed here. Perhaps the making of the wrong choice is something like exposure learning, exploratory behavior. It would seem to be foolhardy to attribute curiosity to the lowly flatworm. Maybe nature has added a little adventurousness to the behavioral repertory of most animal species. "The grass is always greener . . . ." Higher animals, rats, and chimpanzees, however, do not display the behavioral idiosyncracies of the planarian herein described. The problem of planarian behavior is particularly intriguing because it defies our current knowledge.

Memory certainly lies close to the heart of learning. How is learning incorporated into the behavioral devices of an animal organism? Since it is almost impossible to separate form and function, it would seem that learning must have an anatomic or biochemical basis in addition to its functional manifestations. This is one of the last and perhaps the greatest questions of biology. Chemical inhibitors of RNA and protein synthesis have blocked learning and the establishment of long-term memory in a number of experimental animals. An older theory of memory proposed the establishment of new nervous circuits.

At the beginning of the twentieth century, many biologists and a few psychologists believed that intelligence was derived from instinctive behavior. This view has been almost completely abandoned in favor of the idea that the animal kingdom evolved in two major behavioral directions. In the arthropod line, instincts predominated; in the other, which culminates in the mammals, learning came to play the most important role.

The misconception that the lower animals are merely robots is almost three centuries old. Learning is widespread in the animal world, but it is limited in the invertebrate line. Even the lower vertebrates possess a restricted learning potential. Among the higher vertebrates, man has developed the greatest intelligence. He, perhaps, alone, seems to be free of the chains of instinct. It is often said that the insects are in contest with man for supremacy on the land, and that the war between instinct and intellect is not yet ended. There are those who predict that insects will inherit the earth. Who then shall speak in praise of learning?

DISCUSSION QUESTIONS

1. *Are there human instincts?*

2. *Does aggression have a social function? Is aggression pathologic?*

3. *Does any type of behavior have a genetic basis? If so, which?*

4. *What proof is there that RNA may be involved in memory?*

5. *What are the advantages and disadvantages of instinctive behavior? of learning?*

6. *What is the definition of ethology? What are the meanings of the roots from which the word is derived?*

7. *Is nature "red in tooth and claw"?*

REFERENCES

ARDREY, ROBERT, *The Territorial Imperative,* Atheneum Publishers, New York, 1966.

ARDREY, ROBERT, *The Social Contract,* Atheneum Publishers, New York, 1970.

BARNETT, S. A., *Instinct and Intelligence,* Prentice-Hall, Inc., Englewood Cliffs, N. J., 1967.

BIGELOW, ROBERT, *The Dawn Warriors,* Little, Brown and Company, Boston, 1968.

CARR, ARCHIE, "The Navigation of the Green Turtle," *Sci. Am.,* May, 1965.

DE VORE, IRVEN, *Primate Behavior: Field Studies of Monkeys and Apes,* Holt, Rinehart & Winston, Inc., New York, 1965.

EIBL-EIBESFELDT, IRENAUS, "The Fighting Behavior of Animals," *Sci. Am.,* Dec., 1961.

EIBL-EIBESFELDT, IRENAUS, *Ethology: The Biology of Behavior,* Holt, Rinehart & Winston, Inc., New York, 1969.

GILLIARD, E. T., "The Evolution of Bowerbirds," *Sci. Am.,* Aug., 1963.

JOHNSGARD, PAUL A., *Animal Behavior,* William C. Brown Company, Publishers, Dubuque, Iowa, 1967.

LORENZ, KONRAD, *On Aggression,* Harcourt, Brace & World, Inc., New York, 1966.

SCHALLER, GEORGE B., *The Year of the Gorilla,* University of Chicago Press, Chicago, 1964.

TINBERGEN, N., *The Study of Instinct,* Oxford University Press, Inc., New York, 1951.

TINBERGEN, N., *Social Behavior in Animals,* Methuen & Co., Ltd., London, 1953.

VAN DER KLOOT, W. G., *Behavior,* Holt, Rinehart & Winston, Inc., New York, 1968.

OBJECTIVES

1. List five common characteristics of the primates.

2. Draw a tree of primate evolution.

3. Compare the anatomies of typical quadrupeds, prosimians, monkeys, great apes, and men.

4. Give the names of the four great apes and list several characteristics of each.

5. Compare the econiches and ecosystems of the great apes, fossil men, and modern man.

6. Chronologically list five extinct members of the family of man and compare their skull characteristics with those of modern man; describe the evidence and speculations concerning their ways of life.

7. Contrast call systems and language.

8. Describe several stages in the evolution of tool-making.

9. Define race biologically.

10. List the seven characteristics most commonly used in racial classification.

11. List and describe the four most commonly recognized races.

12. Explain the theory of the adaptive significance of skin pigmentation.

Risen Apes or
Fallen Angels

Sᴏᴍᴇᴡʜᴇʀᴇ ᴡᴇsᴛ ᴏꜰ ᴇᴅᴇɴ it seems man was born laboriously. It was a labor of eons in which mother earth outdid herself, for once creating a foresighted organism, destined like the Titan Cronus of Greek myth to overthrow his father the sky. Ill-begot he was, descended from man-apes.

Man did not evolve from modern apes, as was popularly supposed when Darwinism broke upon the Victorian world, which preferred French romance to Huxley's *Man's Place in Nature* (1864). Incredibly, this superficial misconception persists today. Darwinists said that man and the modern apes had a common ancestor, an oft-repeated theme of evolutionary histories. If the fossil record of the modern bears is traced back far enough, a dog-bear finally emerges. The dogs and the bears both belong to the order Carnivora. Man, the apes, the monkeys, and even the near monkeys are all members of the order Primates. Their ancestries converge upon a branch of the tree of life, "whose trunk descends to the black earth and whose roots point to the water."

Man has the strong impulse to impose human form and behavior upon animal and god alike. This feature of human personality is called anthropomorphism, and it is hardly separable from anthropocentrism, man-centered thought, a viewpoint that considers man the central fact and final goal of the universe. Walt Disney made a fortune on the human weakness of anthropomorphism. Few people are interested in organisms that defy attempts to humanize them. It is a rare day at the zoo when the old stand-bys (or stand-ins?) the apes and the monkeys, fail to attract the greatest number of spectators for the longest time. Any fool can see that an ape is close to man. A child sees that a zebra is a horse before he learns the separate name from adults, who very often cannot see the horse beneath the stripes.

About 70 million years ago, the first primates appeared. Primates are placental mammals with long limbs and large hands and feet with five fingers or toes (digits). The primate digits have flattened nails rather than claws. Usually the thumb and great toe can be opposed to the other digits. A well-developed collarbone is another primate feature. The eyes and their sockets are directed forward, and the fields of vision overlap. This arrangement makes for a depth-perceiving vision (binocular and stereoscopic). The stereoscope is an instrument with two eyepieces, through which two slightly different views of the same image are viewed side-by-side, giving a three-dimensional effect to photos seen through it. Vision and hearing are the sharpest senses of the primate; the sense of smell is usually reduced, along with the snout between the eyes. Most primates are adapted for life in the trees.

The most primitive primates anatomically resembled the living tree shrew (Fig. 9-1). Although tree shrews seem superficially to be squirrels, their incisor teeth are quite different from those of the squirrels. The tree shrew has claws instead of nails, but his eyes and brain have begun to enlarge. Such a creature was on the line of evolution of moles, hedgehogs, near monkeys, monkeys, apes, and men.

Fig. 9-1   A tree shrew, the lowest branch on the family tree of primate evolution. This squirrel-like creature is an insect-eater.

NEAR MONKEYS AND HALF-APES

Twenty million years after the emergence of the primordial primates, prosimians or near monkeys appear in the fossil record of Africa. The present-day distribution of the prosimians coincides with the locations of the great apes and fossil men. The near monkeys survive in East Africa and Southeast Asia, where they arose or migrated, and where the primeval tropical habitat persists. The prosimians are unfamiliar creatures with strange, often native names: potos, lorises, galagos, tarsiers, and lemurs. Most of them are twigs on the branch of primate evolution, insignificant creatures of the trees and the night, with grasping hands and feet and large eyes shifted around to the front of the head, where the snout used to be. Only two of these creatures seem to be near the line of monkey evolution.

The Eocene was an early epoch of the modern era, and its Greek roots literally mean "the new dawn." *Eohippus,* the "dawn horse," roamed the Eocene forests of western North America. It was in this time (50 million years ago) that the lemurs (Fig. 9-2) crept silently onto the evolutionary stage. The name of these prosimians is taken from the Latin *Lemures,* which, in Roman mythology, referred to the night-walking spirits of the dead. This term is related to the Greek *lamos,* the abyss, symbolized by a gaping mouth. The monkey and the ape are less mystically named; perhaps they did not suggest such a staggering step in evolution. However, the orangutan was named by Europeans, who used the Malay *oran utan,* which meant "man of the forest." One can only wonder if the sight of the lemur brought the European face-to-face with the chaotic abyss from which evolution says man was haphazardly derived.

The German word for the lemur is *Halbaffe,* which means "half-ape." It seems he is indeed nearly that. The lemur looks like a dog transforming into an ape. The doglike appearance is due to his elongated moist muzzle and large movable ears, but after that the similarity ends. In all other respects, he is typically primate with grasping hands and an imperfectly grasping tail. Most of the lemur's fingers and toes bear nails instead of claws. His brain is simple; the cerebral cortex is smooth. His vision is not binocular and stereoscopic, because his eyes have not moved far enough around to the front of his head, but they are on their way. Unfortunately the snout, although reduced, is still a barricade to eye migration. There is a small compensation for this: the lemur's sense of smell is better than that of other primates.

Fourteen species of lemurs live today on an island off the east coast of Africa, Madagascar, which is in many ways a "floating Eocene museum." At his worst, the lemur is only half an experiment, or as the

Fig. 9-2 Ring-tailed lemur, a prosimian, or near monkey. The German word for lemur is *Halbaffe*, half-ape. (a) Walking in quadrupedal fashion. Note snout and location of eyes. (b) Note grasping hands. (c) Eocene landscape with lemurs and dawn horses. At this time (60 million years ago), there were neither apes nor men.

(c) from *History of Life on Earth Charts,* Denoyer-Geppert, Co., Chicago.

(a)

(b)

(c)

Germans call him, *Halbaffe*. At his best, he remains an evil figure of European poetry, summoning the devil and the nightmare of evolution.

In the waters off the shore of Madagascar, a lobe fin swims slowly toward extinction. Three hundred million years ago his ancestors gave rise to the first amphibians. Man was a long time coming. He might not have come at all.

FROM MONKEYS TO WHOLE APES

Monkeys evolved from lemuroid or tarsioid stock in the dawn epoch. They became and remain biological successes. Many species with large populations are found today in Africa, Southeast Asia, and Central and South America. Their muzzles are gone, and their hands and feet are more flexible, grasping devices than those of the prosimians. Still, monkeys are essentially. quadrupeds, who walk on their palms as well as their soles (Fig. 9-3). Their intestines and stomachs are attached to the back body wall in a hanging fashion much as in dogs, cats, and other quadrupeds. The dog's face is at right angles to the line of his backbone, and the hole through which the spinal cord enters the brain cavity (the

Fig. 9-3   **The baboon is essentially a monkey on the ground. Note that the animal walks "on all fours," much like a dog.**

foramen magnum) is at the very back of the skull. Man's face is parallel to the plane of the spine, and his foramen magnum is on the underside of the skull. The monkey and the ape foramina magna move progressively toward the human location. The canine extreme correlates with complete quadrupedalism, and the human condition reflects absolute bipedalism, the ability to walk erect on two feet.

American monkeys, in contrast to the uselessly tailed Old World monkeys, possess a long, muscular tail that serves as such an effective fifth hand that they can hang by it alone. Their nostrils open on the sides of their flat noses, whereas the nostrils of African and Asian monkeys (also apes and man) open downward. The skull and teeth of the two monkey groups differ in addition. The American and Old World monkeys seem to have arisen separately at different times from the prosimian line. Their similarities in form and habit may be the result of adapting to a common way of life in the trees. Neither living nor fossil apes are found in the Americas. Fossil men are likewise missing; therefore, New World monkeys played no role in human evolution.

The Old World monkeys are not completely bound to the trees; some have come down onto the ground, e.g., the baboons and mandrills. Few monkeys are much larger than medium-sized dogs. Although they possess sharp, moderate canine teeth, arboreal monkeys depend upon flight as a defense mechanism. In consequence of their descent to the earth, baboons were enlarged and equipped with virtual "saber-teeth," long curved canines (Fig. 9-4). The baboon is perhaps the most ferocious primate, and the soldiers of the baboon troop are a match for nearly any predator. Baboons have developed a new muzzle, lower on their faces, since a keen sense of smell is advantageous to the ground forager. They are said to be "dog-headed." Nature has allowed some of these powerful beasts to boast; the mandrill of West Africa is larger than most other baboons and has defiantly decorated his snout with red, white, blue, or purple "war paint" (warning coloration). The apes that came down upon the ground followed baboon evolution closely in many ways.

When the fossil history of the apes is traced back, a monkey-ape finally emerges. How do the experts decide whether the fossil is monkey or ape or both? And how do the living monkeys and apes differ?

Monkeys are rather completely committed to life on all fours. The apes swing in the trees usually by means of their arms alone. The broad, flat chest of apes is strongly muscled (pectorals), and the shoulders are firmly propped away from the breastbone of the body's midline. Apes' arms are longer and more muscular than the "front legs" of monkeys; the monkey's front and rear limbs are almost equal in size and degree of muscular development, and his collarbone and chest muscles are proportionately smaller than those of the ape. How many times do you re-

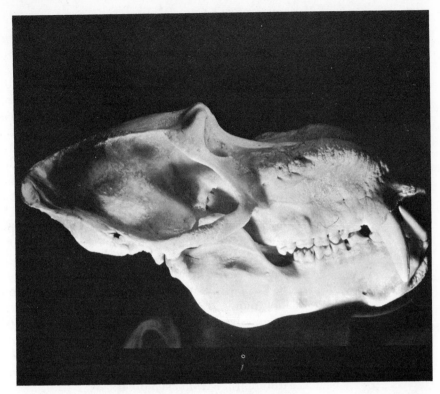

Fig. 9-4 **Mandrill skull. Note the great canines, "saber-teeth."**
From M. D. Nicklanovich and M. J. Zbar, *The Anatomy of Human Evolution*, Saturn
Scientific, Inc., Fort Lauderdale, Fla., 1971.

call seeing a monkey swinging or supporting his whole body weight with
just his arms? The shoulder joint of apes is freer than that of monkeys
and allows wider circular movements of the arm. As a result, the ape can
swing his arm hand-over-hand. In view of their anatomy and function,
the ape forelimbs are fully entitled to be called "arms." They are primarily anatomic tools for swinging the entire body weight without the
aids of grasping feet or tail. The Latin word for arm is *brachium,* and the
term for locomotion by swinging from the arms in the trees is brachiation. The posture of brachiation is erect rather than horizontal as in
quadrupedal locomotion. The spinal column of apes is short and stiffened, compared with the long, flexible backbone of monkeys or other
quadrupeds. The ape's foramen magnum has moved further forward on
the underside of the skull than it has in the monkey.

Many of the living apes are wholly or partly ground dwellers, a situation that has caused natural selection to modify the formerly insignificant legs and hips. The pelvis (hip and lower back bones) of the monkey is about the same size as his shoulder girdles (shoulder blades and collarbones that bind the arm to the body skeleton). Even though no modern apes stand erect or walk on their legs (bipedalism) for any length of time, more relative weight is placed upon their legs than is the case with the monkey. The pelvis of the ape is widened and enlarged, presumably to support the body weight. The ape's stomach and intestines are closely attached to the back body wall and do not sag into a belly of hanging bowels, such as they do in the quadrupedal dog and monkey. The arrangement and attachment of the internal organs of apes is quite similar to that of man.

The shapes and sizes of suspected fossil monkey or ape bones can be compared with what is known from the description and measurements of the anatomy of living monkeys and apes to determine if the fossil skeletal parts are more apelike or monkeylike. The value of a single fossil tooth should not be underestimated. Old World monkeys, like the baboon, have molar teeth with four cusps (points) linked by two ridges. Apes and men have five-cusped molars, and these points are separated by shallow valleys. A 30 million-year-old fossil Egyptian monkey, *Propliopithecus*, had five-cusped molars. He was more apelike than monkeylike, which is to say he was closer to man. Monkeys and apes had begun to differentiate from each other.

THE LIVING APES VERSUS MAN

For 14 centuries, the anatomy of the Barbary Ape, a monkey, served as the guide to the structure of the human body. The correction of the errors resulting therefrom is often hailed as a revolutionary advance in the knowledge of man. Yet the anatomy of this monkey, so studied by the Greco-Roman physician Galen, was with few exceptions an excellent model of human anatomy. Similarities in anatomical structures (homologies) between the Barbary Ape and man are more abundant than differences. Creatures thus related through homologies are close in "blood" or evolutionary relationship, since common genes determine the development of similar anatomies. The possession of like genes is the major criterion of kinship. The study of blood proteins also reveals a close relation between men, chimpanzees, and gorillas. Still, there are many anatomical differences, and the contrast between these structural features has provided the means of separating fossil apes, premen, and men.

The maximal cranial capacity (brain size) of apes is about 600 cubic centimeters (cc). The greatest brain size of man rarely exceeds 1600 cc. The foramen magnum of man is located directly beneath his brain case, evidently an anatomical feature of constant upright posture. The same hole in ape skulls is farther back, most likely because of the ape's inability to stand erect on two legs for any length of time (Fig. 9-5). The bony roof of the mouth (palate) in apes is longer than it is in man (Fig. 9-6); the canine teeth of apes are large, whereas those of man are much reduced (Fig. 9-7). The ape's upper jaw has gaps between the upper lateral incisors and canines for the incorporation of the great lower canines in the bite. This simian diastema ("ape gap") is not present between comparable human teeth. The precursors of the grinding molars, the premolars, are strong and sharp in apes. Man's premolar teeth (bicuspids) are small with two cusps. Ape molars and other teeth are generally large compared with those of man. The arch of the teeth is broader in men than in apes. Modern man has a prominent chin, but the ape's lower jaw recedes: the ape mandible (lower jaw) is much more massive than the human, and the halves of the ape jaw are buttressed where they come together by a ridge of bone, the "simian shelf." Ape skulls present strong eyebrow ridges and foreheads that slant back; man's brow ridges

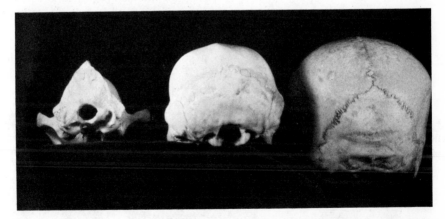

Fig. 9-5   **The position of the foramen magnum in quadrupeds and bipeds. Note that in proceeding from the dog (left) through the chimpanzee (middle) to man (right), the great hole (foramen magnum) moves progressively from the back of the skull to its underside.**

From M. D. Nicklanovich and M. J. Zbar, *The Anatomy of Human Evolution*, Saturn Scientific, Inc., Fort Lauderdale, Fla., 1971.

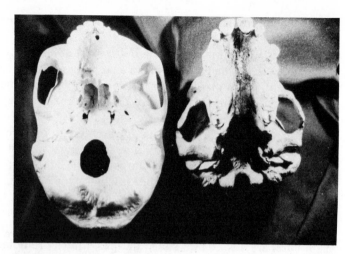

Fig. 9-6   The undersides of human (left) and chimp (right) skulls (lower jaw removed).
                Compare the positions of foramina magna, the lengths of palates, the shapes
                of dental arches, and the sizes of canine teeth. Note the wide gap (simian
                diastema, "monkey space") between the lateral incisors and canines of the ape.
                From M. D. Nicklanovich and M. J. Zbar, *The Anatomy of Human Evolution*, Saturn
                Scientific, Inc., Fort Lauderdale, Fla., 1971.

Fig. 9-7   Lower jaws of chimpanzee (left) and man (right). (a) Side view. Note the mas-
                siveness of the ape jaw and its "chinless" state in front. (b) From above, note
                the greater size of the ape teeth, canines in particular, and the shapes of the
                dental arches once again. Note also the buttressing shelf of bone (the simian
                shelf), where the two halves of the ape jaw join in front. This shelf lies just
                behind the lower incisors.
                Courtesy of Saturn Scientific, Inc., Fort Lauderdale, Fla.

(a)   (b)

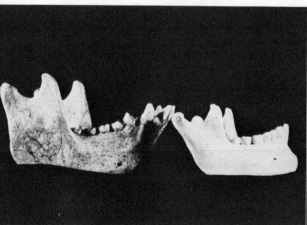

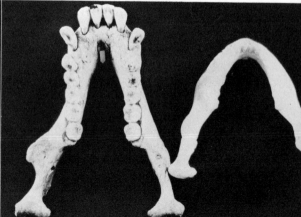

are reduced, and his forehead is nearly vertical. The ape's brain case and cheek bones are markedly ridged for the attachment of the muscles of chewing. These muscles and their bony attachments on human skulls are much reduced. The mastoid process (the bony eminence behind the ear lobe) is poorly developed in apes but prominent in man (Fig. 9-8).

The hip bones of the ape are long, and the pelvis is narrow; man's hip bones are shorter, flanged, and his pelvis is wider (Fig. 9-9). Also, man's hip bones have strong, rough elevations in the back for the attachment of muscles involved in walking and standing erect. Buttocks are particularly human anatomic features. The rear muscles that attach the hips to the thigh bones are poorly developed in apes. The position of the hip socket is more toward the front in man. The socket faces sideways in apes and forward in man. Ape leg bones are very short, compared with those of man. The ape thigh and shin bones are curved, whereas those of man are straight. The ape foot is a loose structure like the hand and has a big toe that is opposable like the thumb. All these structural characteristics are related to the ability or inability to walk on two legs.

The distribution of modern apes coincides almost exactly with the fossil sites of premen. Great apes are found today in Southeast Asia and

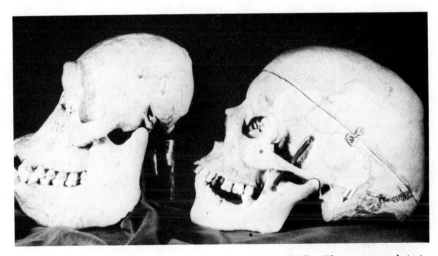

Fig. 9-8   **Side view of chimpanzee (left) and human (right) skulls. The arrow points to the mastoid region of the temporal bone. Note that the mastoid process (so prominent in man) is absent in apes. This view also shows the lower chimp canine fitting into the simian diastema of the upper jaw. Observe that the ape skull is propped up due to the small size of the cranium.**

From M. D. Nicklanovich and M. J. Zbar, *The Anatomy of Human Evolution*, Saturn Scientific, Inc., Fort Lauderdale, Fla., 1971.

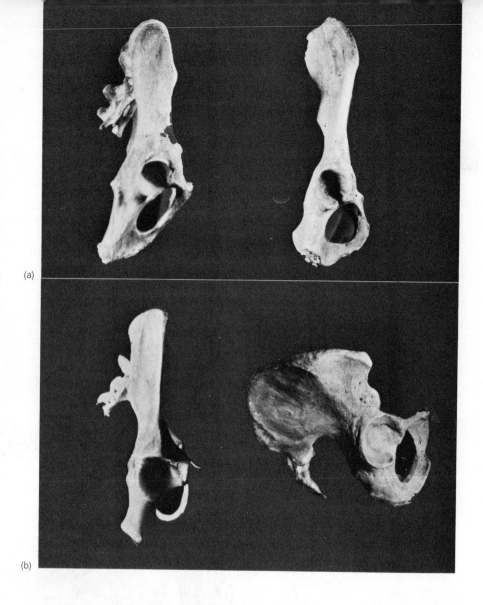

(a)

(b)

Fig. 9-9   The pelvises of quadrupeds, apes, and man. Pelvis is a collective term for the structure formed by the hip bones and the sacrum (made up of fused lower backbones). Aside from the skull, it is the structure most modified in the course of human evolution. In fact, the pelvis was modified and man stood erect long before his brain began to enlarge. Each hip bone is made up of three parts: an upper, bladelike portion, the ilium; a lower, rear part, the "haunch bone," or ischium; a front part, the pubic bone. All three articulate at the hip socket. (a) Side view of cat pelvis (left) and chimpanzee pelvis (right). Note the great resemblances in hip bones. It is not surprising: the great apes are quadrupeds on the ground. (b) Side view of baboon pelvis (left) and human pelvis (right). Baboons are terrestrial monkeys, but their pelvic anatomy resembles that of the

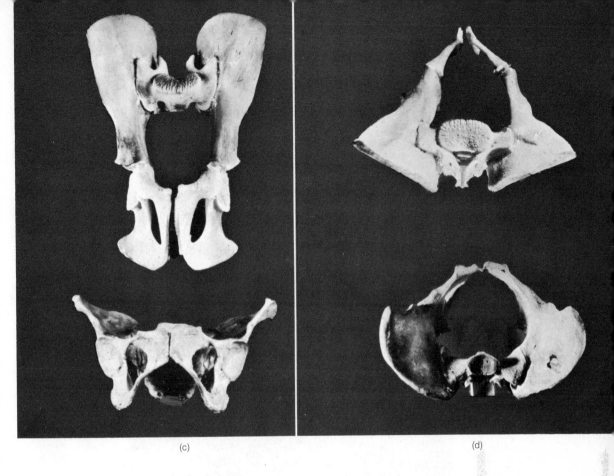

(c)   (d)

great apes considerably. In side view, the similarity to the quadruped pelvis is obvious. Note the great height and narrow surface area of the iliac blades. The human pelvis is shortened; the iliac blade is expanded, prolonged rearward. From it arise two important muscles of walking, the gluteus medius and minimus, which prevent pelvic tilt when support is removed from one leg. The great crest atop the ilium, its tubercle or bump and the buttressing ridge, running from the above across the face of the iliac blade to the hip socket, are all uniquely human features. The anterior inferior iliac spine, the nipple-like bump above the hip socket and below the forward projection of the ilium, is absent in apes. The ape hip socket faces sideways; the human considerably forward. Note the nearness of the sacrum and ischial tuberosity (haunch bone's lower bump) to the hip socket in man, in contrast to their distances from that point in apes. (c) Front view of baboon pelvis (above) and human pelvis (below). The comparative shortness of the human pelvis is even more evident here. Contrast the orientation of the hip sockets and the length of the pubic joint. (d) Superior view of baboon pelvis (above) and human pelvis (below). Note the bowl or basket-like structure formed by the human iliac blades, in contrast with the flat sides of the ape ilia.

From M. D. Nicklanovich and M. J. Zbar, *The Anatomy of Human Evolution,* Saturn Scientific, Inc., Fort Lauderdale, Florida, 1972.

Africa. Like our common ancestors, they dwell in the tropics. Our near relatives may disappear entirely, unless they soon become a protected species. They are indeed worthy of preservation if only to remind ourselves of our comparatively excellent evolutionary fortune.

THE MOST PERFECT APE

The gibbons of Southeast Asia are the smallest apes, only 3 ft high, but they are without doubt the best in the trees (Fig. 9-10). Their long arms reach the ground when these animals stand erect, as they occasionally do. Most of the time, however, they glide through the trees from branch to branch with an incredibly fluid grace, their long arms swinging underhandedly in perfect semicircles. At the zoo their trapeze artistry is a major attraction. Their hands are used as stiff hooks of curled fingers.

Fig. 9-10   **The gibbon, the most advanced brachiator of the great apes.**
From M. D. Nicklanovich and M. J. Zbar, *The Anatomy of Human Evolution,* Saturn Scientific, Inc., Fort Lauderdale, Fla., 1971.

The index, middle, and third fingers are longest, and the thumb and little fingers are quite short. The thumb can oppose the palm, but the joints of the fingers with the hand bones are only slightly flexible when compared with those of other apes. The gibbon hand is poorly constructed for manipulation, but well designed for brachiation. The monkey is a jerky, clumsy amateur in tree swinging compared with the gibbon, who has apparently perfected his arms and hands for this method of movement, much more so than his or man's ancestors, who also swung through the trees, or brachiated. The gibbon is the brachiator par excellence, and his anatomy of brachiation is noteworthy because other great apes and man exhibit the same structure in lesser degrees.

The collarbone is a strut between the breastbone and the shoulder. It is well developed in present-day tree swingers, such as the gibbon, and former brachiators, such as man. In animals that go on all fours, the quadrupeds, the collarbone is greatly reduced, almost vestigial, for the arms must be held close to the body in four-legged running. Such an animal with large clavicles would nearly be running on his face. In cats, collarbones are mere toothpicks imbedded in muscle, hardly attached to any portion of the skeleton. In great apes and man, collarbones are large and firmly attached by ligaments to the shoulder and breastbone. Gibbon legs—their muscles and bones—and buttocks are the most poorly developed of any great ape. This might be expected, since the gibbon relies on his arms; nevertheless, he can stand fully erect briefly, an unusual ability for an ape.

The gibbon depends upon his tree-swinging skill not only for movement but also for defense. No animal can keep up with him in the trees. Unlike monkeys, he seldom falls prey to the great cats. Other defensive features of the great ape body are correspondingly reduced in the gibbon. In general, his teeth are proportionately small, and the canines, so well developed in ground-dwelling primates, such as the gorilla and the baboon, are especially fragile in the gibbon. There is safety in the trees, particulary in many trees.

The canine teeth of the male gibbon are larger than those of the female, and, in contrast to hers, protrude from his mouth over his lips. He is the defender of the territory and the family. Gibbons are monogamous, a single family consisting of an adult male and female with one or more young. That famous student of primate behavior, C. R. Carpenter, studied the habits of the gibbon in Siam. He found that each family occupied a definitely sized but movable territory, and that when two families approached their territorial borders, aggressive behavior was exhibited, short, however, of actual combat. Aggression in the gibbon takes the form of threat displays and vocalizations. Carpenter found that the gibbon could make 11 distinctly different sounds and that more than half were related to territorial behavior.

Of the several great apes, the gibbon is least like man. Nevertheless there are still several species of gibbons, and they are in many ways the best adapted and most successful of the great apes. The gibbon was a logical experiment in brachiation carried to conclusion. Long before man or the other great apes began, the gibbon line appeared. Fossils of the earliest proto-ape, *Pliopithecus,* are more than 23 million years old. His arms were not so long and specialized for brachiation, but, on the basis of skull and teeth similarities, *Pliopithecus* is considered an ancestor of the modern gibbon. Let us leave these fleeing ghosts of the forest with the thought that their evolutionary energies were spent in perfecting arms rather than in developing intellect.

THE OLD MAN OF THE WOODS

The orangutan (Fig. 9-11) of Sumatra and Borneo is so unlike his near continental neighbor, the gibbon, that it is hard to imagine they had a common ancestor. Massive and clumsy, the orang remains in the trees. Small body weight like the gibbon's is an advantage in tree life. The male orang weighs 165 lb, and the female is slightly smaller. Their legs are rather thickly muscled and their feet are quite flexible, large "hands" with very opposable big toes. Although the orang's arms are relatively larger than his legs, his hands and thumbs are similar to his feet. This is not surprising, in view of the fact that he must clutch with all his might to support and transport his bulk in the trees. The orangutan reportedly moves through the trees with great caution.

The incisors and canines of the orang are almost as large as the ground-dwelling gorilla. He certainly cannot rely on defensive flight.

The orang is unsociable yet hardly territorial. Males seem to be solitary, but groups of two or three females and their young often live together. Whether the orangutan is monogamous or not remains unknown, but it seems unlikely. His intelligence has not been carefully studied. Perhaps the orang compensates here. Otherwise, he has the appearance of a dumb ox, bound for the gentle slaughter of millenia, evolutionary extinction.

The orang is almost as unrelated to man as the gibbon. Once orangs ranged all the way to China, but today only a single species remains. Of him Robert Ardrey writes, "The orang stayed in the trees and dies today, as it were, on the vine. The gorilla came down from the trees. . . ."[1] So did man, never to return to those safe, but hypnotic, boughs. Alternatively, he would remain there, dumbly dying or swinging machine-like.

---

[1]Robert Ardrey, *African Genesis* (New York: Dell Publishing Co., Inc., 1961), p. 115.

(a)   (b)

Fig. 9-11   **Orangutans. (a) Female. (b) Male with great cheek pads.**
From M. D. Nicklanovich and M. J. Zbar, *The Anatomy of Human Evolution*, Saturn Scientific, Inc., Fort Lauderdale, Fla., 1971.

A HALF-HEARTED MOVE
TO THE GROUND

Not all tree dwellers are small, light, and agile, depending upon quick escape for protection from predators or members of their own species. The largest, heaviest, and strongest of tree-dwelling monkeys are better able to defend themselves from attack. This disruptive natural selection favors the extremists and tends to eliminate the intermediates, who are neither fast enough to flee successfully nor heavy enough to win in mortal or territorial combat. Perhaps this type of selection played a role in the evolution of the orang and the gorilla. A male gorilla weighs almost a quarter of a ton in nature and more in zoos (Fig. 9-12).

Fig. 9-12   **A female lowland gorilla.**

It seems likely that natural selection operated upon the gorilla follow-ing his descent from the trees to produce a greater defensive size. This evolutionary movement was probably carried to its extreme in the ex-tinct cousin of the gorilla, *Gigantopithecus,* who stood 9 ft tall and is conservatively estimated to have weighed 600 lb. The ground-dwelling monkeys, the baboons, are unusually large and strong as monkeys go, and, like the baboon, the gorilla possesses large canines for defense. The gorilla is probably every bit as ferocious appearing to potential preda-tors, such as the leopard, as he is to man—so much so, in fact, that this greatest of the apes remains essentially unmolested. Possibly as a result of this state of affairs, the gorilla seems to have lost much of his aggres-sive behavior and territorial instinct somewhere along the littered evo-lutionary wayside, where not only parts but also behavioral charac-teristics are discarded. Gorilla groups range over areas of about 15 sq mi, but this range is not defended against intruders, of course. The leaders of meeting troops fight with stares. All that muscle is most often merely for threatening display or show, poker bluffing.

Nature seems to have lost interest in the brute after it protected him with a bully's build—bare compensation for leaving the cradle of the boughs. There he stands today, trying to look up from the earth unsuccessfully, since he ridiculously poises upon his knuckles. There is no satisfaction for him in possessing arms better developed than the orang's, because their added dimensions are barely adequate to support additional weight above. His small legs and loose feet-hands help little. Both feet and hands are unquestionably designed for life in the trees, but there he crouches on the ground as much on his hands as on his tree-feet. George Schaller, famed scholar of the mountain gorilla, never saw one swing on a branch in a year of watching. Most gorillas won't even climb into a tree crotch to spend the night, but instead build nests upon the ground. The gorilla is a vegetarian with a low-calorie diet and a huge mouth to feed. He spends most of his waking hours from dawn to dusk scouring the mountainsides for bamboo shoots and celery stalks, incessantly munching and browsing like an unfillable cow. His large molars are indeed reminiscent of those of grazing animals (Fig. 9-13).

Nature should have converted his flexible, clutching hand-feet into tight, true feet, developed his hip and leg bones and muscles, stood him up, bade him walk, and last, but not least, made of him a carnivore. Alas, nobody dictates a course of action to faithless nature with its personality of chance. Often at the very moment that a margin of success—survival at least—is achieved, the work is abandoned. Sometimes evolution carries on relentlessly in a rare display of renewed energy, what the scien-

Fig. 9-13   **Male gorilla skull, front (a) and side (b) views. The crest atop the skull and the prominent cheek arches serve as origins for great chewing muscles.**
From M. D. Nicklanovich and M. J. Zbar, *The Anatomy of Human Evolution*, Saturn Scientific, Inc., Fort Lauderdale, Fla., 1971.

(a)   (b)

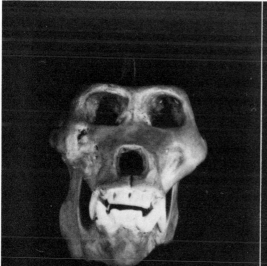
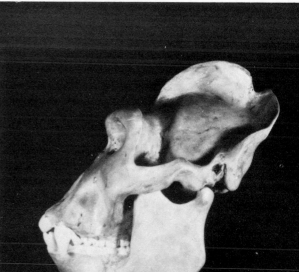

tist calls positive feedback, to fabulous success. So it would be with man. Even then the road twisted and turned, and the way was strewn with discarded works, the old species that failed. Nature gave the gorilla a heel, small amends for eventual evolutionary mortality.

Only two types of gorilla remain: a mountain type near Lake Edward, Africa, and a lowland form in the rain forests of the Congo. Gorilla intelligence is being studied intensively at the present time. Although his brain is about the size of the chimpanzee's, one must note that it serves a greater muscle mass and is therefore relatively or proportionately smaller than the chimp's. The gorilla is closer to man than the other great apes discussed, but he is not at all on the line of human evolution. The gorilla and the chimpanzee seem to be descended from a well-known early ape, *Proconsul,* whose fossils are more than 20 million years old.

THE PENALTY OF NONCOMMITAL

Blood proteins of chimpanzees are more similar to man's than to those of the other living apes. Once again, this is not to say that man descended from the chimpanzee. (Likewise, gorilla blood proteins are quite near man's serum proteins.) The chimpanzee is often said to be the most manlike in appearance of the great apes (Fig. 9-14); he seems to be the most intelligent. He has been the object of numerous psychological studies. The chimpanzee uses natural weapons and tools, sticks and stones, for defense and food gathering. He has considerable capacity for learning. In captivity, he has been taught a number of complicated activities. All attempts to teach chimpanzees human language have failed—at least they fell short of producing new sounds. Still they have learned to respond to a variety of speech commands. Recently, psychologists have taught chimps sign language, and they have been trained to "speak" with metal symbols. With these, they have constructed simple sentences for rewards.

Chimpanzees spend most of their lives in the trees, although they descend to the ground more than any other great ape with the exception of the gorilla. Their overall body build is essentially one of a brachiator with great arms, broad shoulders, a flattened chest, poorly developed legs, and grasping feet. They seldom walk without the aid of their forelimbs, and they are not as successful at tree swinging as gibbons.

Nature seems to have experimented somewhat with the chimpanzee intellect. He has the largest relative brain capacity of any of the living apes. The chimpanzee brain is actually almost as large as the gorilla's,

Fig. 9-14   **Chimpanzee, adult male "walking." Note the oblique quadrupedal or "semi-erect" posture, characteristic of the great apes.**
Courtesy of Saturn Scientific, Inc., Fort Lauderdale, Fla.

even though the chimp has less than one-fourth the muscled body mass. He moves faster on the ground than the gorilla, and has no difficulty escaping to the trees, where he is certainly more agile than the orang, since he weighs less.

The chimps are polygamous, and their social organization is flexible. Several males sometimes roam together, and groups of mothers and infants are often without males. It has also been reported that gangs of chimpanzees range over an area of about 8 sq mi, but their groupings change daily.

The territorial behavior of the chimpanzee is not strong. The lighter weight chimpanzee moves better in the trees than the gorilla does and relies upon flight as the final defense. Although a cousin of the gorilla, the chimp ventured only partially to the ground, and has therefore not evolved brawn. Neither did he perfect a way of life in the trees, as the gibbon did. His mind is superior to either that of the tree ape or the ground ape. Wolfgang Kohler, noted student of chimpanzee mentality, says, "The time in which a chimpanzee lives is limited to past and future."[2] The chimpanzee is generalized, not specialized for anything;

---

[2] Ibid., p. 350.

still, that seems to be good enough. There are several varieties of chimpanzees. They will probably hold their own better than the orang or their brethren gorillas upon the ground. A few ounces of intellect is more precious than a ton of brawn.

DOWN FROM THE TREES AND UP FROM THE EARTH

By chance, about 20 million years ago a small ape climbed down from the African trees to the grasslands below, or at least to the forest's edge. He must have come hesitantly at first, like the chimpanzee, spending only a fraction of his time in the hostile environment for which he was poorly equipped physically, and then more and more, generation after generation, as he became just barely fit to occupy this formerly unfilled niche. The forest's edge was then indeed "no man's land," but the trees were always nearby. Still there was the grassland beyond, beckoning, even daring nature to send forth this tree creature upon the African plain, a seeming sacrifice, the end of an insane experiment.

The place where the grasslands meet the forest is another example of an ecotone, where two ecosystems meet, a zone of gradual evolutionary trial, the site most frequently of the modification of the species. An ecotone often abounds with a greater variety of organisms than can be found in the ecosystems on either hand. This is most likely so since creatures from both extremes can tolerate the gradual transition briefly as well as give rise to new varieties better fit to dwell more permanently in the place of change itself.

It is often stated that man's ancestors left the trees as a drought of millenia brought about a recession of the forest and the rise of the grasslands as the major terrestrial biome. Many scientists no longer hold this view. They feel instead that proto*hominoids* (preman-apes) were "lured" to the ground and the forest's edge by the wealth of this ecotone. Here, the shrubs that do not reach too far from the earth find it unnecessary to compete for the meager sunlight coming down through the forest canopy inside, or rootholds among the dense stand of grasses on the plain. Also, the trees of the forest's edge do provide some shade and help to break the dry winds that take the water from the soil of the parching grasslands. Berries are numerous on the shrubs, and edible roots and grubs are abundant in this transitional zone. Probably several species of protohominoids existed. It is unlikely that some would not become adapted to a way of life in the forest's edge. It is not difficult to imagine one such not too conscious, imperfect creature peering forth from the shrubbery onto the sea of grass, sometimes deathly still and then waving in the wind.

One evolutionary day (1 million years), his descendants would come gradually out into the savannah, finally freed from even the forest's edge, and the journey would be attended by the death of whole species. The ancestors of the gorilla and the chimpanzee would remain behind. This twilight creature himself would be modified, disappearing as a type in the process, and even types derived from types would be extinguished by the slow, heavy hand of nature.

The curtain went up from the forest's edge, and the dramas that followed upon the stage of the plains seem to have been played by ignorant actors, protoapes, man-apes, and premen, without playwrights, scripts, or directors. Yet the themes of conflict, death, and survival were constant. Nature herself cannot qualify as a dramatist. In fact, "nature" is a playwright created by the imagination of the only sophisticated spectators, still millions of years away from these times and events. Now the story will be more difficult to tell. Is it "a tale told by an idiot signifying nothing"?

An ape was made to walk, and the hands first bound to the boughs were now freed from the earth. The beast was, nevertheless, still an ape that walked. The brain was a long time coming, and races of walking apes would become extinct without developing minds superior to that of the chimpanzee. The hands of the walking apes were often empty, but the hands of his descendants would hold the world, in the palms of the mind.

The ancient ground apes were poorly adapted in many ways, but the brachiating arms and grasping hands were "preadapted" for tool using. Sometimes anatomical devices of great value in one environment are surprisingly useful in another. Neither the lungs of lungfishes nor the muscled fins of lobe-fin fishes were intended for the invasion of the land; rather, they were adaptations of survival value in the fresh water, which is subjected to periodic drought. The lobe fins were used to clamber from one drying mudhole to the next, in an attempt to *stay* in water rather than leave it. Yet these devices were combined in the land-going amphibian. Natural selection was not foresighted. Preexisting materials were modified. Little alteration of the brachiating arm and grasping hand was required; the tree-leg and tree-foot had to be changed.

The pelvis had to be shortened, widened, broadened, and buttressed to support the weight above without the help of hands. The small muscles of the buttocks had to be enlarged. Hand-in-hand with their magnification went increases in the bony ridges of the pelvis and thigh bone to which they were attached. The long bones of the leg were lengthened and thickened at the same time that muscles of the thigh and calf were further developed. The flexible, grasping tree-foot was stiffened, for hands are ill designed to walk upon. An ankle bone was turned into a heel. Recall that a heel is one of the gorilla's few evolutionary gifts. At one time it was suggested that man's ape ancestors "learned" to walk to make their way more efficiently from one stand of trees to another, crossing the grasslands between. Usually it is said that tool using logically followed the evolution of walking and the liberation of the hands, but it has been recently hypothesized that natural tool using preceded walking and might have inclined natural selection toward additional freeing of the hands through the development of bipedal locomotion. One way or another, tree-grasping hands were already adapted for tool using. Yet the greatest tool of all would be the word.

THE BONES OF ADAM

The oldest fossilized remains of a suspected member of the family of man were recently discovered in northern India and shortly thereafter in Africa. A separate genus, *Ramapithecus,* was made for a single, incomplete upper jaw. The canines are smaller than those of the apes. The palate is arched and wider at the back like man's. Radioactive dating of the fossil revealed an age of about 14 million years, beyond our wildest imagination of the antiquity of man. Few manlike fossils are found until about 2 million years ago, 12 million years after *Ramapithecus.* Just a few years ago, a lower jaw to match the upper was found in a museum collection. *Ramapithecus* was a contemporary of *primitive* apes yet seems on the way to man. Even though no other fossilized skeletal parts of *Ramapithecus* have been unearthed, no such fossil jaws have ever been found with fossilized parts of the skeletons of nonhuman beings. The tree of primate evolution is diagrammed in Fig. 9-15.

About 2 million years ago, two types of near men were present in Africa. Both walked on two legs, and shared other anatomic similarities. The largest of the two, *Paranthropus robustus,* stood about 5 ft high and weighed nearly 140 lb. *Paranthropus* had a bony crest running from front to back along the midline of his skull roof, scooped out temple bones, prominent cheek bones, massive lower jaws, and huge molar and premolar teeth (Fig. 9-16). Microscopic examination of the surface of these fossilized grinding teeth revealed sand scratches, usually characteristic of root-eating vegetarians. The ridges of *Paranthropus'* skull are

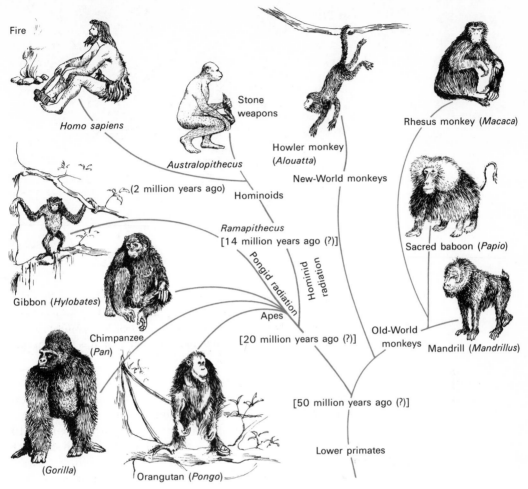

Fire

Homo sapiens

Stone
weapons

Australopithecus

(2 million years ago)

Howler monkey
(*Alouatta*)

New-World monkeys

Rhesus monkey (*Macaca*)

Hominoids

Ramapithecus
[14 million years ago (?)]

Sacred baboon (*Papio*)

Gibbon (*Hylobates*)

Pongid radiation

Hominid
radiation

Chimpanzee
(*Pan*)

Apes

[20 million years ago (?)]

Old-World
monkeys   Mandrill (*Mandrillus*)

[50 million years ago (?)]

(*Gorilla*)

Lower primates

Orangutan (*Pongo*)

Fig. 9-15   **A tree of primate evolution.**
From N. M. Jessop, *Biosphere*, Prentice-Hall, Inc., Englewood Cliffs,
N. J., 1970.

decidedly apelike. Similar elevations of the gorilla skull serve as attachment sites for chewing muscles, which are well developed in this ape,
the literal "cow of the primates."

Scientists are certain that *Paranthropus* walked upright, because the
great hole of his skull is directly below the brain case and his pelvis is
decidedly human. His brain size was about 500 cc. Although there is no
evidence to suggest that he manufactured tools, *Paranthropus* was quite

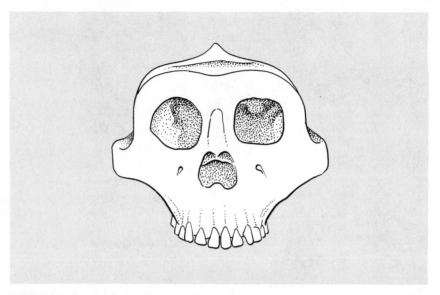

Fig. 9-16   **The skull of this creature from the family of man shows several apelike features. The crest, prominent cheek bones, and large teeth are all apelike characteristics. Yet *Paranthropus* ("alongside man"), or *Australopithecus robustus*, possessed enough human traits to be placed in the family of man. It is thought that he was a vegetarian.**

likely a natural tool user. He is placed in the family of man but not on the main line of human evolution. Although several of his characteristics are strongly apelike (the ridged skull, vegetarian diet, small brain size to body ratio), *Paranthropus* must have been much brighter than modern or ancient gorillas, whose body weights are three to four times as great as this man-ape's, at the same time that their brain capacities are nearly equal. From the time of his appearance 1.8 million years ago, until his extinction 600,000 years ago, *Paranthropus robustus* evolved slightly if at all. It has been speculated that his brain size did not change throughout his million years because of his failure to advance into tool manufacture. *Paranthropus* was a branch of man's tree that broke. In many ways, he was nearly an equal mixture of human and ape traits; at best, he was a man-ape. At least he was a brighter walking ape.

Prehumans, probably on the line of man's descent, seem to have appeared before *Paranthropus*. The South African ape-men stood "intermediate between anthropoids and man," according to their discoverer, anatomist Raymond Dart. They were classified in a genus apart from modern man. Dart dubbed his find *Australopithecus africanus*.

Australopithecus walked erect. His pelvis was wide and flanged, essentially like modern man's. The ape-man's hip sockets were further back on the pelvis than modern man's. Because of this arrangement, *Australopithecus* must have been somewhat clumsy when running and walking. That wedge of fused lower backbones, the sacrum, that joins the hip bones in the rear, was slighter in the South African ape-man than in modern man. Man-apes must have suffered from a great deal of "back trouble."

The brain size of *Australopithecus* was not relatively very much bigger than that of *Paranthropus*. The brain volume of the South African ape-man ranged from 450 to 600 cc, still less than half the average brain size of modern man but slightly larger than the chimpanzee brain, whose normal range is from 350 to 450 cc. *Australopithecus* was only 4 ft tall and weighed 70 to 100 lb. His brain size to body weight ratio is therefore plainly the largest discussed so far. The face of the ape-man was low browed and chinless. Although his lower jaw was massive, his canines were small. A mastoid process was present. The skull of *Australopithecus* lacked the bony ridge on top, and his cheek bones, jaws, and teeth are all much reduced compared with those of *Paranthropus*. See Figure 9-17.

It seems that *Australopithecus* was a carnivore, since the broken bones of antelopes and baboons are found with his fossilized remains. Meat, fat, and blood make up a very high-calorie diet. The switch from a herbivorous diet (at best slightly omnivorous) to a largely carnivorous diet must have been a huge step in human evolution. The ape-men were no longer fated to feed continually upon largely indigestible plants, and their energies could be devoted to other ends. Not all primates are exclusively vegetarians. Baboons often devour fawns that they find hidden in the grass; gorillas and chimpanzees frequently eat insects and grubs, although they feed upon fruits, nuts, shoots, and tubers most of the time. *Australopithecus* was probably omnivorous, but he ate more meat than any primate before him. It seems logical to suppose that early man was first a scavenger of the kills of more successful predators, and it has been proposed that man's early association with wild dogs might have developed in this fashion.

The body of this ape-man was poorly equipped for hunting or even defending himself. At least *Paranthropus* had some size; *Australopithecus* was a lightweight. Neither had large canine tusks. It has been suggested that the reduction of these defensive teeth was correlated with the use of natural weapons near at hand as extrasomatic (beyond the body) defensive devices. The use of such articles would tend to reduce the selection pressure for the maintenance of large canines. The one bone most frequently associated with australopithecine fossil sites is the

Fig. 9-17 **The South African man-ape (*Australopithecus prometheus*) skull (center) compared with those of chimpanzee (left) and modern man (right). Note that the man-ape skull possesses a mastoid process and lacks large canines and tooth gaps. These are human traits. The size of his lower jaw and his cranial capacity (near that of the chimpanzee) are apelike features. His foramen magnum is fully on the underside of his skull, and his jaws protrude to a lesser degree than those of the ape.**

From M. D. Nicklanovich and M. J. Zbar, *The Anatomy of Human Evolution,* Saturn Scientific, Inc., Fort Lauderdale, Fla., 1971.

antelope upper arm bone, the humerus. The double-headed elbow end fits perfectly into the double-indented fractures of fossilized baboon skulls from the ape-men's kitchens. Many other animal remains suggest that the ape-man hunted often and successfully. In Olduvai Gorge of East Africa, Louis Leakey found primitive stone tools associated with advanced australopithecine fossil layers. Crudely chipped choppers and more advanced hand axes were probably manufactured and used to kill and skin small animals. The extent of the use of fire or language by the South African preman remains unknown.

In contrast to the immutability of *Paranthropus, Australopithecus* changed. The first small primitive australopithecines appeared more than 2 million years ago. They were natural tool users. Several hundred thousand years later, about 1.6 million years ago, a larger, more advanced *Australopithecus* shows up in the fossil record with apparently manufactured tools. Thus, it seems that tool making appeared in human evolution before the brain had enlarged to any great extent.

The evolution of bipedalism completely freed the hands for manipulation. Larger and larger sensory and motor areas were devoted to the hands and fingers, and increasing associational fiber tracts (and centers) coordinated the special sense of vision with manipulation. The cerebral cortex expanded largely with the increased manipulative capacity of the hand. Hand and mind developed together in lockstep fashion, "hand-in-hand," one preceding and pulling the other, like muscle and bone and way of life in positive feedback. As nature selected for greater and greater tool making and designing ability, more and more abstract, somewhere along the line the greatest human tool of all appeared, language. The evolution of language will be considered as a separate topic later.

Many authorities consider *Paranthropus* to be an australopithecine. Some experts have renamed *Australopithecus*, placing the variety of creatures hereunder within the genus of modern man, but with a separate species designation: *Homo transvaalensis* (Transvaal is a province of South Africa). Transvaal man or *Australopithecus* was a contemporary of *Paranthropus*. They seem to have coexisted or competed in the same ecosystem. *Paranthropus* probably favored the more lush habitats, since we know his econiche was largely that of a vegetarian. He might have stayed closer to the forest's edge, while *Australopithecus* must have ventured farther out onto the plain, eating anything he could. During the wet season or shortly after the rains, they might have competed for fruit and berries. *Paranthropus* made the fatal mistake of *not* learning to make tools, but competition with Transvaal preman certainly did not extinguish his species. *Paranthropus* lived in Africa and Java until 600,000 years ago. *Australopithecus* (*Homo transvaalensis*) disappeared about 800,000 years ago. Most scientists agree that he is a true ancestor of man and feel that *Paranthropus* was a dead end, a blind evolutionary alley.

Half a million years ago, the first completely human being was probably derived from the most advanced of the australopithecines. Racial types of *Homo erectus* ("erect man") were discovered at several sites, one in Java in 1894, a second near Peking, China in 1927. Since that time, fossils of *Homo erectus* have also been found in Europe and Africa. In Olduvai Gorge, Louis and Mary Leakey uncovered typical *Homo erectus* skeletal fossils quite like those of Peking man. Parts of skulls and teeth midway between *Australopithecus* and *Homo erectus* were unearthed in intermediate strata.

Homo erectus was unquestionably a man. He was a superior walker to *Australopithecus*, owing to improvements in his hip and leg anatomy. *Homo erectus* was an advanced tool maker, which fact alone suggests a brain advance.

The brain size of Java-Peking man ranged from 750 to 1300 cc. His average capacity was slightly less than 1000 cc. This is almost twice the average australopithecine brain volume (500 cc), and the maximum cranial capacity of *Homo erectus* (1300 cc) falls well within the range of modern man, whose average brain capacity varies between 1200 and 1500 cc (Fig. 9-18). Although the average brain size of Java-Peking man is less than that of modern man, the difference is not as great as the distance between *Australopithecus* and *Homo erectus*.

Homo erectus possessed great eyebrow ridges and was relatively chinless. His skull lacked the large frontal swelling and prominent vertical forehead of modern man, indicating that his frontal lobes and forebrain were smaller than modern man's. Since the forebrain is an area of thought, the thinking ability of *Homo erectus* must have been less than that of modern man. Still, the brain of Java-Peking man seems large enough to have done most of the things a modern brain can do. The increase in brain size reflects a sharp rise in brain cell numbers and the complexity of their linkages. Could *Homo erectus* speak?

He apparently hunted in groups, since cooperative effort was required to kill the large beasts with which *Homo erectus'* tools are frequently found. It seems probable that he talked. *Homo erectus* knew the use of fire. Since many human long bones and skulls were split open in the Peking caves, Peking man was probably a cannibal.

The races of *Homo erectus* died out about 400,000 years ago. At one time, about 900,000 years ago, *Paranthropus, Australopithecus,* and *Homo erectus* were alive and evolving at the same time. *Homo erectus* succeeded *Australopithecus* along the evolutionary lineage of modern man.

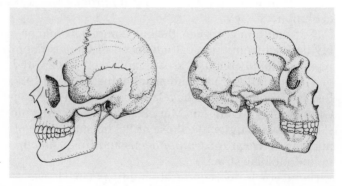

Fig. 9-18   **The skull of modern man (left) and a reconstructed skull of**
Homo erectus (Java man) on the right.
From M. Bates, *Man in Nature,* Prentice-Hall, Inc., Englewood Cliffs, N. J., 1964.

TO MODERN MAN

About 300,000 years ago, nearly modern men appear in the fossil record. Fossil skulls from Swanscombe, England, and Steinheim, Germany, have a larger average brain capacity and smaller eyebrow ridges than *Homo erectus.* The curvature and size of the individual bones of the roof and back of the skull of these specimens fall within the range of *Homo sapiens,* modern man. The Steinheim skull has heavy brow ridges and a lower forehead than that of modern man. These ancient premodern men seem to link *Homo erectus* and the famed Neanderthal man.

Neanderthal man, a subspecies of modern man, spread wide over Europe, the Mediterranean, and perhaps even East Asia and Africa between 100,000 and 35,000 years ago. He was 5.5 ft tall and weighed approximately 140 lb. He was a stocky welterweight. His brain was as big as modern man's, sometimes bigger because of a long backward projection of his brain case. His forehead sloped away from prominent brow ridges and his chin receded (Fig. 9-19). Neanderthal was an advanced tool maker. Since he often buried his dead, it has been speculated that he was "capable of abstract and religious thought." There is really no reason to suppose otherwise.

Neanderthal man used to be placed in a separate species, *Homo neanderthalensis.* Modern biologists consider Neanderthal at least a modern human subspecies, *Homo sapiens neanderthalensis.* In 1932, a fossil skeleton of a Neanderthal woman was found in a cave on Mt. Carmel, Palestine. Her eyebrow ridges were less massive than the classic Neanderthaler of Europe, and the back of her skull was more rounded. Her body was unquestionably Neanderthaloid. This Tabun woman was a first clue that Neanderthal was not very far from modern man. Other Neanderthaloid fossil specimens from this site also exhibit a mixture of modern and primitive features suggesting interbreeding or hybridization. In 1957, a complete male Neanderthal skeleton was found in a cave near Shanidar, Iraq. His brow ridge was likewise reduced, and the other skeletons found with him all had modernized faces. It seems more and more that Neanderthal was just a race of essentially modern man. Solo man of Southeast Asia and Rhodesian man of Africa are today considered close relatives of Neanderthal. Modern men from North Africa and the Middle East invaded Europe and either absorbed (through interbreeding) or eliminated the classic Neanderthal race of Europe.

Modern man inherited the earth. Meek he had never been; wise he had yet to become. He is called *Homo sapiens sapiens,* which can be literally translated as "wise, wise man." The word was with him.

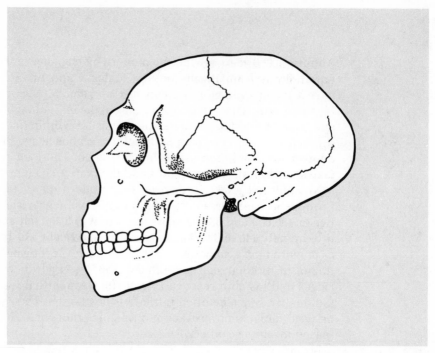

Fig. 9-19   A common type of Neanderthal skull. Except for the brow ridges and a few
other minor features, it is entirely modern. The cranial capacity is easily within
the range of modern man, and the Neanderthalers are now considered an extinct race of modern man.

THE EVOLUTION OF LANGUAGE

Unfortunately, fossils cannot speak. The story of man's evolution would
be incomplete without an attempt to account for the development of that
unique human asset, language, without which it is safe to say there
would be no man. The best that can be done is to examine vocal communication in our near relatives, the living apes, assuming that their primitive vocalizations are similar to those of the man-apes.

Ape call systems usually consist of a half-dozen or more vocal signals
or responses to a biologically important stimuli. The gibbons, for instance, have one call for food discovery and another for danger detec-
tion. A third call indicates friendly interest, and a fourth just tells the
location of the caller to the members of his band, keeping them from
spreading out too far as they move through the trees. Other calls indicate

sexual interest, need for maternal care, and pain. Mating calls and the announcements of territorial possession are closely related.

Gibbons can vary the loudness, length, and repetitions of a call to communicate the strength of the stimulus. When a gibbon is confronted with both food and danger, he emits only one sound or the other, never both. Nor does he utter a mixture of the two sounds. The call system is thus mutually exclusive. It is said to be closed. Language is an open system, because of the greater variety of sounds available and the larger number of choices of response open to the animal with the more advanced nervous apparatus.

Gibbons never give the food call unless they have actually found food. They never make the call away from the stimulus. The users of language can talk of food in its absence, when they are displaced from the stimulus, just as they can "see" a tool in unworked stone. They can even speak of nonexistent, abstract symbols. In the open system of language, there are also meaningless sounds used to link meaningful utterances; the closed call system has only unlinked, meaningful calls.

Call systems are largely inherited genetically, not transmitted to any great extent by tradition. A language of a given culture is, of course, wholly learned by traditional transmission, even though the mechanical ability to speak is genetically inherited. The failure to teach chimps language has been discussed.

The call system must have been opened even before *Australopithecus*. Group hunting necessitated increased flexibility in communication, and nature undoubtedly selected for it in premen. The ability to blend sounds to indicate shades of meaning in a situation is an obvious advantage. Planning for a cooperative hunt requires the capacity to displace sounds from their immediately meaningful context to the future.

Prelanguage, a stage of communication between the closed call system and open language, must have evolved right along with tool making and group hunting. Just as the hand areas of the cortex expanded with tool making, so must have the speech areas and association centers enlarged with hunting, planning, and other activities of the "civilizing" band. This was undoubtedly another positive feedback situation, a chain reaction in evolution. Advanced australopithecines must have had at least prelanguage; *Homo erectus* probably had language, and the road to *Homo sapiens* was traveled by creatures with an ever-expanding language.

RACE

What is the biological basis of race? From where did the races come? A species is defined in sexually reproducing organisms as a population (or populations) of similar individuals that actually or potentially interbreed. All the races of man are potential interbreeders. It is most likely that all races have a common ancestry.

It is very improbable that the major steps in the evolution of man occurred nearly identically and simultaneously in a number of widely separated geographic areas. It has been noted that the former species of man were far ranging, cast over most of three continents. Races of *Homo erectus* and Neanderthal have been mentioned. Java and Peking man represent races of *Homo erectus*. Solo man and Rhodesian man are probably Neanderthaloid races. Neanderthal himself can be considered a race of modern man. Thus the evolution of human races has occurred before. What is the definition of race?

The term race can be used to refer to a subspecies or variety, levels of classification beneath the species level. Sometimes biologists use the word race to describe varieties of subspecies. Popularly, race has been used as a synonym of nation, culture, and ethnic group; many of these "psychological races" are biologically indistinguishable. Unfortunately, social controversy has clouded the issue of race. Sometimes we tend to think of race as the figment of a bigot's imagination, but it has a definite biological basis. What harm is there in this fact? Of what value is this knowledge?

The UNESCO statements on race made in 1950 and 1952 are worthy of intensive study, but we can only include highlights here. The social scientists recommended that the term "race" be dropped from popular parlance and the phrase "ethnic group" be substituted for it. Yet their biologic definitions of race in the technical sense were excellent:

> . . . *From the biological standpoint the species* Homo sapiens *is made up of a number of populations, each one of which differs from the others in the frequency of one or more genes. . . .*
> . . . *In short, the term "race" designates a group or population characterized by some distribution, of hereditary particles (genes) or physical characters, which appear, fluctuate, and often disappear in the course of time by reason of geographic and/or cultural isolation.* [3]

---

[3] "The UNESCO Statement by Experts on Race Problems, July 18, 1950," UNESCO House, Paris.

Physical anthropologists and geneticists hardly improved upon this statement. They defined race as "a classification device providing a zoological frame within which the various groups of mankind may be arranged and by means of which studies of evolutionary processes can be facilitated."[4]

Sometime in our third thousand recorded years, today's races will probably have disappeared in the melting pot. None of the races of man has remained in geographic and reproductive isolation long enough to approach subspeciation, but it is obvious that the tendency to evolve variety was at work and play here. Some physical anthropologists recognize 30 or more races of human physical variation. A physical anthropologist can almost tell where a man comes from by the shape of his face and skull, provided that he is descended from emigrants of regions stable for centuries. There are no foolproof rules, however. As one moves from north to south in Europe, then into the Middle East and Africa, progressively darker races are encountered. From western Central Asia toward the east, the frequency of "slant eyes" increases. These gradual geographic changes in the appearance of men are reflections of a racial cline. Only in long and widely separated populations are the changes relatively abrupt.

The characters most used in the classification of races are (1) texture, form, and distribution of hair; (2) skull dimensions; (3) facial features; (4) color and form of the eyes; (5) average size of adults; (6) percentage of different blood groups; (7) skin color. Most racial traits have a genetic basis (frequently multiple genes), and the minor differences between the races are reflections of the differing gene frequency that nature has wrought sometimes for a purpose, sometimes randomly.

The Negroid races have in general wooly, kinky, coarse hair, sparsely distributed over their face and bodies. The nose is frequently broad and flat, and the lips are thick. The forehead is prominent—even bulging. Skin color varies from light brown to black, but eye color is always dark brown or black. There is considerable variation in size and other traits. The pygmies of the Congo average 55 inches in height, whereas the nearby Watussi almost always approach 7 ft. The Bushman of the Kalahari Desert is so distinct in appearance as to be easily distinguished from all other Africans except the Hottentots. The buttocks of Bushman females are seats of immense fat deposits forming great cones, a condition called steatopygia ("fat tail"). The Rh positive blood factor has a 100% occurrence in unmixed Negroid races.

---

[4] "Statement on the Nature of Race and Race Differences—by Physical Anthropologists and Geneticists, September, 1952," UNESCO House, Paris.

The Caucasoid races have wavy or straight hair, often abundantly distributed over the face and body, prominent and projecting noses, and thin lips. Although skin, hair, and eye pigmentation vary widely, there is a great frequency of blue-eyed blonds and brunettes. Yet some Caucasoid East Indian populations are darker than many Negroids. The total percentage of Rh positive persons among a typical white population averages approximately 87%. Only two-thirds of the Basque population is Rh positive; in this trait they differ significantly from other Europeans. The Basques are a very ancient people living in the mountains of northern Spain and speaking a non-Indo-European language.

The Australian aborigines have both Negroid and Caucasoid traits. Their skin is medium brown. Their head hair is wavy or curly and dark brown to auburn. The facial and body hair is like that of Caucasians in amount and distribution. The nose is broad, but more prominent than flat. The forehead retreats, and the chin recedes. They have well-developed brow ridges and long heads. Many of these characteristics are considered to be primitive features. Although the average aboriginal cranial capacity is only 1280 cc (low range for modern man), there is no indication that they are less intelligent than other races; Rh positivity is 100%. The Australian natives have probably been isolated longer than any other race.

The Mongoloid races have straight, black, coarse head hair, and their facial and body hair is the least of any human type. The eyes are brown or black, and the skin pigmentation varies from waxy white to golden brown. Cheekbones are prominent and give the face a "squarish" appearance. The upper eyelid is folded into an epicanthic fold that arches outward from just inside the low-bridged nose. (This Mongoloid or epicanthic fold is also found in Bushmen.) Mongoloid incisors differ from the Caucasoid. The classic Mongoloid type is medium to tall in height and is found in northern China, Mongolia, and Siberia.

As one moves westward from White Russia toward Mongolia, a great mixture of Caucasoid and Mongoloid traits is encountered. Many of the nomadic "Soviets" of the steppes of Central Asia are Mongoloid peoples. Alexander Borodin, the Russian chemist who composed the entrancing "On the Steppes of Central Asia," is said to have had a decidedly Mongolian caste to his features. The Mongol face is no stranger in Russia, and many Slavic faces bear Mongoloid traces. Americans are surprised by the height and build of northern Chinese, since they are accustomed to the appearance of the shorter, southern Chinese, who have emigrated most frequently and resemble the Southeast Asian Mongoloids more so than the northern Chinese. The Japanese also are more similar to the Thais, Indo-Chinese, and Malaysians. Eskimos and American Indians are usually classified as distinct Mongoloid subraces.

THE EVOLUTION OF RACE

What is the significance of racial features? Are all racial characteristics adaptive, or are some the results of random genetic drifting?

It was suggested that darkly pigmented skin was naturally selected by the tropical forest, since dark individuals are much less visible than light ones. Recently a more probable theory has been advanced.

There can be "too much of a good thing"; for example, too much vitamin D is as bad as too little. Vitamin D is required for calcium absorption from the gut; deficiency results in rickets in children and poor mineralization of bone tissue in adults. Excessive vitamin D is likewise damaging, causing excess mineralization and demineralization at different sites. Kidney damage is a common complication of hypervitaminosis (excess) D. Besides dietary sources, a type of vitamin D is synthesized in the tissue beneath the skin, following exposure to the ultraviolet portion of sunlight. Aside from the role of ultraviolet in vitamin synthesis, this radiation is mutagenic (mutation causing).

Tanning, the increase in pigmentation following exposure to the ultraviolet wavelengths of sunlight, is an individual physiologic response intended to decrease harmful ultraviolet penetration. Racial pigmentation is perhaps an adaptive evolutionary response of generations to excess or lack of sunlight and the vitamin D it generates. Northern Europeans who receive little sunlight during much of the year, are minimally pigmented to allow maximal penetration of ultraviolet. Nature selected for genes for light pigmentation, increasing their frequency in European populations. Negroes, who are continually bathed in bright sunlight, are maximally pigmented to prevent even the minimal penetration of ultraviolet. Nature selected for genes for dark pigmentation, increasing their frequency in African populations.

Some anthropologists have attempted to explain the adaptiveness of other human racial traits. They have argued for a correlation between body build and environmental temperature: stocky builds would be most appropriate for cold climates and thin types would be best for warm climates. Which has the greater surface area to volume ratio? What does this have to do with heat loss or gain? It has been proposed that the epicanthic fold of the "slant eyes" of northern Mongoloids would protect them against the glare of snowfields. The lack of facial hair has been interpreted as an advantage for life in the cold, since ice crystals tend to form on beard hairs. According to these hypotheses, the Mongoloid Eskimo would be ideally suited for life in the Arctic; however, the Eskimo's adaptation to arctic life is largely cultural.

Many racial traits may be the results of nonselective random change.

The skull dimensions of the various races, length, and length to breadth ratios seem to have had little influence in developing differing intelligences. The Yugoslav Dinarics have a short broad skull, flat in back, the "most advanced" skull type in all of Europe. The function of the conical buttocks in Bushwomen is unknown; perhaps there is no adaptive significance.

Too often the adaptive nature of evolutionary change is exemplified; frequently, however, evolutionary change has little meaning. Change for change's sake is not a rare event in the living world. Possibly in the races we can see some of the unconscious, goalless gropings of "nature."

Alpha (α) and omega (ω) are the first and last letters of the Greek alphabet. We have used them to symbolize the beginning and the end, foreign concepts to "nature's mind." We forgot our origins, never learned our fate. We made the calculus and deduced evolution. Delta (δ) is the fourth letter of the Greek alphabet, and it literally means "door." We have used it to symbolize a change in numerical values. A delta is also a triangular deposit of sand or soil in the river's sea mouth that divides the stream. Delta separates the old and new worlds of thought. The only constant is change. Man is neither alpha nor omega, but delta.

DISCUSSION QUESTIONS

1. How does the anatomy of a brachiator differ from that of a quadruped? What brachiational anatomy does man retain? How does the anatomy of man differ from that of the gibbon?

2. How is the position of the foramen magnum correlated with posture and locomotion?

3. Is the baboon an ape? What features do baboons share with ground apes?

4. What pelvic and lower limb modifications were made in the evolution of bipedal locomotion?

5. Were the australopithecines apes or men? How so?

6. How did language evolve?

7. How did the races evolve?

8. What role did warfare play in the evolution of the human species?

REFERENCES

ARDREY, ROBERT, *African Genesis*, Atheneum Publishers, New York, 1961.

BATES, MARSTON, *Man in Nature*, Prentice-Hall, Inc., Englewood Cliffs, N. J., 1964.

CLARK, W. E. LeGROS, *The Antecedents of Man*, Quadrangle Books, Inc., Chicago, 1960.

CLARK, W. E. LeGROS, *Man-Apes or Ape-Men?*, Holt, Rinehart & Winston, Inc., New York, 1967.

COON, CARLETON S., *The Origin of Races*, Alfred A. Knopf, Inc., New York, 1962.

DOBZHANSKY, THEODOSIUS, *Mankind Evolving*, Yale University Press, New Haven, Conn., 1962.

HOWELL, F. CLARK, *Early Man*, Time-Life Books, New York, 1965.

HUXLEY, THOMAS H., *Man's Place in Nature*, University of Michigan Press, Ann Arbor, Mich., 1959.

LEAKEY, LOUIS S., *Adam's Ancestors*, Harper & Row, Publishers, New York, 1953.

LEAKEY, LOUIS S., *Olduvai Gorge*, Cambridge University Press, New York, 1965.

LEAKEY, LOUIS S., *The Progress and Evolution of Man in Africa*, Oxford University Press, New York, 1965.

MONTAGU, M. F. ASHLEY, *Physical Anthropology*, Charles C. Thomas, Publisher, Springfield, Ill., 1960.

MORRIS, DESMOND, *Naked Ape*, McGraw-Hill Book Company, New York, 1967.

NAPIER, J. R., "The Evolution of the Human Hand," *Sci. Am.*, Dec., 1962.

SIMONS, ELWYN L., "The Early Relatives of Man," *Sci. Am.*, July, 1964.

VON BONIN, GERHARDT, *The Evolution of the Human Brain*, University of Chicago Press, Chicago, 1963.

OBJECTIVES

1. Compare the perspectives of vitalism, mechanism, and organicism.

2. Briefly summarize the biophilosophies of René Descartes, Herbert Spencer, T. H. Huxley, Charles Darwin, Teilhard de Chardin, Homer Smith, Jacques Monod, and Van Rensselaer Potter.

3. Contrast gnosticism, agnosticism, and atheism.

4. Distinguish between naturalism, theism, and natural theology.

5. Define, defend, and criticize social Darwinism and bioethics.

6. Compare the attitudes of the various faiths toward evolution.

7. Contrast orthogenesis and most biologists' interpretation of evolution.

8. Contrast the bioethics of Spencer, Huxley, and Van Rensselaer Potter.

9. Distinguish between teleology and teleonomy.

Biophilosophy

IF WE BOILED THE 5 BILLION YEARS of the earth's age down to a compre-
hensible 3 days, then "One minute ago, flint-chipping, fire-kindling,
death-foreseeing, a tragic animal appeared—called Man."[1] So spoke
novelist Updike's central character, the teacher Caldwell, in *The Centaur*. The centaur, a creature of Greek myth, half-man, half-horse, "a
man above the beast below," perhaps symbolizes the struggle between
the spirit of man and his animal body. The animal nature of man has
been vehemently denied in the past and violently advanced in the present. More than 2000 years ago, the Greek philosopher Empedocles suggested that the four elements, earth, air, water, and fire, had separated
or united under the influence of the two forces love and hate, producing
either monsters destined to die or a few normal survivors. Here was the
germ of the idea of the mechanism of natural selection. With the added
life dimension of almost boundless consciousness, man became the animal "most alive." He was the only animal that foresaw his death and had
need of a philosophy of life. French priest and biologist Pierre Teilhard
de Chardin said that man is the only creature who knows that he knows,
reflects upon his reflections. In the process of natural selection, death is
a necessity. The old must give way to the new varieties that ever so
slowly meet the gradually changing times.

It is often said that the quest for wisdom is an ever-humbling en-

---

[1]John Updike, *The Centaur* (New York: Alfred A. Knopf, Inc., 1963), p. 40.

deavor. The intellectual history of mankind seems to support this rule. The first great blow to man-centered thought (anthropocentrism) was Copernicus' proposal that the earth was not the center of the solar system. The Darwinian revolution dwarfed the Copernican, since the central position of man in his own world was questioned. The physical anthropologist and writer Loren Eiseley called the nineteenth century "Darwin's Century." To whom will the twentieth century belong? And the twenty-first? Will the advance of the philosophy of reality destroy the spirit of man? Should it? Evolution and religion, biology and philosophy, are topics that are difficult to separate.

PHILOSOPHIES OF LIFE

The dramatic historian portrays the story of man as the struggle between opposite philosophies. The histories of both biology and philosophy have as their central themes the conflict between the philosophies of life, *vitalism* and *mechanism*. Vitalism is the belief that life can never be explained in terms of physicochemical or natural law. According to this philosophy, the vital force of life is undefinable. This viewpoint is often criticized as unscientific, since it proposes that the investigations of life are doomed to ultimate failure. Vitalism is idealistic, supernatural, and related to faith, a belief in the unknowable. Note, however, that the modern vitalist accepts that science will explain many of the mechanisms around the edge of life. The old vitalists believed that hardly any of life's activities could be explained scientifically. Ancient vitalism yielded century after century to its opponents. Even today, however, a modified vitalism remains, often inside the scientific community itself.

Mechanism is the faith that all of life will be explained by the natural laws of physics and chemistry, and the "vital force" will be shown to be a figment of the imagination. Mechanism is definitely materialistic, even atheistic, some have said. Mechanism cannot be separated from biologic experimentation, and mechanists have won much of the territory formerly occupied by vitalists. Yet life has not been completely explained in this fashion. Life scientists are still engaged in taking living organisms apart, the method of analysis, and the final test will be life synthesis. Although some viruses have been dismantled and reassembled, they are a far cry from the living organism of the vitalist. It seems safe to say that the virus has no soul; also, it is extremely difficult to attribute "life" as we use the word to viruses. Because of this failure to synthesize even a simple living organism and the distortion produced by the study of the isolated parts of complex organisms, a third viewpoint has emerged. This philosophy has probably been with us a long time, although it was unnamed.

Organicism is defined as the theory of biology that an organism's own functional system is life. "The whole is greater than its parts," according to this philosophy. Life activities are seldom reduced to functional parts, and circular control systems interact to produce the steady state. The puzzle of the organicist is like the surprised realization of the analyst that "life in a test tube" is very often different from living activities in the natural organism. The histories of biology and philosophy are often told in terms of the struggle between the mechanists and vitalists. Any story is dull without the tension of conflict, but moderate organicists can be found in all ages. Organicism is not a mere compromise between mechanism and vitalism, but rather a recognition of the "holiness" of the whole organism.

Before recorded history, when biology was a practical science of survival, primitive man gave his religions a naturalistic design. He was bound by the natural world and attributed its incomprehensible and uncontrollable machinations to the hands of the "gods of nature." Man must have become a dualist, separating body and soul, quite early. Knowledge of death must have been among the first cultural transmissions. Studies of primitive societies in recent times suggest that early man populated the spiritual world with the souls of departed men and animals. These creatures were said to be the vital forces that controlled the dynamics of nature and human destiny. Religion probably came into existence before magic, but their births were close in time. They had in common the attempt to establish a relationship with the supernatural. The famous scholar of mythology, Sir James Fraser, author of *The Golden Bough,* proposed that magic had more in common with science than religion, since primitive man felt that he could control the supernatural through incantation and ritual, whereas prayer laid man before the will of uncontrollable gods.

THE GREEKS

The tendency toward the formation of "two cultures" began in the world of Greek scholars. The philosopher Plato could be accused of "literary intellectualism." He was an idealist rather than a materialist. Plato's prize pupil Aristotle was a "vulgar philosopher" and the first scholar of recorded history to leave behind a massive work of biology.

Aristotle's natural history, *Historia Animalium,* was an admirable, ambitious—although inaccurate—attempt to classify animals on the basis of such characteristics as type of reproduction (egg laying, live bearing, etc.), method of locomotion, and lack or possession of blood. Although these are inadequate foundations for classification, Aristotle did recognize the proper relationships between many organisms. He allied birds with reptiles on the basis of structural similarities and placed monkeys between man and four-legged beasts. Most important for posterity, however, was his arrangement of organisms in a graded series from the imperfect to the perfect (man). This "ladder of perfection," the *Scala naturae* (scale of nature), was not meant to outline a progressive evolution with organisms climbing from rung to rung but implied it nevertheless. Aristotle stated that "nature makes so gradual a transition from the inanimate to the animate kingdom that the boundary lines which separate them are indistinct and doubtful." He believed in the spontaneous generation of small animals and even large creatures, such as eels. Here was the step onto the first rung of the ladder. Today, the most accepted theory of the formation of life is a modification of this older spontaneous generation.

Despite his belief in spontaneous generation and the design of his ladder, Aristotle did not arrive at a theory of evolution, but his scholarly heirs in future generations did. The scale of perfection was later expanded to include the angels, archangels, and God above man, and it was this "tower of Babel" that became one of the four great principles that led Darwin and others to evolution.

All science begins with description, the observation of things of the world and the creation of language for recording. Once enough is described, experimentation begins. So it was in the history of biology, and it seems as if we have only entered the stage of biologic experiment. Recorded biological observation really began with Aristotle.

He spent days in small boats off the island of Lesbos studying the reproduction and development of octopus and cuttlefish in the clear tidal pools. Aristotle observed and recorded the nest-defending (territorial) behavior of a male catfish of the River Achelous. Two thousand years later, the first great American biologist, Louis Agassiz of Harvard, confirmed his description. This was probably the first recorded study of territorial behavior in animals. Aristotle discovered the yolk sac placenta in certain sharks that give birth to live young. He was the father of embryology, and his descriptions of the development of the chick were hardly improved until the seventeenth century. Not until Darwin, perhaps, do we encounter a comparably painstaking observer with such wide curiosity. In our jet-paced world it is difficult to recall the near proverb that "new eyes" (insight) are given to the patient and the curious.

Aristotle's god was not of the flesh, but a "prime mover," the source of motion, and quite like his "formative principle," which, imposed upon matter, produced the variety of nature. Aristotle was a vitalist; he believed in vital force. There was an internal guidance, "an entelechy." To Aristotle, man was an animal first, but second *the rational* animal. Here is a reflection of the dualism of his philosophy.

Although Aristotle was primarily a vitalist, his writings reveal a mechanistic streak. He analogized the "prime mover" with energy; he placed "form" and matter in constant interaction. Nevertheless, he built a purposeful internal guidance system into the organization of nature. Aristotle never really arrived at mechanism in biology. Perhaps he approaches organicism, however.

Very few of the Greek biological theories can withstand our knowledgeable examination. Our reverence for the ancients has often resulted in rather flattering interpretations of their naïve ideas. Some of their theories were strikingly foreshadowing. Anaxagoras suggested that man's intelligence was the product of the use of his hands for manipulation rather than locomotion, which is reminiscent of several features of today's theory of human evolution. It is said that the descent from the trees followed by the evolution of bipedal locomotion freed the hands, and the increase in brain size and manipulative ability went "hand-in-hand" thereafter. (The descent of man from the apes was not a Greek belief.) Aristotle inverted Anaxagoras' proposition. He thought instead that man used his hands because he was intelligent. Empedocles' "idea of evolution," was previously mentioned. Recall that he proposed that the parts of organisms were united haphazardly by the force of love, and selections made from the resultant combinations. Only by a considerable stretch of the imagination could this be called "the idea of natural selection." More likely, it is a reflection of the Greek mythology, which was filled with centaurs and satyrs.

In summary, Greek biology was qualitative, descriptive, observational. There were few observers of Aristotle's caliber. He was the only "bioscientist-philosopher." His study of the real world led him to attempt to synthesize what he had seen and that which he had not into a harmonious whole. Aristotle was a "vulgar philosopher," who attempted to viscerate his speculative philosophy. The roots of vitalism, mechanism, organicism, and "two cultures" (the scientific and the humanistic) can all be traced to Greek thought.

THE ROMAN PERIOD, MIDDLE AGES, AND PRE-RENAISSANCE

In the Roman period and the Middle Ages thereafter, little was done in biology, and vitalism was the triumphant philosophy. The conflict between science and the church was settled in favor of the church. Just before the Renaissance, scholars returned to the study of Greek learning, which the church had considered pagan. This revival of the study of the ancients has been called the Scholastic Period. Between the tenth and fifteenth centuries, Aristotle's writings were ingested, digested, absorbed, and assimilated. Biology continued to be the encyclopedic endeavor of collecting information about living things. The most outstanding scholastic philosopher was St. Thomas Aquinas.

Before Aquinas, St. Paul had distinguished between revealed truth and knowledge won by worldly means. "Where is the wise man? Where is the scribe? Has not God turned to foolishness the wisdom of this world? For since, in God's wisdom, the world did not come to know God by wisdom, it pleased God by the foolishness of our preaching, to save those who believe. For the Jews ask for signs, and the Greeks look for wisdom. . . .?"[2]

Aquinas argued that man could arrive at the knowledge of God and His characteristics by natural reason. It is often said that he tried to prove the necessity of the existence of God. Aquinas was a student of Aristotle and through his writings made the study of the ancients more acceptable to the church. Aquinas wrote that a Prime Mover was a necessary postulate to account for the movement in the universe. He added that a divine intelligence was the only explanation of the order of the universe. He said that it was necessary to presuppose a supremely perfect being as the cause of the scale of perfection. In all these assertions, the influence of Aristotle's writings is recognizable. Aquinas believed in potential creation rather than special, yet he felt that a guiding hand was evident in the development of nature. St. Thomas proposed that there were two sources of truth: that divinely revealed to the faithful and that discovered through the senses and reason. Like Plato, he believed there was a reality beyond the senses.

Thomas Aquinas remains a very influential philosopher in modern Catholicism. Although he is often credited with reconciling Christian dogma and reason, many scholars argue that he failed in his resort to faith rather than pure reason. In the history of influential thought, however, Aquinas did pass on a strong leaning toward natural theology, the

---

[2]New Testament, *Corinthians*, I, 20–22.

philosophy that God is revealed in the works of nature. This philosophy reached its peak strength just before Darwin's time and was a major ingredient of the environment in which evolution was proposed. At that time, natural theology was used as an argument against and an alternative to Darwinism. Today, this philosophy is used by many theologians to mellow the harsher aspects of evolution.

The first half of the Middle Ages is sometimes referred to as the Dark Ages, for in retrospect it seemed that the flame of Greek reason, held aloft even by the Romans, faltered and went out, leaving Europe in blackness. Then the Scholastic philosophers lit the torch again. The seeds of science, central to Greek philosophy and newly sown in the Scholastic mind, germinated in the Renaissance soul, and the "rational animal" progressed again, this time almost relentlessly to the present. Mechanism reared its pagan head.

REVIVAL

The Belgian Andre Vesal took apart the body of man and in 1543 described the formerly forbidden machine with a thoroughness unknown before. In the sixteenth century William Harvey measured the cardiac output in animals. Quantitative biology was born. Without measurement there could be no meaningful experiments. With a tourniquet about the upper arm of a man, Harvey demonstrated experimentally that the blood in the veins flowed in one direction: toward the heart. He said that the heart was a pump, and the blood circulated.

DESCARTES

In the seventeenth and eighteenth centuries, the natural philosophers tried to explain the universe and some biological matters in terms of natural law. Foremost among them was René Descartes (Fig. 10-1), who is generally credited with introducing the philosophy of mechanism into biology. Unfortunately, he is more remembered for his famous quotation: "I think, therefore I am." Homer Smith said of him: "I find it paradoxical that his greatest positive contribution to philosophy, the doctrine of mechanism, should remain unnamed and unrecognized by those

Fig. 10-1   **René Descartes (1596–1650), French mathematician, engineer, philosopher, and father of mechanism in biology.**
Bettmann Archive.

who write on the history of philosophy."[3] Here, again, is a symptom of the two cultures.

Descartes was a mathematician and engineer. The axes that lie at the heart of trigonometry are named after him—the Cartesian coordinates. In reference to them, numerical points are plotted in space, the sets of numbers acquiring a dynamic, functional significance. Descartes showed that algebraic equations could be graphed and displayed characteristic curves. Thus was born analytic geometry, the forerunner of calculus, the mathematics of motion, the mathematics of change, an evolutionary mathematics. Constants were to become the relations between variables.

Harvey had shown the heart to be a mechanical pump forcing blood through the conduits of the arteries and veins. With this in mind, Descartes went further. To him the whole animal body was a machine. The muscles worked the limbs according to the principles of the lever. He believed that the nerves were hollow channels with valves at strategic

---

[3]Herbert Chasis and William Goldring, *Homer William Smith* (New York: New York University Press, 1965), p. 159.

locations and that they transported a "vapor," which, when released into the muscles, caused the latter to swell in width and contract in length.

To Descartes, humans were more than animal machines, because they had immortal souls. "The greatest of all prejudices we have retained from infancy," he said, "is that of believing that the beasts think." [4] Only to man did he assign the thinking substance. Descartes believed that the soul was immortal substance, not part of the perishable body, and interacted with the body by way of the pineal gland in the dead center of the brain. This foolish hypothesis is surprisingly out of Cartesian character.

Although Descartes was a founder of mechanism in biology, he was an idealist, not unlike Plato, in his speculative philosophy. The common ground of idealism and vitalism has been mentioned before. According to Descartes, lower animals were robots, and he reserved the thinking substance and an immortal soul for man alone. He was at least a human vitalist. His famous quotation ("I think, therefore I am") is a perfect expression of the idealism that led a whole school of thought in the eighteenth century to the indefensible position that the existence of real things depends upon the workings of the mind.

Descartes was a dualist. He believed that man had both a physical and a spiritual nature, and he separated mind from matter. These views are unremarkable. Descartes' greatest contribution to biology and philosophy was his bold assertion that the lower animals were machines. Subsequent philosophers preferred to study his idealistic writings. Fortunately, because of his widespread fame as an all-around scholar, biologists at least heard his cry for mechanism.

Julien Offrey de La Mettrie (1709–1751) was trained for the priesthood but switched to the study of medicine. After his mechanistic writings on natural history had aroused the anger of churchmen, he fled from France to Holland. In the Netherlands he wrote *L'Homme Machine* (*The Human Machine*), which brought the wrath of the Dutch down upon him. This time his flight took him to the liberal court of Frederick the Great of Prussia.

Unlike dualistic Descartes, La Mettrie was a complete mechanist and materialist. He completely rejected the immortal soul of theologians and the vital force of life. All organisms including man were simply machines. Nature, he said, rather than supernatural religion, defined good and evil. La Mettrie argued that nature was supremely good, and that natural instincts should serve as the guide to healthy and virtuous behavior. It is said that La Mettrie gorged himself on truffle pastry and died, thereby dying a materialistic death as he had lived—with a mechanistic philosophy.

---

[4] Ibid., p. 162.

A PHILOSOPHER OF EVOLUTION

Perhaps the greatest biophilosopher, Herbert Spencer, lived and worked in Darwin's century (Fig. 10-2). No philosopher ever took so much of his philosophy from biology. Spencer's biography would cause the most unconcerned to gasp in admiration. In our times of educational criticism, it is fascinating to note that he undertook at age 40, following a "nervous breakdown," probably the last nearly successful attempt to construct a philosophy which included all fields of knowledge and had as its central theme evolution. All this Spencer did, even though he was formally "uneducated." The historian Will Durant says of him, "He is the clearest expositer of complex subjects that modern history can show; he wrote of difficult problems in terms so lucid that for a generation all the world was interested in philosophy."[5]

[5] Will Durant, *The Story of Philosophy* (New York: Washington Square Press, Inc., 1963), pp. 359–360.

Fig. 10-2   **Herbert Spencer (1820–1903), English philosopher who coined the historic phrases "struggle for existence" and "survival of the fittest," and who used evolution as the theme of his psychology, ethics, and sociology.**
Bettmann Archive.

Spencer had been a surveyor and civil engineer, a builder of railways and bridges, and an inventor. In those days, civil engineers were informally educated. The coproposer of the theory of natural selection, Alfred Russell Wallace, had also been a surveyor, and his education was quite like Spencer's. At any rate, Spencer was a "practical scientist," "beneath," it might be added, "the pure scientist" of today's two cultures. He was more of a journalist than a poet, and so fell far below the ivory towers of "literary intellectualism." It was a boon for all mankind that he was an expert at nothing; for 40 years he worked upon his Synthetic Philosophy.

Spencer was not a biologist, yet he influenced the thinking of the bioscientists of his day. In his 1852 essay on "The Theory of Population," he restated the "dismal theorem" of Malthus, which so influenced Darwin and Wallace, and coined historic phrases when he said the "struggle for existence" led to a "survival of the fittest." In the "Development Hypothesis" of the same year, Spencer became the first to use the word "evolution" in the modern way—i.e., the origin of new species by the adaptive modification of older species. Remember that Darwin's *Origin of the Species* was not published until 1859. The subtitle of *The Origin of Species* is *The Preservation of Favoured Races in the Struggle for Life*. Unfortunately, Spencer believed in Lamarckian teleology rather than Darwinian mechanism.

The times were ripe for the man of the age, Charles Darwin, and he would do more than hypothetically propose the theory of evolution by means of natural selection. It has been said that Spencer wrote his *Principles of Biology*, the second and third volumes of the *Synthetic Philosophy*, after having read a single biological source: Carpenter's *Comparative Physiology*. Spencer's *Biology* appeared in 1872, and he supposedly had not read the *Origin of the Species*; however, he had "picked the brains" of T. H. Huxley ("Darwin's bulldog") and other biological experts at the Athenaeum Club.

In 1863, Spencer had assumed that the hereditary material was in the form of "physiological units." The definition of the gene is still evolving, but the most sophisticated, modern description of it hardly improves upon Spencer's keen phrase. This was quite remarkable for any mind in 1863. Mendel was unpublished. Even when he was, Mendel would not give such a functional definition to his hereditary factors. Darwin's genetics was like the ancient Greek misconceptions about the subject. Prominent biologists of the time (and even until 1900) would describe the hereditary material with mystic phrases like "idioplasm" and "stirp." Reading that one book on physiology, in addition to lacking the professional biologist's terminological entrenchment, allowed Spencer to cut through to the core of the matter with his typical journalistic skill.

Karl Ernst von Baer, the father of modern embryology, had said in 1828 that "the development of every organism is a change from homogeneity to heterogeneity." That is to say that the adult, composed of many specialized cells, develops from a single unspecialized cell, the fertilized egg. Herbert Spencer used this dictum as the theme of his evolutionary philosophy. Spencer combined von Baer's rule of embryological development with the physicist's law of conservation of energy when he proposed that "Evolution is an integration of matter and a concomitant dissipation of motion, during which the matter passes from an indefinite, incoherent homogeneity to a definite, coherent heterogeneity, and during which the retained motion undergoes a parallel transformation." [5] Development and evolution are the two major themes that recur throughout Spencer's *Synthetic Philosophy*. His evolutionary idea had its greatest impact on social science. Let us now examine other aspects of Spencer's philosophy.

In a later edition of his *Biology,* Spencer vitalistically stated, "We are obliged to confess that life in its essence cannot be conceived in physicochemical terms." [7] Nowhere can a more perfect expression of vitalism be found. This is not to say that Spencer was religious. He felt that even the scientist "knows that in its ultimate nature nothing can be known." This statement of frustration reflects the modern physicist's difficulty in visualizing the electron. The electron has wave and particulate characteristics at the same time. Its position in space cannot be known with certainty. Only a probability map, which has the form of an "electron cloud," can be constructed to show "where it is most often." Spencer's realization of the impossibility of obtaining this kind of final knowledge is what the German physicist Heisenberg called the uncertainty principle and T. H. Huxley called agnosticism. Spencer's God was thus unknowable. The Plantonic philosopher's favorite field of endeavor, metaphysics, which seeks to explain the ultimate nature of reality and being, was to Spencer a fool's quest.

In his *Principles of Psychology* (1873), Spencer argued for a gradual evolution of consciousness, a scale of awareness and learning ability. He recognized simpler responses than the reflex and asserted that instinct had evolved by the compounding of reflexes, transmitted thereafter by the inheritance of the acquired behavioral characteristics. (Here, again, his unfortunate Lamarckianism is obvious.) Memory and imagination succeeded instinct and preceded intellect and reason, the highest activity of the fully conscious mind. Spencer noted the expansion of adaptive ability in this progression. Long before Jung, the twentieth-century psychoanalytic successor of Freud, he emphasized the importance of the

---

[6] Ibid., p. 365.
[7] Ibid., p. 371.

experience of the race in the evolution of the mind. Spencer postulated the evolution of threatening aggressive behavior from actual physical violence.

Of the mind he said, "Were we compelled to choose between the alternatives of translating mental phenomena into physical phenomena, or of translating physical phenomena into mental phenomena, the latter alternative would seem the more acceptable of the two."[8] This statement is quite like the problem we face today, as scientists investigate the chemistry of learning. If they succeed in reducing mental activities to biochemical reactions, one of the last strongholds of vitalism, the mind, will have yielded to the forces of mechanism.

It was in the newborn field of sociology that Spencer was most influential. Eight of the largest volumes of his *Synthetic Philosophy* are devoted to this subject. Even in sociology his biological theme of evolution is obvious. He compared society with the organism. Society, he said, had organs of nutrition, circulation, coordination, and reproduction. The evolution of cities with specialized trades and interdependence was like the differentiation of the multicellular organism. He praised the industrial state, and it was to be in political and economic philosophy that his ideas would live longest. Herbert Spencer more than any other figure was responsible for the theory that has come to be known as "social Darwinism."

Like La Mettrie, Spencer called for a new morality to be built on a biological foundation: "Acceptance of the doctrine of organic evolution determines certain ethical considerations."[9] He felt that the new code of morality should follow the design of nature and meet the test of natural selection. "Good" behavior would be adaptive and "bad" behavior maladaptive. With some exceptions, pleasure was the indicator of biologically sound conduct, and pain identified biologically dangerous behavior. It should be mentioned that the Darwinist T. H. Huxley argued to the contrary, that biological design had no place in human ethics but was as brutal as the poet Tennyson's "nature red in tooth and claw."

Spencer believed that fierce economic competition between men was natural, and that the resulting survival of the fittest contributed directly to progress. Each individual should prosper according to his ability and his work. Every man was free to do as he willed, provided that he did not infringe upon the "equal freedom" of others. Spencer was firmly opposed to any interference by the state with the "natural development of society." In his view, the purpose of the state was the insurance of the freedom of the individual, who would become perfect if left alone through enough generations. Spencer advocated unchained business

---

[8] Ibid., pp. 374–375.
[9] Ibid., p. 385.

competition and criticized state aid to the poor. "The ill-fitted must suffer the evils of unfitness, and the well-fitted profit by their fitness. . . . if, among adults, benefit were proportional to inefficiency, the species would disappear forthwith. . . ." [10] The economically unfit should and would be eliminated naturally. Spencer felt that charitable behavior would become instinctive through natural selection for its social practicality.

He was revolted by any thought of rule by the working class. Spencer believed that socialism would limit the development of healthy, perfected individuals and would instead produce dependent, unconscious slaves.

Unfortunately, Spencer argued that the progress of humanity resulted from the conflict of the races. He held that the more powerful or better adapted races continuously overran the less powerful or poorly adapted races, driving the latter to undesirable habitats, and occasionally exterminating them. Darwin's writings actually supported this view. Darwin did not distinguish between biology and culture; he felt that an evolutionary progression connected the most primitive living people and the most civilized. He believed that savages were close to apes and Europeans were far removed from them. The conception of racial inferiority and the equation of power and adaptedness were to be used as rationalizations for some of the sorriest pages in the history book of man. Largely because of these features of his social evolution, Spencer is held in low regard by sociologists today.

On the subject of social Darwinism, modern historians have said, "The pseudoscientific application of a biological theory to politics, whereby a nation is regarded as an organism, constituted the most perverted form of social Darwinism . . . led to racism and anti-Semitism and was used to show that only 'superior' nationalities and races were fit to survive." [11] Social Darwinism is far from dead, and its assertions are easily recognized in the platforms of major political parties, not only in the United States but all over the world. This philosophy still forms much of the naturalistic justification for free enterprise. Nations are still given individual, racial personalities, and their behavior in international relations often resembles primitive or feudal man's violent conduct. At this level, it seems that the sublimation of aggression fails, and the daily newspapers are filled with records of violence and threats, basic tactics of international politics. Spencer himself would probably be sickened by this failure in human progress.

The emphasis that some animal behaviorists place upon the role of

[10] Ibid., p. 389.
[11] T. Walker Wallbank, Alastair M. Taylor, Nels M. Bailkey, *Civilization: Past and Present* (Chicago, Scott, Foresman & Company, 1962), p. 581.

territorial behavior in human competition and aggression is reminiscent of social Darwinism, but perhaps there are positive aspects of the theory as it applies to human behavior and economic philosophy. Spencer's naïvete can be forgiven, considering the size of his task. If the proliferation of information was staggering in Spencer's day, then it is safe to say that the explosive growth of knowledge in the twentieth century is overwhelming. No one man has undertaken the construction of a synthetic philosophy in view of it. Maybe none will ever again be forthcoming, unless it is the work of teams of men. Amazingly, well-rounded scholars still appear now and then.

DARWIN

In Darwin's time and for almost two centuries before, a natural theology reigned. Many famous biologists were both ministers and natural historians. John Ray, who is ranked as a father of modern biological classification, along with Linnaeus, was both a biologist and a clergyman. His interbreeding definition of the species still stands, but he did not believe in the mutability of the species. Ray said that organisms were "the Works created by God at first, and by him conserved to this Day in the same State and Condition in which they were first made." In *The Wisdom of God Manifested in the Works of the Creation* (1691), Ray argued that the widespread adaptation of structure and function in nature was evidence for the wisdom of the Creator. This is a perfect expression of natural theology.

Seventeenth-century science had left an ideal, clockwork universe to the world of eighteenth-century thought. Isaac Newton had synthesized the work of previous astronomers with his law of gravitation into the unifying principle of a mechanical universe. The world came to regard God as the Great Mechanic and the universe as his perfect machine. This argument for divine design was carried into biology by the natural theologians. Archdeacon William Paley compared the adaptation of structure and function in nature with the machinery of a clock in his famous work *Natural Theology* (1802). As a student, Darwin had read Paley's *View of the Evidences of Christianity*. Like Aquinas before him, Paley tried to put the faith on a reasonable basis. First Darwin and then Einstein would shake the natural theology or physicotheology of the Old World at its foundations.

Darwin himself was trained for the Unitarian ministry. Unitarianism is a branch of the Protestant faith that accepts the teachings of Christ but denies his divinity and the concept of the Trinity. Many Christians would say that he had therefore already given up most of the faith. On the other hand, Darwin has been described as a "reasonably orthodox Christian."

The Darwinian theory of natural selection was a mechanism that stood in direct contrast to Special Creation and Grand Design. Organic evolution has often been portrayed as a history of imperfection, a story of unconscious, plodding experimentation, met with failure as often as success, which frequently was only momentary in the vast span of geologic time. The central theme is chance, probability, the accident of adaptive mutation.

Yet even Darwin could not escape the times in which he lived. In the final sentence of *The Origin of Species* he wrote, "There is grandeur in this view of life, with its several powers, having been originally breathed by the Creator into a few forms or into one; and that whilst this planet has gone cycling on according to the fixed law of gravity, from so simple a beginning endless forms most beautiful and most wonderful have been, and are being evolved." Here Darwin expresses faith and suggests that evolution should be substituted for Special Creation as the Grand Design. Note also his faith in a universe governed by Newtonian mechanics.

The origin of life from nonliving matter supports the position of mechanism. Darwin was unusually foresighted here. He wrote, "It is often said that all the conditions for the first production of a living organism are now present, which could ever have been present. But if (and oh! what a big if) we could conceive in some warm little pond, with all sorts of ammonia and phosphoric salts, light, heat, electricity, etc., present, that a protein compound was chemically formed ready to undergo still more complex changes, at the present day, such matter would be instantly devoured or absorbed, which would not be the case before living creatures were formed." [12] This statement almost designs the Urey-Miller experiment that was discussed previously. The acceptance of potential creation in place of special creation has been considered a surrender of the fortress of church philosophy by many. By contrasting these two Darwinian quotations, we can see that Darwin was uncertain, ambivalent, or had mixed feelings. In his unabridged autobiography we read, ". . . I gradually came to disbelieve in Christianity as a divine revelation. . . ." This is not to say that he was without faith in God. He also said, ". . . I had gradually come, by this time, to see that the Old Testament from its manifestly false history of the world and from its attributing to God the feelings of a revengeful tyrant, was no more to be trusted

---

[12]Francis Darwin, *The Life and Letters of Charles Darwin* (2 vols.) (New York: D. Appleton & Company, 1887), p. 202.

than the sacred books of the Hindoos, or the beliefs of any barbarian."[13]

Darwin had little inclination to approach the subject of the evolution of man, and his *Origin* dealt mainly with the evolution of the lower animals. In the last chapter of his famous work, he had said, "Much light will be thrown on the origin of man and his history." Only this and nothing more. Perhaps Darwin tended toward a Cartesian separation of man and the animals. If Wallace had not pressured him into a publication of his theory, it is likely that *The Origin* would have been published *after* Darwin's death. He had no desire to venture into public controversy. It has been suggested that Darwin suffered from a neurotic psychological illness and that a factor in this sickness was his fear of making public his theory with its sacrilegious overtones. When he finally wrote the *Descent of Man and Selection in Relation to Sex* (1871), only the first few pages were given to the topic of the evolution of man. If man was haphazardly evolved, was he without purpose? Equally important, what should be the definition of evolved man's death? We have seen that death is an essential feature of the mechanism of natural selection, the old making way for the new varieties to meet perchance the changing times.

Darwin cast natural theology aside: "The old argument of design in nature, as given by Paley, which formerly seemed to me so conclusive, fails, now that the law of natural selection has been discovered. We can no longer argue that, for instance, the beautiful hinge of a bivalve shell must have been made by an intelligent being, like the hinge of a door by man. There seems to be no more design in the variability of organic beings and in the action of natural selection, than in the course which the wind blows."[14] Darwin was neither gnostic (knowing or faithful) nor atheist. He wrote: "I cannot pretend to throw light on such abstruse problems. The mystery of the beginning of all things is insoluble by us; and I for one must be content to remain an Agnostic."[15]

The American anthropologist and writer, Loren Eiseley, called the nineteenth century "Darwin's Century." In many ways he has not surrendered the world of thought of the twentieth century, for even the Einsteinian revolution, dealing with the inanimate world, does not completely threaten the centrality of man in "God's world." Although Darwin was a master of the English language, he refused to enter public debate with the churchmen of his time. It was Thomas Henry Huxley who publicly argued for Darwinism and spoke first of its implication for human evolution.

---

[13] Charles Darwin, *Autobiography* (New York: W. W. Norton & Company, Inc., 1969), p. 85.
[14] Ibid., p. 87.
[15] Ibid., p. 94.

THE FIRST AGNOSTIC

Thomas Huxley (Fig. 10-3) is chiefly remembered as the public defender of evolution and the founder of an intellectual dynasty whose descendants are still with us. This family of genius has been much studied with regard to the inheritance of intelligence. Aldous Huxley was an outstanding figure in the culture of literary intellectuals. He wrote extensively on Hindu philosophy and mysticism, and, as the author of the famed novel *Brave New World* (1932), he expressed fear for the future of mankind, based upon his distrust of the then current trends in politics and applied science. This is a strange position for a grandson of Thomas Huxley. Another grandson, Julian Huxley, is more of a member of both cultures. Through his great writing skill, he has been a leading figure in relating science to social life and religion. His views will be examined later. Andrew Fielding Huxley recently won the Nobel prize in medicine

Fig. 10-3 **Thomas Henry Huxley (1825–1895), "Darwin's bulldog," the first to write extensively upon the evolution of man, coiner of the word "agnostic."**
Bettmann Archive.

for his monumental studies of the electrophysiology of the nerve impulse. Thomas H. Huxley was as versatile as his grandsons.

Thomas Huxley early expressed a great desire to become a mechanical engineer but took instead a degree in medicine. When he was quite young, he published original papers on the microscopic anatomy of man. It was not at all uncommon in those times for a physician to be a naturalist as well. Just as Darwin voyaged on the H. M. S. Beagle, so T. H. Huxley sailed on board the H. M. S. Rattlesnake, bound on a mapping expedition of the Torres Strait between northern Australia and Papua. In addition to his duties as ship surgeon, Huxley found time to make an historic study of the surface life of tropical seas. His important paper on the anatomy and relationships of jellyfishes and their relatives appeared in the *Philosophical Transactions of the Royal Society* in 1849. After he left the naval service, he became a lecturer in natural history at the School of Mines, a post that he held for more than 20 years.

Huxley destroyed the archetypal theory of Lorenz Oken and the German poet Goethe. These men had tried to define ideal anatomical forms that fitted animals for various ways of life, swimming, flying, etc., and toward which all organisms were striving. Huxley proved that the facts would not allow such a sweeping generalization. Until 1859, he was more concerned with specific organisms and the differences existing between them than he was with the theme of similarity. Herbert Spencer, whom Huxley met in 1852, was unable to convert him to a belief in universal evolution. At about the same time, Huxley met Darwin, and told the future prophet of evolution that a sharp line of demarcation separated the natural groups. It is said that Darwin smiled. Huxley would become the most fanatic convert to Darwinism, pick up the sword of evolution, and slay the Old World.

In a discussion of Darwin's theory before the 1860 meeting of the British Association, one Bishop Samuel Wilberforce (whose name, except for this one famous incident, would be forgotten) asked T. H. Huxley whether he had traced his ape ancestry from his mother's or father's side of the family. Huxley is rumored to have arisen in pale anger and replied that he would rather have an ape for an ancestor than an intellectual prostitute. According to his own account, Huxley rose without passion and replied with impeccable Victorian wit: "If then the question is put to me would I rather have a miserable ape for a grandfather or a man highly endowed by nature and possessing great means and influence and yet who employs those faculties and that influence for the mere purpose of introducing ridicule into a grave scientific discussion—I unhesitatingly affirm my preference for the ape." [16]

---

[16] T. H. Huxley, "The Huxley Papers," Imperial College, 15: 117–118.

It was inevitable that the Darwinists would be charged with atheism. Since the debate has not yet cooled, it is not difficult to imagine the heat of the times. It was a false accusation; Darwin's mixed feelings have been examined. Here, Huxley performed another invaluable service when he coined the word "agnostic." The dictionary defines an agnostic as "a person who thinks it is impossible to know whether there is a god or a future life or anything beyond material phenomena." "Atheist" is given as a synonym. The meanings of words change with usage, and that is forgivable. Everything changes, evolves. In this case, the original meaning intended by the coiner of this word is very important. Atheism is synonymous with materialism in common usage, and, unfortunately, agnosticism is a popular synonym of the two. Huxley said, "The one thing in which most . . . good people were agreed was the one thing in which I differed from them. They were quite sure they had . . . solved the problem of existence; while I was quite sure that I had not, and had a pretty strong conviction that the problem was insoluble. . . ." [17] "My fundamental axiom of speculative philosophy is that materialism and spiritualism are opposite poles of the same absurdity—the absurdity of imagining that we know anything about either spirit or matter." [18] Agnosticism was not as many theologians insisted (and still insist) an intellectual evasion of the question nor a compromise between atheism and faith. Agnosticism is the genuine position of not knowing (with certainty). This is the position of science itself with regard to the supernatural.

Huxley defined the word protoplasm, coined by the Czech scientist Purkinje, as "the stuff of life." The term was used to refer first to the then unanalyzed substances of the cell and later to the fluid portions of the cell, such as the cytoplasm, where little structure had been observed. With the advance of biochemistry in the clarification of the enzymatic reactions occurring there and the progressive elucidation of the electron microscopic anatomy of this region, revealing definite fine structure where none had been suspected before, the word protoplasm became a meaningless term except in its historic context. Huxley's definition seems vitalistic, mystic, and it receded before the advance of mechanistic investigations. The term protoplasm died a slow death, and there was a hesitancy to bury it, even when it was no longer useful. Max Schultze had defined protoplasm as "the physical basis of life." This definition reveals a mechanistic hope. Still, today, the term has not disappeared from basic biology textbooks, even though it is less and less frequently encountered as every year goes by and new texts are pub-

[17] T. H. Huxley, *Collected Essays*, v, pp. 237–238.
[18] Leonard Huxley, *The Life and Letters of T. H. Huxley* (vol. 1), (New York: D. Appleton & Company, 1913), p. 262.

lished. Huxley would probably be the first to fling a fistful of dirt on the coffin of this word.

Before the meeting of the British Association in Belfast in 1874, Huxley lectured "On the Hypothesis that Animals are Automata and its History." His views differed from Descartes' importantly. Like Spencer, Huxley did not believe that man could have come to consciousness without there being a gradual evolution of consciousness or gradations of it in other "brutes." He felt that nature had experimented with degrees of this ability in other organisms before man. He spoke both of the behavior of a decerebrated frog and of a medical case involving a man with brain damage. The famous English comparative anatomist Richard Owen claimed that man was sharply marked off from apes by the anatomical structure of his brain. Huxley said that it was merely a matter of the degree of development of certain areas. He stated that, "The soul stands related to the body as the bell of a clock to the works," and assumed that "molecular changes in the brain are the causes of all the states of consciousness. . . ."[19]

Yet this was the same Huxley who said, "In the 8th century B.C. in the heart of a world of idolatrous polytheists, the Hebrew prophets put forth a conception of religion which appears to be as wonderful an inspiration of genius as the art of Pheidias or the science of Aristotle."[20] He argued that orthodox Christianity was a "varying compound of some of the best and some of the worst elements of Paganism and Judaism."[21] Still he said, ". . . there is no evidence of the existence of such a being as the God of theologians."[22] Huxley said that morals were a practical social check on the struggle for existence; he could not envision moral purpose in nature or in the universe. Ethics were manufactured by man. Two years later he died having done his reasonable best. His son was a scholar.

FOSSILS AND GOD

Pierre Teilhard de Chardin (Fig. 10-4) was a man of the cloth and a biologist. He was both paleontologist and priest, a rare blend, indeed. Few men have looked at the graveyard of the past and the whirlpool of life and come away with their faith unshaken. Fewer still are collared evo-

---

[19] T. H. Huxley, *Collected Essays*, i, 199.
[20] Ibid., iv, 162.
[21] Ibid., v, 142.
[22] Leonard Huxley, *The Life and Letters of T. H. Huxley* (New York: D. Appleton & Company, 1900), vol. ii, p. 162.

Fig. 10-4   **Pierre Teilhard de Chardin (1890–1955), priest, geologist, evolutionist, and philosopher, "twentieth-century Aquinas."**
Photo by Philippe Halsman.

lutionary scholars. Chardin was one such, whose studies included research into the origins of man. He came from the studious order of Jesuits. In 1937, he was awarded the Mendel medal, which commemorates the intellectual achievement of that other great biologist and priest, Gregor Mendel, the father of genetics.

Teilhard de Chardin was a member of the team of scientists that excavated the Peking site and reconstructed the anatomy and way of life of Peking man. Later, he visited the African sites of Sterkfontein, the Taung, and Olduvai, where Dart, Broom, and Leakey had unearthed australopithecine fossils. Chardin recovered crude tools and fossilized bones in Southeast Asia, where Dubois had found Java man, and where the gibbon and the orang still swing through the trees.

In 1925, conservatives tried to discredit evolutionists like Chardin within the Church and asked the Holy Office to denounce evolutionary doctrines. They failed, but in September of 1937 Rome forbade Chardin to publish anything philosophic or theologic. Yet he had said: "Truth is simply the complete coherence of the universe in relation to every point contained within it. Why should we be suspicious of some sort of anthropocentric illusion contrasted with some sort of objective reality? In fact, there is no such distinction." [23] Here Chardin criticizes Kant's criticism of pure reason, which was based upon the proposal that man could not escape from his subjective senses. Chardin called for a new faith: "What all of us, more or less, are lacking at present is a new formula to express what is meant by holiness." [24] He wrote *The Phenomenon of Man* (1955), *Man's Place in Nature* (1963), and, prophetically, *The Future of Man* (1959), in which he expressed the utmost hope for an "ultraman," evolved from the new level of consciousness, even transcendant. Chardin had seen the face and amoral culture of Peking man, and still refused to embrace agnosticism. Because of his denial of the faith of doubt, even though he was a knowledgeable evolutionary scientist, Chardin is often exemplified by the faithful as their intellectual soldier, and his philosophy as a model of the enlightened compromise possible in the conflict of evolution and religion. As Darwin offered evolution as the Grand Design in the last sentence of the *Origin*, Chardin proposed, "Evolution, which offers a passage to something that escapes total death, is the hand of God drawing us to himself." [25] Chardin saw evolution as "a grand orthogenesis of everything living towards a higher degree of immanent spontaneity." [26] Orthogenesis has been discussed in the chapter on evolution.

Most biologists frequently view evolution as a dismal, haphazard mechanism. Julian Huxley, the grandson of Thomas, said that natural selection never plans ahead or works to complete a design. "Further-

---

[23] Jeanne Mortier and Marie-Louise Aboux, *Teilhard de Chardin Album* (New York: Harper & Row, Publishers, Inc., 1966), p. 138.

[24] Ibid., p. 140.

[25] Ibid., p. 185.

[26] Pierre Teilhard de Chardin, *The Phenomenon of Man*, trans. B. Wall (Harper & Row, Publishers, Inc., 1959), p. 151.

more, it often leads life into blind alleys, from which there is no evolutionary escape."[27] Still he speaks of progress in evolution: "There is sometimes a path out of the impasse, but it is generally a devious one; it is through its twists and turns that life finds its way into a new field of maneuver; and this marks the beginning of another distinct step in progress."[28] Evolution, Huxley says, is opportunistic, relativistic, more hindsighted than foresighted. The historian John C. Greene says that Julian Huxley tries to "draw moral inspiration from the progress of nature."[29] Huxley is not alone in this attempt.

The famous geneticist Theodosius Dobzhansky says, "Man is a zoological species. But this species has evolved properties so unique and unprecedented on the animal level that in man biological evolution has transcended itself."[30] This statement sounds quite like the philosophy of Chardin. However, it is not the expression of theistic evolution but rather a statement of naturalistic humanism.

George Wald, one of the greatest American biology teachers and winner of the Nobel prize for his studies into the chemistry of vision, states the following: "When speaking for myself, I do not tend to make sentences containing the word God; but what do those persons mean who make such statements? . . . What I have learned is that many educated persons now tend to equate their concept of God with their concept of the order of nature. This is not a new idea; I think it is firmly grounded in the philosophy of Spinoza."[31] Spinoza did not remove God from nature. George Bernard Shaw, in his play *The Shewing Up of Blanco Posnett,* expressed a modified Spinozan philosophy. Nature and God were portrayed as inseparable, imperfect, and evolving. A modest optimism is evident in this definition.

Although a personal friend of Chardin, George Gaylord Simpson, one of the greatest evolutionary biologists, branded Teilhard's evolutionary theology as mysticism:

Teilhard's beliefs as to the course and the causes of evolution are not scientifically acceptable, because they are not in truth based on scientific premises and because to the moderate extent that they are subject to scientific tests they fail those tests. Teilhard's mystic vision is not

[27]Julian Huxley, *Evolution in Action* (New American Library of World Literature, New York: Harper & Brothers, 1957), pp. 47–48.
[28]Ibid., p. 101.
[29]John C. Greene, *Darwin and the Modern World View* (New York: New American Library of World Literature, Inc., 1963), p. 75.
[30]Theodosius Dobzhansky, "Human Nature as a Product of Evolution," in *New Knowledge in Human Values,* ed. A. H. Maslow (New York: Harper & Brothers, 1959), p. 75.
[31]George Wald, "Innovation in Biology," *Scientific American,* Vol. 199, No. 3 (Sept., 1958), p. 101.

thereby invalidated, because it does not in truth derive from his beliefs on evolution—quite the contrary. There is no possible way of validating or of testing Teilhard's mystic vision of Omega. Any assurance about it must itself be an unsupported act of mystic faith. [32]

P. B. Medawar, Nobel laureate, was less kind to Teilhard. In a review of *The Phenomenon of Man,* Medawar accused Teilhard of importing the ancient vital force into the Mendelian machine. Medawar found the work "antiscientific in temper. . . . written in an all but totally unintelligible style, . . . construed as prima facie evidence of profundity." [33] Last but not least, he called the Phenomenon "a bag of tricks."

Most evolutionists take issue with Teilhard's belief that man, the epitome of evolution, can know where he is bound. While Teilhard believes that science and religion will show the way, contemporary evolutionists hold that the final destiny of man is unknown and unpredictable. Teilhard's willful evolution sounds remarkably Lamarckian.

AN AUGURY OF LUNGFISH GUTS

It seems like a long way from kidneys to agnosticism, but maybe it's not so far after all. Homer William Smith (Fig. 10-5) made the journey. In 1930, to "relieve the boredom of a Pacific crossing en route to Siam," Homer Smith wrote *Kamongo, or The Lungfish and the Padre.* He used the lungfish as "a springboard from which to take the reader on an exploration of the cosmos" as a biologist saw it. The padre of *Kamongo* was the Right Reverend Frank O. Thorne, Bishop of Nyasaland. The lungfishes, you will recall, are said to be relicts from an experiment of nature in an attempt to cope with drought and the accompanying stagnation of fresh water. This same supposedly, marginally successful trial is believed to have eventually led to crude limbs and the unconscious conquest of the land by vertebrates. *Kamongo* records the debates of a young biologist and a clergyman aboard a steamer bound up a Conradian tropical river, one to his ministry of primitive people, the other to his studies of primitive animals.

French physiologist Claude Bernard said that the maintenance of the constant composition of the "internal sea," what Walter Cannon called homeostasis, was a prerequisite to the invasion of the land. Foremost

---

[32] George Gaylord Simpson, *This View of Life* (New York: Harcourt, Brace & World, Inc., 1964), p. 232.
[33] P. B. Medawar, "Review of *The Phenomenon of Man,*" *Mind,* LXX (1961), pp. 99–106.

Fig. 10-5   **Homer William Smith (1895–1962), kidney expert, novelist and philosopher, author of *Kamongo, Man and his Gods, From Fish to Philosopher*, and numerous other books on biology and philosophy.**
Photo by Bachrach.

among homeostatic organs are the kidneys. Homer Smith was at the time of his death (1962) the foremost authority on the kidney in the world. He was the chairman of the Department of Physiology at the New York University School of Medicine, and his books on the kidney included *Lec-*

tures on the *Kidney* (1943), *The Kidney: Structure and Function in Health and Disease* (1951), and *The Principles of Renal Physiology* (1956), in addition to numerous articles and hundreds of published researches.

From Fish to Philosopher (1953) is one of the finest volumes on evolution in its entirety ever written, and yet it is primarily the story of kidney evolution. Smith's *The End of Illusion* went to press in 1935. The book he would undoubtedly like most to be remembered by is *Man and His Gods* (1952), which has now been reprinted eight times and continues to be a near best seller.

This scholarly work is generally ranked close to Frazer's *Golden Bough*, the anthropologists' bible of mythology and religion. *Man and His Gods* is not so wide in scope. The primary religions considered are Egyptian, Judaic, and Christian. Of this book Albert Einstein said, "The work is a broadly conceived attempt to portray man's fear-induced animistic and mythic ideas with all their far-flung transformations and interrelations. It relates the impact of these phantasmagorias on human destiny and the causal relationships by which they have come to be crystallized into organized religion. This is a biologist speaking, whose scientific training has disciplined him in a grim objectivity rarely found in the pure historian." [34]

Here is this grim biologist-poet's parody of Darwin's last lines in the *Origin*:

"There is grandeur in this view"—that living organisms warred perpetually to satisfy an appetite, and died only to make way for others to war a little differently, and perhaps a little more successfully. That the sun of heaven shone to sustain a continuing holocaust where love was primarily an impulse to copulation and a fuel to unselfishness only through an impermanent cultural pattern, where the most elaborate instinct was blind egocentric mechanism, where mercy, tolerance and charity but variants of behavior bearing an egotistically pleasing hue. That at every moment unnumbered living creatures subsisted by devouring other living creatures, the whole of animate creation a pyramid of murderers feeding upon the murdered—"teeth and talons whetted for slaughter, hooks and suckers moulded for torment—everywhere a reign of terror, hunger, and sickness, with oozing blood and quivering limbs, with grasping breath and eyes of innocence that dimly close in deaths of brutal torture"—until the most clever creature of all, "who fats all other creatures to fat himself," could sit down three times a day to his foul repast and pride himself upon being the highest animal, the Lord

---

[34]Homer W. Smith, *Man and His Gods* (New York: Grosset & Dunlap, 1957), Foreword by Albert Einstein.

of Creation the Very Ultimate Goal of a cosmic process that had required untold billions of years and the whole of the astronomical universe to achieve this magnificent result. Darwin can be forgiven his phrase because it expressed, not an assessment of the universe, but the inexpressible joy which the explorers of that universe feel when they penetrate one of its mysteries and discover one of its truths. The evolution that Darwin discovered was process, not progress.

If man, who shaded by degrees back into ape, into mute and insentient beast, wanted to call himself a "higher" animal, rather than just a more clever one, there was no other species vocal enough to gainsay his choice of adjective or his conceit. Endless forms most beautiful and wonderful have been evolved, but beauty and wonder are too much in the eye of the beholder to afford reliable standards of cosmic progress, and, taken in such vein, man might better consider his existence modestly, or else report it as a "tale told by an idiot." [35]

From the particulars of the kidney, by the process of reasoning called induction by that other "vulgar philosopher" Aristotle, Homer Smith had arrived at the general conclusion of agnosticism.

JUST YESTERDAY, TODAY, AND TOMORROW

In 1925, John Scopes was brought into a Tennessee courtroom for teaching evolution in a classroom. Clarence Darrow successfully defended the intellect, but his client was found guilty. Evolution is still most frequently taught as theory. Gerald Sykes shrewdly observes, "Evolution, though accepted now in theory when it confines itself to African monkeys or Tennessee courtrooms, is still not enjoyed as a factor in our own lives. We like to think we were born yesterday. It makes us feel younger." [36] Evolution is one of the Four Horsemen of the modern philosophic Apocalypse, and it rides a grey horse. What are the present views of the Christian churches on evolution?

Except for the fundamentalists, who interpret the bible literally, particularly Genesis, the Christian modernists mostly admit the probable fact of evolution and consider the biblical accounts of Creation "poetic

[35] Ibid., p. 434.
[36] Gerald Sykes, *The Cool Millenium* (Englewood Cliffs, N. J.: Prentice-Hall, Inc., 1967), p. 27.

interpretations." Modernists feel that evolution and faith are not in conflict and that a compromise is possible. An evolution *of sorts* is considered certain fact. Nevertheless, it would be a distortion of the truth to say that *Darwinian* evolution is accepted. Darwin's relenting hope expressed in the last lines of the *Origin* is a reality. *An* evolution has replaced Special Creation as the Grand Design in modern Christian thought. A "new" natural theology has been "born." The evolution to which most modern Christians subscribe, however, differs importantly from the Darwinian. There is a strong tendency toward the orthogenesis of Chardin. The biographer of Chardin, Canon C. E. Raven, sums up the predominant position of the modern churchman: "It remains as true as ever that the world is so ordered that life develops and the evolution of higher types takes place upon it; and it is as absurd as it ever was to suppose that it was 'merely accidental.'" [37] The canon praises John Ray, that great natural theologian of the eighteenth century. Raven sees "real progress" in evolution, but equates progress with "travail." Recall that Julian Huxley also writes of progress in evolution but firmly believes in the role of chance or accident in the process. Canon Raven continues: "There was in my own young days a period of sheer materialism when it looked as if the scientific and religious outlooks were wholly contradictory, and when some of us had to say how hard it is to be a Christian not only on moral and spiritual but intellectual grounds. Indeed, it is still true that we have not yet realized the full consequences of Darwin's work; or seen all its effect upon our traditional ideas." [38]

Beyond puny man and his trivial planet there was always the eternal, perfect universe. Darwin expressed his faith in the order of the Newtonian mechanical universe. Then in the twentieth century, Einstein's theory of relativity twisted the Cartesian coordinates and distorted the clockwork universe. The universe was expanding; now, it seems to be contracting. The universe itself is thus evolving. Perhaps there is a cycle of universes. Herbert Spencer noted that all theories of the origin of the universe drove man to inconceivabilities. When the priest says, "God made the world," the child asks, "Who made God?" How shall the mind of man ever imagine anything uncaused or beginningless? Infinity is just a word. It seems that one would need the mind of God to comprehend it. These revelations in the physical sciences have contributed *not* to the restoration of faith in order and progress, but rather to a confusion of the popular mind. Darwinism, with its central theme of chance, and the Einsteinian relative universe have helped to drive even educated men to either philosophies of despair or redoubled faith in "the other world."

---

[37] *A Short History of Science* (Garden City, N. Y.: Doubleday & Company, Inc., 1951), p. 106.
[38] Ibid., p. 108.

The rare atheist sees neither cause nor origin, while the gnostic sees both beginning and end.

The pessimistic philosophy of existentialism holds that each man is cast down alone into a purposeless universe. Although the growth of this philosophy of despair is often attributed to the tragic political and social events of the World Wars, the scientific developments of Darwinism and relativity are probably equally important factors in the evolution of existentialism. Dictionaries define existentialism as a philosophy of nihilism, and many scholars have seen in it the death of philosophy, an anti-philosophy, the absence of philosophy, emptiness. According to existentialism, every man is "on his own," responsible to himself for his own actions. Existentialism need not be viewed negatively. Although many existentialists are literary intellectuals, the philosophy is not without followers from the ranks of science; it is from this quarter that a positive existentialism has emerged.

The Nobel laureate Jacques Monod is a molecular geneticist whose studies clarified the mechanism by which substances in the external and internal environments of cells regulate genetic activity. In his latest book, *Chance and Necessity: An Essay on the Natural Philosophy of Modern Biology,* he advocates the substitution of science for the animistic tendencies still present in human belief. (Animism is just another term for vitalism.) The title of his essay is taken from a statement attributed to the ancient Greek Democritus: "Everything existing in the Universe is the fruit of chance and of necessity."

Historians of evolution have puzzled over the fact that, while evolutionists reject the idea of purpose or design in nature, they nevertheless resort frequently to the metaphor of progress in describing the course of evolution. It is just this seeming paradox that Monod masterfully resolves. Evolutionists do not hold with original causality or ultimate purpose, seemingly necessary accompaniments of vitalism. Teleology is a term applied to the study of final causes or the state of being directed toward a definite end. Teleology also refers to the vitalistic belief that natural events are determined not only by mechanical causes but by an overall design or purpose in nature. Teleology therefore opposes mechanism. To most natural scientists teleology is anathema, to be avoided at all costs, particularly with regard to life and its evolution.

Yet, living beings without exception are endowed with a purpose, which they show in their structures and functions. Monod maintains that the recognition of this meaningful quality is essential to the definition of living things. He calls this characteristic property of living beings *teleonomy.* The machines of man are wrought in accordance with the designs of extrinsic thought. They are not born by chance. The designs of living machines, however, are inherent, within themselves, and they have been theoretically conceived by the meetings of chance variations with environmental circumstances. Monod points out that living things are self-reproducing machines at the same time that they are self-constructing machines, that they are conservative yet capable of change, a seeming paradox. The genetic code replicates itself faithfully. DNA is conserved, but it can and does mutate, rarely—but enough.

Monod is a bacterial geneticist and has encountered in his field of study at least two examples which have led him to propose the teleonomic apparatus. A consideration of these will perhaps illustrate what he means by teleonomy.

One subject of his studies was the inducible enzyme system of the common intestinal bacterium, *Escherichia coli*. The normal microbe of this type neither takes up the double sugar lactose (milk sugar) nor produces the enzyme lactase, necessary for the digestion of lactose, a step essential to its use as a food source. The bacteria do not grow for some time after inoculation into a lactose broth, but they then shoot up in a characteristic growth curve. The lag period is, however, much longer than normal. Analysis of the growing bacteria reveals that they are synthesizing significant quantities of the digestive enzyme lactase. Apparently the lactose has somehow called forth or induced the manufacture of the enzyme required for its digestion. Enzymes of this type are designated inducible in contrast to the constitutive enzymes, which are normally present and manufactured regularly. Monod and his coworkers discovered that the genes for lactase and permease (a factor essential to lactose uptake) were usually switched off by a repressor substance from a regulatory gene. They proposed that the inducer (lactose) combined with the repressor to inactivate it and thus allow the switching on of the genes directing the enzyme syntheses. Thus, a substance in the environment regulates genetic activity. Note that the gene for lactase is already present. It does not arise *de novo*. The "lac" gene is not a mutant. Mutations of the regulatory gene are known that fail to repress the lactase gene, resulting in the enzyme becoming constitutive, normally present and synthesized.

Perhaps the greatest seeming flaw in the armor of evolutionary theory for most students is the fact that mutations play a central role in it, yet most mutations are random (nondirected), recessive, meaningless garblings of purposeful information, or meaningful in a negative sense, harmful and even lethal in double dosage. It is here also that Monod's microbial background comes to our aid. Due to the conservative near perfection of the replicative apparatus (DNA), he points out, any mutation considered alone is a rare event. The probability of a given bacterial gene undergoing a mutation which would affect the function of the determined protein is between one chance in a million and one chance in a hundred million per cellular generation. A few drops of water can harbor several billion bacteria. In such a large population any given mutation is probably carried by ten to a thousand bacteria. The total number of mutants of all kinds in this population is estimated to be one hundred thousand to one million. This is a literal "sea of chance."

In the billions upon billions of sperms produced by an average male in his reproductive life, millions of mutants must likewise occur. Certain human genes have a higher mutation rate than others. Some producing recognizable defects occur as frequently as once in every hundred thousand to once in every million samples. These figures do not take into account those individually undetectable mutations which, combined in the shuffling of sexual reproduction, can produce significant effects. Monod points out that mutations of the latter type have probably played a greater role in evolution than those whose individual effects are easily detected. He estimates that in today's human population of three billion that a hundred billion to one trillion mutations occur with each new generation. If for some reason of environmental circumstance a mutant is favored (natural selection), the incidence of it in a given population can rise in a display of positive feedback. Monod compares this unconscious game to a "gigantic lottery," from which nature draws the numbers.

The best example which Monod cites to illustrate the teleonomic process is the mechanism of antibody formation. Antibodies are proteins with the ability to recognize, by specific shape association, foreign substances that have invaded the organism. These alien substances are called antigens (antibody generators), and they include bacteria and viruses. The antibody selectively recognizes a foreign substance after the first encounter. The organism is capable of forming antibodies designed to deal with almost any natural or synthetic configuration. The allergies—drug and otherwise—are considered to be imperfect immune reactions. The possibilities for antibody formation seem infinite.

For most of this century it was supposed that the source of information for the manufacture of the antibody's specific associative pattern was the shape of the antigen itself. It was as though the lymphocytes and plasma cells (antibody manufacturers) possessed an intellect, first sizing up the specifications of the antigen and then making a mirror image antibody or complement. This was a teleological explanation. It has recently been proved that just the reverse is the case: antibody structure is already determined *before* the antigen is seen!

Within the animal the antibody-manufacturing cells (which exist in great numbers) have certain "hot" or highly mutating genetic segments that determine antibody structures. Mutations occur at random, in complete ignorance of the antigen. The antigen plays the role of natural selector, favoring the reproduction of those cells which have produced the complementary antibody by *accident*. This is natural selection at the cellular level, the evolution of direction or purpose through the coincidence of chance and circumstance. It is something like the pre-adaptation discussed previously. The teleologic hypothesis of antibody formation is destroyed. It is a shock for teleologic thinkers to find a game of chance at the root of the highly adaptive mechanism of immunity, yet Monod points out that only chance would supply an adequate variety to underly the animal's extremely wide range of chemical defense. The immune response is probably the most plastic teleonomic performance.

Any mutation that results in new teleonomic action (adaptive response) insures the survival and reproduction of its bearer. Mutations are stable. The same replicative apparatus (DNA) that conserves the old preserves the new. On a grand scale, evolution is something like the antigen–antibody response. The evironment is the antigen selecting the individuals producing the proper antibody, not by design or with forethought, but by chance mutation. Teleologic evolution, orthogenesis, is unlikely if there is a logical, mechanistic alternative accounting for the development of purpose. The reason why comes after the fact of undirected mutation, not before. It might be weakly argued that the absence of foresighted design is the grandest design itself, but that would be logical suicide, since a teleologic apparatus would be immeasurably more adaptive than the natural selection of chance variation. What philosophic possibilities are open to a man accepting this vision? Must the Prime Mover be moved from the natural world, which according to the above mechanism includes the living? Natural theology becomes an untenable position. Only a supernatural theology is permissible, not, however, essential.

Jacques Monod is an interesting figure in modern philosophy because he is both an existentialist and a scientific humanist. Although he believes in the essential loneliness of man in the universe, Monod does not despair but confirms his faith in the metaphor at least of the human spirit. He condemns, however, the persistence of vitalistic philosophy, belief in the separability of body and spirit, faith that the soul is the guiding, vital principle of organic evolution, the attribution of conscious life to nature and natural objects. Monod recommends that the ethic of man should be the ethic of knowledge discovered, not the possibility of supernatural secrets, questionably revealed occasionally. He feels that "scientific socialist humanism" is the proper philosophy for man and stresses that the scientific method is indispensable to its practice: "It is the conclusion to which the search for authenticity necessarily leads. The ancient covenant is in pieces; man knows at last that he is alone in the universe's unfeeling immensity, out of which he emerged only by chance. His destiny is nowhere spelled out, nor is his duty. The kingdom above or the darkness below: it is for him to choose." [39]

The author's narrative poem of evolution, "Conquests of the Galapagos," concludes with similar lines: "I am the bastard child of chance and time, yet owned and never grown. I accept the lot of my imperfect birth and perfect death without an afterthought." The dedication page of *Chance and Necessity* bears a quotation from the existential novelist Camus' *Myth of Sisyphus*. In Greek myth, Sisyphus was a cruel king of Corinth sentenced forever in hell to roll up a hill a great stone which perpetually rolled down again, a frustrating fate, to work to no avail, in addition to being in hell. Camus reinterprets the myth: "I leave Sisyphus at the foot of the mountain! One always finds one's burden again. But Sisyphus teaches the higher fidelity that negates the gods and raises rocks. He too concludes that all is well. This universe henceforth without a master seems to him neither sterile nor futile. Each atom of that stone, each mineral flake of that night-filled mountain, in itself forms a world. The struggle itself toward the heights is enough to fill a man's heart. One must imagine Sisyphus happy." [40] Like Camus' Sisyphus, Monod accepts the natural boundaries of the human world, but not in defeat. Many of the problems of the current human predicament can be solved through the application of the time-proven scientific method.

There are several definitions of humanism. It can refer to either the study of the humanities or the belief in natural man and the value of the scientific method in realizing his potential, rejecting supernaturalism. The English physicist and novelist C. P. Snow coined the phrase "two cultures" to refer to the split of the intellectual community into two camps, the scientists and the humanities' scholars, whom Snow dubs "literary intellectuals." The latter frequently harbor antiscientific sentiments and hold the scientific community to be unprincipled, uncultured, while scientists often characterize the scholars of the humanities as foppish, uneducated, "foggy thinkers." The philosophy of science is usually taught as a subject separate from courses in general philosophy, which are frequently obituaries of dead philosophers as well as their dead philosophies.

Originally, philosophy meant the love of wisdom or knowledge, and

---

[39]Jacques Monod, *Chance and Necessity* (New York: Alfred A. Knopf, Inc., 1971), p. 180.
[40]Albert Camus, *The Myth of Sisyphys and Other Essays*, trans. J. O'Brien (New York: Alfred A. Knopf, Inc., 1967), p. 123.

the mathematicians of ancient Greece considered themselves philosophers. Unfortunately, we live in an age of narrow specialization, which has too often created prejudiced, crippled thinkers, rather than balanced philosophers. At the same time, the explosive growth of knowledge has almost made it impossible to be universal men, capable of creating in several widely separated fields. Fortunately, the human mind still tries to synthesize its culture, to fuse the fragments. Scientists have been more successful at assimilating the subject matter of the humanities than the humanists have been in absorbing the sciences. Calls have gone out for the recreation of one culture. Science is as much a part of the human culture as the arts, and rather than mirroring an age the scientific method has the potential to change the times.

Van Rensselaer Potter urges the creation of a new discipline of *bioethics* to be a bridge between the two cultures: "In the past *ethics* has been considered the special province of the humanities in a liberal arts college curriculum. It has been taught along with logic, esthetics, and metaphysics as a branch of Philosophy. Ethics constitutes the study of human values, the ideal human character, morals, actions, and goals in largely historical terms, but above all *ethics implies action* according to moral standards. What we must now face up to is the fact that human ethics cannot be separated from a realistic understanding of ecology in the broadest sense. *Ethical values* cannot be separated from *biological facts.* We are in great need of a Land Ethic, a Wildlife Ethic, a Population Ethic, a Consumption Ethic, an Urban Ethic, an International Ethic, a Geriatric Ethic, and so on."[41] He further states: "Man's natural environment is not limitless. . . . Man is considered as an errorprone cybernetic machine. . . ."[42]

Although science is closer to magic than to religion, science is neither magic nor religion. Scientists recognize the limitations of their methods. At one time popular belief held that science would find a way to support the "growing population." Scientists were quick to puncture this myth. Any attempt to break natural law is doomed to failure. The solution to overpopulation is simply population control.

It is crucial to the survival of civilization that we resynthesize our culture, educate ourselves to obliterate the greatest cultural lag, the failure of social institutions to keep pace with the rapid advances in science. The frustrations of this discrepancy have rekindled popular interest in the occult and have produced an antiscience reaction, sentiments with catastrophic implications. It is not time to "hoe one's own garden." The

---

[41]Van Rensselaer Potter, *Bioethics* (Englewood Cliffs, N. J.: Prentice-Hall, Inc., 1971), p. vii.
[42]Ibid., p. 1.

problems of the twentieth century cannot be solved by a retreat to romantic naturalism. It is time for more not less science. It is time for a teleologic performance with man as the vital force and science as the method.

"If man is a living thing, and biology is the study of living things, then the study of any and all aspects of man should fall within the realm of biology in the broad sense."[43] Biology is just a nearsighted term for one huge compartment of our knowledge, the totality of which we used to call philosophy. In our ecocrisis, it is time for a living philosophy, a biophilosophy.

---

[43] Robert Bigelow, *The Dawn Warriors* (Boston: Little, Brown and Company, 1969), p. 10.

DISCUSSION QUESTIONS

1. Will the triumph of mechanism justify atheism?
2. Is natural theology presently justifiable?
3. Is there any relation between evolution and existentialism?
4. What is bioethics?
5. What role should the facts of biology play in personal, social, and international philosophy?

REFERENCES

CHASSIS, HERBERT, and WILLIAM GOLDRING, Homer William Smith, New York University Press, New York, 1965.

DURANT, WILL, The Story of Philosophy, Washington Square Press, New York, 1952.

EISELEY, LOREN, The Firmament of Time, Atheneum Publishers, New York, 1960.

GREENE, JOHN C., Darwin and the Modern World View, Louisiana State University Press, Baton Rouge, 1961.

HOFSTADTER, RICHARD, Social Darwinism in American Thought, George Braziller, Inc., New York, 1959.

JONAS, HANS, The Phenomenon of Life, Dell Publishing Co., Inc., New York, 1966.

MONOD, JACQUES, Chance and Necessity, Alfred A. Knopf, Inc., New York, 1971.

POTTER, VAN RENSSELAER, Bioethics, Prentice-Hall, Inc., Englewood Cliffs, New Jersey, 1971.

RENSCH, B., Biophilosophy, Columbia University Press, New York, 1971.

SIRKS, M. J., and CONWAY ZIRKLE, The Evolution of Biology, The Ronald Press Company, New York, 1964.

SMITH, HOMER, Man and His Gods, Grosset & Dunlap, Inc., New York, 1952.

SPENCER, HERBERT, The Evolution of Society, University of Chicago Press, Chicago, 1967.

INDEX

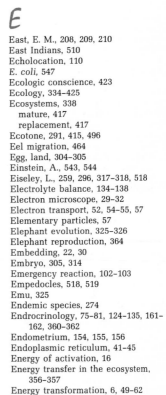

R

Rabbit plague, 361, 420
Race, biologic, 275
Races, human, 503, 506, 508–512
Radial symmetry, 105
Radioautography, 236–238
Ramapithecus, 498, 499
Range, 356
Rassenkreis, 276
Rat reproduction, 108–111, 154–157, 364
Rat social behavior, 363
Rattlesnakes, 378, 379, 382, 442, 443
Raven, C. E., 545
Ray-fins, 291
Ray, John, 274
Receptor end organs, 90, 92, 93
Recessive, 194
Recon, 247
Red-green color blindness, 218
Red tide, 400
Reflex:
  chemical, 80
  nervous, 90–99, 454
Regulation, genetic, 248–249, 547
Regulator gene, 248, 249, 547
Reindeer moss, 385
Relativity, 545
Releasers, 434, 439, 440
Relict species, 291
Religion, 259–260, 505, 519, 522, 533–541, 544, 545, 549
Renal corpuscle, 119, 122
Renal papilla, 121, 142
Renal tubule, 119
Renin, 136, 140
Replication, 235
Repolarization, 85, 86
Repressors, 249, 547
Reproduction, 6, 150–187, 433–439
Reproductive ability, 369
Reproductive potential, 364
Reptiles, 300–312, 439, 442, 443, 464
Resolving power, 29
Respiration, 6, 49–57, 345, 346, 347
Resting membrane potential, 84
Rh factor, 216–217, 509–510
Rhea, 325
Rhinoceros, 443
Rhodesian man, 505
Ribosomes, 42–44, 241, 246
Ritual feeding, 433
Ritualization of aggression, 431, 438–446, 529
Rivers, 407, 408, 421
RNA, 10–12, 41–44, 228–249, 469
Rocky Mountains, 386–388
Rocky Mountain Spotted Fever, 418
Rods, 381
Romer, A. S., 290
Root–nodule bacteria, 347

Ruminants, 396
Rut, 154

S

Sagebrush-grasslands, 388, 397
Sagegrouse, 436–437
Saguaro, 374, 376
Salamanders, 168, 297, 463–464
Saltatory conduction, 87
Salt glands, 147, 148
Salt lakes, 342, 372
Salt water hypothesis of vertebrate origin, 146, 290–291
Santa Gertrudis, 250
Savannah, 394
Sawgrass, 411
Scala naturae, 268, 520
Scallop gonads, 183
Scavengers, 355
Schaller, G., 493
Scholastic period, 522, 523
Schooling behavior, 449
Schrödinger, E., 259
Scopes, J., 544
Screwworm, 420
Scrotum, 171
Sea birds, 147
Seal, 146, 306
Sea mammals, 146
Sea urchin, 402
Secondary sex characters, 163
Secretion, 41–46
Secretory cells, 41–44
Secretory granules, 23
Seed, 312–313, 314, 315
Seed plants, 300
Segregation, 194
Selection:
  artificial, 271
  natural, 267, 271–272, 491
  sexual, 280–282, 436
Selective permeability, 37
Seminal vesicles, 169
Seminiferous tubules, 166
Semipermeability, 33
Sensitivity, 7
Sensory neuron, 89, 91, 93
Sewage, 344, 421
Sex attractants, 420
Sex chromatin, 180, 181
Sex chromosomal abnormalities, 180
Sex chromosomes, 177, 218
Sex determination, 177–186
Sex hormones, 13, 14, 161–162, 361, 454
Sex-influenced genes, 220
Sex-linkage, 217–219
Sexual development, 182, 185
Sexual dimorphism, 433
Sexual reproduction, 172
Sexual selection, 280–282, 436

Seymouria, 298
Shade tolerance, 358, 391
Sharks, 291
  osmoregulation in, 145–146
Shaw, G. B., 540
Siamese fighting fish, 438
Sickle-cell anemia, 247, 330, 331
Sierra Nevada, 386
Sign stimulus, 432, 434
Simian diastema, 483, 484
Simian shelf, 483, 484
Simpson, G. G., 540–541
Siren (mud-eel), 299, 300
Skeletal muscle, 414
Skin, 304
Skin color, 210–212, 511
Slash pine, 387
Sleeping sickness, 352
Sloth, 393
Smith, Homer, 117, 290, 523, 541–544
Smith, William, 269
Snake origin, 309
Snow algae, 341
Snow, C. P., 550
Snowshoe-hare, 358–360
Snowy owl, 359
Social Darwinism, 529–531
Social insects, 451, 456–465
Societies:
  insect, 451, 456–465
  primate, 445, 450, 529
Sodium pump, 38, 83, 125
Soils, 339
Solo man, 505
Soul, 525, 537
South Africa, 374, 375, 500–503
Southern coniferous forests, 387
Spawning, 167
Special creation, 259
Specialization, evolutionary, 311
Species, 273–279, 508
Spencer, Herbert, 526–531, 545
Spermatogenesis, 165–166
Sperm banks, 254
Sperms, 166, 177–179, 288–290
Sphenodon, 312
Spider webs, 463
Spinal animals, 96–97
Spinal cord, 93
Spinal reflex arc, 93
Spinoza, B., 540
Spiny anteater, 318
Spontaneous generation, 283, 520
Spores, 176, 177
Sporophyte, 176
Spurge "cactus," 374
Squid, 104, 105, 288
St. Hilaire, 262
St. Paul, 522
St. Thomas Aquinas, 522
Stagnant water, 344
Stamens, 186, 315, 316